THE FRONTIERS COLLECTION

Series editors

Avshalom C. Elitzur
Unit of Interdisciplinary Studies, Bar-Ilan University, 52900, Gières, France
e-mail: avshalom.elitzur@weizmann.ac.il

Laura Mersini-Houghton
Department of Physics, University of North Carolina, Chapel Hill, NC 27599-3255
USA
e-mail: mersini@physics.unc.edu

Maximilian Schlosshauer
Department of Physics, University of Portland
5000 North Willamette Boulevard Portland, OR 97203, USA
e-mail: schlossh@up.edu

Mark P. Silverman
Department of Physics, Trinity College, Hartford, CT 06106, USA
e-mail: mark.silverman@trincoll.edu

Jack A. Tuszynski
Department of Physics, University of Alberta, Edmonton, AB T6G 1Z2, Canada
e-mail: jtus@phys.ualberta.ca

Rüdiger Vaas
Center for Philosophy and Foundations of Science, University of Giessen, 35394,
Giessen, Germany
e-mail: ruediger.vaas@t-online.de

H. Dieter Zeh
Gaiberger Straße 38, 69151, Waldhilsbach, Germany
e-mail: zeh@uni-heidelberg.de

For further volumes:
http://www.springer.com/series/5342

THE FRONTIERS COLLECTION

Series editors
A. C. Elitzur   L. Mersini-Houghton   M. Schlosshauer
M. P. Silverman   J. A. Tuszynski   R. Vaas   H. D. Zeh

The books in this collection are devoted to challenging and open problems at the forefront of modern science, including related philosophical debates. In contrast to typical research monographs, however, they strive to present their topics in a manner accessible also to scientifically literate non-specialists wishing to gain insight into the deeper implications and fascinating questions involved. Taken as a whole, the series reflects the need for a fundamental and interdisciplinary approach to modern science. Furthermore, it is intended to encourage active scientists in all areas to ponder over important and perhaps controversial issues beyond their own speciality. Extending from quantum physics and relativity to entropy, consciousness and complex systems—the Frontiers Collection will inspire readers to push back the frontiers of their own knowledge.

For a full list of published titles, please see back of book or
http://www.springer.com/series/5342

Dean Rickles

A BRIEF HISTORY
OF STRING THEORY

From Dual Models to M-Theory

 Springer

Dean Rickles
Unit for History and Philosophy of Science
University of Sydney
Sydney
Australia

ISSN 1612-3018   ISSN 2197-6619 (electronic)
ISBN 978-3-662-50183-2 ISBN 978-3-642-45128-7 (eBook)
DOI 10.1007/978-3-642-45128-7
Springer Heidelberg New York Dordrecht London

Printed on acid-free paper

Springer is part of Springer Science+Business Media (www.springer.com)

For Gaia and Sophie,
my charmed quarks!

Preface

> *If superstring theory does turn out to be the TOE, historians*
> *of science will have a hard job explaining why it came*
> *into being.*
>
> Joel Shapiro

String theory seemingly has a very bizarre early history. In his recent book, on the so-called "monster sporadic group" and its relation to physics (including string theory), Terry Gannon writes of string theory that it "is still our best hope for a unified theory of everything, and in particular a consistent theory of quantum gravity," but, he continues, "[i]t goes through periods of boom and periods of bust, not unlike the breathing of a snoring drunk" ([6], p. 277). This book describes some of these periods of boom and bust: the snoring drunk. I aim to reveal aspects of string theory's development in what I hope are more honest terms that the accounts of a pristine, unique, ineluctable structure that form much of the current string theory literature (especially the popular presentations of the theory[1]). It tells the rather volatile story of string theory from just before its conception, toward the end of the 1960s, when it was bound to so-called "dual-resonance models" (themselves the high-point of Geoffrey Chew's 'bootstrap' approach to strong interaction physics), to the advent of *M*-theoretic ideas in the mid-1990s, where it isn't really clear that it is a theory of *strings* at all.

We do face something of a historiographical problem here: in a very real sense, **string theory**, in the very general sense of a physical framework grounded in a fundamental principle (which delivers the dynamics and physical degrees of freedom) along the lines of the equivalence principle of general relativity, say— **does not exist!** Instead, we have a strange inversion of the usual relationship between *principles* and *theories* according to which one derives the theory from the principles, rather than the other way around. So why write a history of string

---

[1] At the same time, it avoids the recent trend of engaging in 'string bashing,' presenting the history in such a way that stands aside from questions of the truth or falsity or the theory—though I do step down from the historian's podium, briefly, in the final chapter.

theory if there is no theory as such to speak of? There are several reasons why I think we should certainly proceed, despite this glaring incompleteness:

- String theory, though not an empirically well-confirmed theory (that is, beyond already known, "old evidence" such as the existence of gravity and various gauge symmetries), has been around for over 40 years. This is a fairly sizeable chunk of recent physics history—it could likely match more chronologically mature theories in physics in terms of the number of physicist-hours that have been devoted to it.

  However, as John Schwarz bemoaned in a recent talk on the early history of string theory [18], there appears to be a complete lack of interest in the subject from within the history of science community.[2] We can perhaps trace this to the unwillingness of historians to invest time (an awful lot of it, given string theory's technical demands) studying a theory that might well be "firing blanks". However, given both the length of time string theory has been on the scene, and, more crucially, given its importance and dominance in physics, and even in pure mathematics, I share Schwarz's belief that "there remains a need for a more scholarly study of the origins and history of string theory" (ibid., p. 1).[3] Even if string theory should prove to be an *empirical* dud,[4] its role in the development of the physical and mathematical sciences mark it out from many empirical duds of the past. It has, after all, been intimately entangled with various key episodes in

---

[2] There are a small handful of notable exceptions [3, 4, 11], though in each case string theory is dealt with only very briefly. This does not include the recent book *The Birth of String Theory* [1], which, strictly speaking, falls outside of 'professional' history of science (however, I have more to say about this below).

[3] This really ought to be expanded to a more general study of the development of quantum gravity research, for which see [15], for a first attempt (dealing, initially, with quantum gravity research up to 1957). This book is a *brief* history, intended to be read by nonspecialists (hopefully) without overwhelming them. As such, it does not aim to be exhaustive, sketching the broad outlines of development and seminal publications, rather than the most intricate details— there is such a wealth of primary literature spanning string theory's 40+ years that the task of writing its history will certainly require the contributions of many authors. Hence, though I do include enough detail to provide a coherent account of the evolution of the theory, it should be remembered, of course, that this involves much selection and filtering on my part that does not necessarily imply that what was filtered out was unworthy of inclusion.

[4] This is already highly unlikely as a general claim, since string theoretic models, and in particular duality symmetries present (or originating) in string theory, have led to several groundbreaking results that have experimental ramifications. In particular in the field of 'quark-gluon plasmas' which extremely difficult problems in particle physics are probed using what is essentially the study of black holes (see [12] for a good review; or [13] for a more elementary account). String theory has more recently found applications to the very difficult problem of explaining high-T_c superconductivity—this was carried out by Mihailo Cubrovic, Jan Zaanen, and Koenraad Schalm: [2, 9]. (This is not to mention certain early empirical successes of the hadronic string theory, such as a physically intuitive mechanism for quark confinement and Regge behaviour, which should not be underestimated.)

the wider history of physics, including the development of quantum chromo-dynamics, supersymmetry, supergravity, black holes, cosmology, statistical physics, and the anthropic principle. It is far richer in such connections than one might expect, given its esoteric reputation. To ignore it is to leave behind a notable gap in the historical record.

- Moreover, there is real value in looking at the earliest history of a subject and exposing it as it was originally presented (regardless of whether the subject can be considered "complete"). Reading the original work that paved the way to the creation of a field of research can prove to be very enlightening: not only can one gain more intuition for the field, but it is often very surprising how much has been *forgotten* in the intervening period, especially in terms of general guiding, physical principles. As Robert Woodhouse elegantly put it in his 1810 study of the early days of the calculus of variations: "the authors who write near the beginnings of science are, in general, the most instructive; they take the reader more along with them, show him the real difficulties and, which is the main point, teach him the subject, the way they themselves learned it" ([21], p. 1). Auguste Comte, in his *Positive Philosophy*, put it more strongly: "To under-stand a science it is necessary to know its history". Though this quotation is over-used, there is certainly some truth to it. Unless all one is doing is number-crunching, one will wish to know what the numbers and structure represent and why they have the form they have. But this is often exactly what is lost as a subject develops momentum, and such details are condensed into a misleading, simplified package. Problems of the theory are tackled for their own sake, as interesting puzzles independent from the fundamental *raison d'etre* of the theory—in fact, there appear to be several such *raisons d'etre* in string theory: it is far from being a unified endeavour. Of course, such puzzle solving often proves very fruitful, and can lead to expansions and generalisations of the theory. However, the point remains, through history comes a greater depth of understanding.[5]

This book is, then, intended to plug the string theoretic hole (or a decent portion of it) in the history of science literature, identified by Schwarz. For better or for worse, string theory is a central part of the landscape of science. Not only that, it is a science that has spread into the public domain because of its obvious popular appeal (theory of everything, the multiverse, colliding branes, hidden spatial dimensions, pre-big bang scenarios, etc.)—not to mention the spirited presentation in Brian Greene's PBS television series, *The Elegant Universe*. It is also a very

---

[5] I might add to this list (to echo the opening quote from Shapiro) the fact that string theory's history is, in many places, just downright strange and for that reason alone makes for an interesting book! However, I would take issue with Shapiro's comment: regardless or whether string theory is true or not has no bearing on how the history is written up to this point.

difficult subject for outsiders (and, I would guess, many insiders!) to understand. A study of this kind is clearly warranted and I think much needed.[6]

In fact, since Schwarz wrote of the dire lack of historical work on string theory, there has been some emerging interest, resulting in two collections of articles: [1] and [7] (though the latter is more of a *festschrift* for Gabriele Veneziano, who we will meet later: nonetheless, it contains some excellent chapters on the early history of string theory). These books have been enormously useful in providing clues as to the various stages of development of the subject (beyond the "internal" published literature), as has John Schwarz's own two volume collection of important papers from the early days of string theory, pre-1985 [16].[7] There remains, however, no monograph on the subject. One notable near-exception to this is James Cushing's masterly monograph (already cited), *Theory Construction and Selection in Modern Physics* [3], in which he treats the development of S-matrix theory from Heisenberg and Wheeler's early work to the emergence of the dual models that were forerunners of superstring theory. I am a great admirer of Cushing's book, and will take the baton from him, following the story from the dual models he concludes with to *M*-Theory.[8]

I take as my starting point, then, the birth of so-called "dual models" in the late 1960s, which have their origins in the attempt to capture patterns in the physics of hadrons. To make sense of this, however, it will be necessary to first say something about the state of particle physics in the 1960s in general, and about the developments in mathematical physics that paved the way for dual models, and ultimately string theory as we know it today. My ending point is the shift to the notion

---

[6] Copernicus wrote on the frontispiece to his *De Revolutionibus* that "mathematics is written for mathematicians". I think there is an element of this attitude in the string theory literature, making it very hard for non-string theorists to penetrate its labyrinthine structure. For example, one finds David Olive writing that "[a]n ideal introduction [to superstrings] could, with some justice, be subtitled 'The string theory prerequisites for mathematics'"([14], p. 1). John Schwarz (writing in 1986) predicted that "[t]he mathematical sophistication required to be a successful string theorist of the future is so much greater than what has been needed until now that there are sociological consequences worth considering" ([17], p. 200). He was quite right. Part of the purpose of this book is to attempt to expose the inner workings of string theory for the benefit of non-string theorists and those who do not possess the skills of a professional mathematician. A particularly good way of achieving this is to trace its 'life story' (again, following Woodhouse), as we shall attempt here.

[7] I must admit, however, encountering the former book, *The Birth of String Theory* [1], considerably slowed down the production of my own book, since it forced me to revisit and reassess very many parts of it in the light of the various 'from the horses mouth' reminiscences contained therein. I urge readers also to consult this book, as a companion, for more information on how the various architects of superstring theory think about the history of their subject, and their own roles in that history—though with the historian's warning that one must be careful, of course, to avoid what Jeff Hughes and Thomas Søderqvist label 'the living scientist syndrome': "memory and personal archives are notoriously selective" ([10], p. 1). However, Hughes and Søderqvist rightly admit that such recollections can be enormously fruitful in pointing out new lines of inquiry for historians to follow, and for cross-checking with other sources.

[8] However, I should point out, I am not grinding Cushing's methodological axe of 'contingency in physics' in this book—but see "Notes to the Philosopher of Science" below.

of *M*-Theory, signalled by the appearance of various non-perturbative clues brought to light by the discovery of duality symmetries linking together the various perturbative definitions of the several consistent superstring theories (and 11-dimensional supergravity). I leave it for other writers (mathematically better equipped) to follow the story further into this still rather *Mysterious* realm. I make no apologies for this somewhat early curtailment: unlike the period of string theory I cover (from around 1967 up until 1995), the development of *M*-Theory is incomplete in ways that make the historical study of the subject genuinely hard even to contemplate.[9] Despite this curtailment of the story at 1995, the biography of string theory between dual models and the beginnings of *M*-Theory provides very many interesting glimpses into the workings of both physics and pure mathematics, and their practitioners. Let me finish this preface by providing some bespoke reader's notes for the various audiences that I hope this book will attract.

Notes to the General Reader. Though this book is inevitably technical, I have tried to write with a wide readership in mind: there are elements that will interest philosophers, sociologists, and historians, in addition to physicists and mathematicians (the more obvious readership). However, given the nature of the subject matter, the book may appeal to a rather more general audience too. In view of this, I have made every attempt to keep this book self-contained, introducing and defining the necessary mathematical and physical concepts *en route*, integrating them within the general narrative. I encourage readers who fall within this domain to attempt the mathematical parts; but failing this, simply skip the formal details and press on. There is enough "informal" material in the text to provide sufficient information to follow the story to the end: the central focus is firmly on the evolution of *ideas*.

Notes to Historians and Sociologists of Science. Historians of science will most likely be disappointed with the strong whiff of Whiggishness in this book.[10] However, I have tried to make sure the book is not infused with too much presentism. The book is intended to be of some use to a wide variety of readers, so an overly vigilant eye on avoiding "future developments" of the theory as a guide to the selection of past sources would be likely to alienate non-

---

[9] I do provide a very brief review of some key aspects of string theory's development since 1995 in the concluding chapter, but this is less history and more a discussion of certain conceptual issues.

[10] Silvan Schweber defines Whiggish history as "the writing of history with the final, culminating event or set of events in focus, with all prior events selected and polarized so as to lead to that climax" ([19], p. 41). Though it can play an important role in the *pedagogy* of science, it ought really, in excessive quantities, to be given a wide berth in the history, philosophy, and sociology of science. However, in some ways, quantum gravity research programmes, such as string theory, provide an ideal laboratory where Whiggish footholds have not yet had time to fully form: quantum gravity is in many ways (despite the *problem's* relatively long life) a revolution still waiting to happen. The contingencies are often an open book, with loose, uncertain principles not yet fit to function as 'final, culminating events' (save, perhaps, in a local sense).

historians.[11] Rather than restrict myself to the published literature, I have
aimed to use as many sources as possible: interviews, archives, reminiscences,
correspondence, and so on. The aim was to put some flesh of the rather skeletal
historical accounts that one finds thrown in the introductory chapters of textbooks on string theory, and also to deepen such accounts—and, if necessary,
correct them. This is by no means intended to be a "definitive history"
(whatever that might mean). Rather, it is simply this author's attempt at providing a coherent story of a very difficult field: a more objective perspective
will, of course, require the patching together of many more such attempts.

Notes to Philosophers of Science. This book surveys the development of string
theory from the early dual models that prefigured string theory to the more
recent (non-perturbative) work on *M*-Theory. I am most concerned with the
earliest work, and end with only a brief examination of *M*-Theory and related
ideas. Though primarily a work of history, I am also deeply concerned with
methodological issues and more general conceptual issues that emerge from
this history. I think a good knowledge of the historical basis of their subject,
along with the methodological and philosophical ramifications that go hand in
hand with this, could prove invaluable to those string theorists who were not
involved with string theory from its beginnings and perhaps have a different
view of its status (clouded by the immutable appearance one finds in modern-
day presentations of the subject). The present book might thus open up some
new channels for dialog between philosophers and string theorists. Of more
direct relevance to philosophers of science, string theory offers up some
methodological novelties concerning "the way science works". Several of
these novelties would make good case studies, and some might be generaliz-
able to other areas of physics—I shall flag several of these explicitly as we
proceed. Finally, though philosophers of physics (even those working on
quantum gravity) have tended to shy away from string theory, I aim to show
how string theory is rich with interesting philosophical projects, especially
concerning the nature of spacetime.

Notes to Mathematicians. As Gomez and Ruiz-Altaba note in the opening of their
review of string theory, "[t]here is much pretty mathematics in string theory"
([8], p. 2). There has, of course, been a close interaction between pure mathematics and string theory since its earliest days, and this book was written with
mathematicians in mind as well as physicists. Indeed, string theory, especially
in its formative years, is especially intriguing as a case-study in the "physics–
mathematics" interaction. I know of no comparable episode in the history of
science in which the flow of ideas between physics and mathematics is so
prolific *in both directions*, with both mathematical physics and physical

---

[11] There is a wealth of published literature from the dual model and post-anomaly cancellation
eras. I have naturally been forced to be very selective, as previously mentioned. Whenever
possible my choices have been guided by citation analysis. However, often there are articles
included purely for their historical interest, rather than their importance in the development of the
theory.

mathematics.[12] The contents of the penultimate chapter will probably be of most relevance, containing discussions of the ubiquitous Calabi-Yau manifolds, as well as the links between string theory and such areas of mathematics as finite group theory—however, I should point out that groups theory and Calabi-Yau manifolds constitute just one small area of the overlap between string theory and pure mathematics.

Notes to Physicists. Finally, I address this book's most likely audience: physicists (still more likely: string theorists themselves). While I have striven for accuracy in terms of the formal presentation of the theory and the unfolding of events in its life story, errors will undoubtedly have crept in both cases: string theory's history is almost as complex as the theory itself. Note that, unlike Lee Smolin's recent book, *The Trouble with Physics* [20], and Peter Woit's book, *Not Even Wrong* [5], this book is *not* a critique of string theory (if anything, quite the opposite). I have nothing invested in any particular approach to quantum gravity, and am interested in it in a more general way, and more so in the *problem* itself rather than specific proposals to resolve it—in any case, string theory goes beyond the problem of quantum gravity, and initially had no connections with it at all. The book is instead an attempt to understand string theory, in numerous senses of "understand": to understand the theory and its claims; to understand its origins; to understand "influences" on its development, and, perhaps, to better understand its current dominant position in the research landscape. Whether this dominance is or is not deserved is, as mentioned above, not a concern of this book; though I do discuss the emergence of the controversy over string theory's status as one of many events in the life story of superstrings that have shaped its present appearance in the public eye.

Ultimately, however, I hope the richness of string theory's historical development, with its many and varied connections to other fields, will convince readers that it is a theory worth pursuing, whether as a practitioner, or in some other capacity (e.g., as an historian or philosopher of physics).

Sydney, October 2013   Dean Rickles

References

1. Cappelli, A., Castellani, E., Colomo, F., & Di Vecchia, P. (Eds.) (2012). *The birth of string theory*. Cambridge: Cambridge University Press.
2. Cubrovic, M., Zaanen, J., & Schalm, K. (2009). String theory, quantum phase transitions, and the emergent Fermi liquid. *Science, 325*(5939), 439–444.
3. Cushing, J. T. (1990). *Theory construction and selection in modern physics: The S-matrix*. Cambridge: Cambridge University Press.

---

[12] It constitutes what Peter Galison calls "a trading zone" ([22], p. 24).

4. Galison, P. (1995). Theory bound and unbound: Superstrings and Experiments. In F. Weinert (Ed.), *Laws of nature: Essays on the philosophical, scientific, and historical dimensions* (pp. 369–408). Berlin: Walter de Gruyter.

5. Galison, P. (2004). Mirror symmetry: Persons, values, and objects, In M. Norton Wise (Ed.), *Growing explanations: Historical perspectives on recent science* (pp. 23–63). Durham: Duke University Press.

6. Gannon, T. (2006). *Moonshine beyond the monster: The bridge connecting algebra, modular forms and physics.* Cambridge: Cambridge University Press.

7. Gasperini, M. & Maharana, J. (Eds.) (2009). *String theory and fundamental interactions. Gabriele Veneziano and theoretical physics: Historical and contemporary perspectives.* Berlin: Springer.

8. Gomez, C., & Ruiz-Altaba, M. (1992). From dual amplitudes to non-critical strings: A brief review. *Rivista del Nuovo Cimento 16*(1), 1–124.

9. Hand, E. (2009). String theory hints at explanation for superconductivity. *Nature 25*(11), 114008–1140021.

10. Hughes, J., & Søderqvist, T. (1999). Why is it so difficult to write the history of contemporary Science? *Endeavour 23*(1), 1–2.

11. Kragh, H. (2011). *Higher speculations: Grand theories and failed revolutions in physics and cosmology.* Oxford: Oxford University Press.

12. Myers, R. C., & Vázquez, S. E. (2008). Quark soup *al dente*: Applied superstring theory. *Classical and Quantum Gravity 25*(11), 114008-11480021.

13. Natsuume, M. (2007). String theory and quark-gluon plasma. *Progress of Theoretical Physics, Supplement 168*, 372–380.

14. Olive, D. I. (1989). Introduction to string theory: Its structure and its uses. *Philosophical Transactions of the Royal Society of London A, 329*(1605), 319–328.

15. Rickles, D. P. (forthcoming). *Covered in Deep Mist. The Development of Quantum Gravity: 1916–1956.* Oxford: Oxford University Press.

16. Schwarz, J. H. (1985). *Superstrings: The first fifteen years of superstring theory.* Singapore: World Scientific.

17. Schwarz, J. H. (1987). The future of string theory. In Brink, L. *et al.* (Eds.), *Unification of fundamental interactions* (pp. 197–201). *Physica Scripta*, The Royal Swedish Academy of Sciences. Singapore: World Scientific.

18. Schwarz, J. H. (2007). *The early history of string theory: A personal perspective.* arXiv:0708.1917v2 http://arxiv-web3.library.cornell.edu/abs/0708.1917v2.

19. Schweber, S. S. (1984). Some chapters for a history of quantum field theory. In B. DeWitt & R. Stora (Eds.), *Relativity, groups, and topology II* (pp. 40–220). Elsevier: North Holland.

20. Smolin, L. (2006). *The trouble with physics: The rise of string theory, the fall of science, and what comes next.* New York: Houghton Mifflin Books.

21. Woodhouse, R. (2004). *A history of the calculus of variations in the eighteenth century.* New York: American Mathematical Society.

22. Woit, P. (2006). *Not even wrong. The failure of string theory and the search for unity in physical law.* New York: Basic Books.

Acknowledgments

This work would not have been possible without the very generous financial support of the Australian Research Council (via discovery project DP0984930: *The Development of Quantum Gravity*)—this book was written as a key part of that project. I also thank my own institution, the University of Sydney, for additional financial support. The American Institute of Physics, The Max Planck Institute (thanks to Jürgen Renn), and the FQXi funded several mini-projects that were used in the writing of this book, including interviews and archive visits providing much raw data. The Caltech archives were especially useful, and I thank them for their assistance and for the generous Maurice Biot grant that enabled me to conduct research there. Don Salisbury deserves a special thanks for joining me with some of the interviews and making them such good fun. I would also like to add my deep thanks to John Schwarz and Feraz Azhar for their detailed comments on the entire manuscript.

Many of the physicists involved in the history of string theory were very helpful in supplying information. In particular, I would like to thank: Lars Brink, Elena Castellani, Stanley Deser, Prashant Kumar, Dieter Lüst, Holger Nielsen, Roland Omnès, Joseph Polchinski, and Mahiko Suzuki. If I have failed to mention others, my apologies. I also thank Angela Lahee at Springer for her limitless patience and vital nudges during the (protracted) completion of this book. Finally, as always, I would like to thank my long-suffering family for putting up with my incurable workaholism.

Contents

Chapter 1
History and Mythology

The superstring theory has perhaps the weirdest history in the annals of science.

Michio Kaku

1.1 Serendipity and Strings

Gabriele Veneziano is widely heralded as the man that 'gave birth' to string theory in 1968, with the publication of his paper "Construction of a Crossing-Symmetric, Regge-Behaved Amplitude for Linearly Rising Trajectories" [27], in which he introduced a formula that included virtually all of the desirable features for describing the behaviour of hadronic scattering then known. He began his PhD research just 2 years earlier, in 1966, at the Weizmann Institute in Rehovot, Israel, and had been inspired into action by a remark of Murray Gell-Mann's while the latter was lecturing at a summer school in Erice in 1967. Gell-Mann happened to mention the (finite energy sum rule: FESR) duality conjecture of his Caltech colleagues, Richard Dolen, David Horn, and Christoph Schmid, relating s-channel resonances to t-channel Regge poles (in the context of pion-nucleon scattering).[1] Veneziano was interested in the possibility that this duality was a general feature of the strongly interacting world, in which case one ought to see it exhibited within meson-meson processes (albeit initially

---

[1] The duality (to be discussed more fully in the next chapter) states that what were seen to be distinct diagrams (at least in the context of orthodox Feynman diagram-based quantum field theory) were really two representations (or rather *approximations*) of one and the same underlying process. The so-called Harari-Rosner duality diagram that represents this equivalence class of diagrams (the dual processes, captured by Veneziano's formula) played a crucial role in the genesis, early development and understanding of string theory *qua* theory of one dimensional objects and their two dimensional worldsheets. Indeed, the original introduction of the worldsheet concept used it to provide an *explanation* of duality (via conformal symmetry and the extreme deformability of worldsheets)—see Sect. 4.1.

D. Rickles, *A Brief History of String Theory*, The Frontiers Collection, DOI: 10.1007/978-3-642-45128-7_1, © Springer-Verlag Berlin Heidelberg 2014

only in thought experiments). This train of thought was leading (in Veneziano's mind) to a 'bootstrap' according to which consistency conditions (on the S-matrix for such processes) would determine what one measures in actual experiments involving mesons.

Veneziano joined together with a team of fellow strong interaction enthusiasts on his journey back to the Weizmann Institute, including Marco Ademollo, from Florence, and Hector Rubinstein and Miguel Virasoro also from the Weizmann Institute.[2] They probed this newly conjectured DHS [Dolen, Horn, Schmid] duality (linking resonances at low energy and Regge poles at high energy) further, and discovered that mesons had to lie on a linearly rising Regge trajectory (for which spin is plotted against mass squared on a so-called 'Chew-Frautschi plot') with the slope given by a universal parameter α' with some particular intercept $\alpha'(0)$ (both of which will be seen to continue to play an extremely important role in the development of string theory). Given this seemingly well-regimented behaviour, Veneziano naturally reasoned that there would likely exist a neat mathematical formula summing up all of the nice physical properties.[3]

Though there has been no scholarly historical study of the origins of string theory, as with most scientific theories, there is a 'mythology' surrounding its discovery, repeated over and over again until it becomes firmly entrenched. The standard story one can find in the brief historical passages that appear in the string theory literature recount how in 1968, Gabriele Veneziano, then a mere slip of a student as I mentioned, found *entirely by accident* a connection between the previously mentioned regularities and desirable properties of hadronic physics, and the Euler beta function. For example, Michio Kaku (himself responsible for some important work in superstring theory) puts it like this:

> Thumbing through old mathematics books, they [Veneziano and Mahiko Suzuki] stumbled by chance on the Beta function, written down in the last century by mathematician Leonhard Euler. To their amazement, they discovered that the Beta function satisfied almost all the stringent requirements of the scattering matrix describing particle interactions. Never in the history of physics has an important scientific discovery been made in quite this random fashion [12, p. 4].

Veneziano has himself, not without some wry amusement, corrected this particular myth.[4] Much like Fleming's discovery of penicillin, one needs to know *what one is*

---

[2] The "gang of four," as Veneziano himself describes them [28, p. 185].

[3] This, in part, reflects the belief, true for most theoretical physicists, in deep links between mathematics and physics: where there is some piece of physics to be described (involving certain lawlike features), there will usually be a piece of mathematics that can do the job, if one only looks hard enough. Eugene Wigner referred in this context to "the unreasonable effectiveness of mathematics in the natural sciences" [29]. In fact, Dirac had earlier drawn attention to this feature, in 1939, pointing to "some *mathematical quality in Nature*" that enables one to infer results "about experiments that have not been performed" [4, p. 122]—note: it is from this basis that Dirac develops his principles of *simplicity* and *beauty* that characterise his work. See [8] for a superstring-relevant discussion of this subject.

[4] See, for example, his talk, "The Beginning of String Theory or: How Nature Deceived us on the Sixties": http://online.itp.ucsb.edu/online/colloq/veneziano1. In [28] he writes, with somewhat less

looking for to know that one has found it! To borrow Pasteur's famous phrase, even the most serendipitous discovery requires a "prepared mind".[5] There was nothing random or accidental about the discovery itself, though there may have been multiple elements of serendipity in both cases. For example, it is indeed rather remarkable that Euler's formula, discovered in entirely independent circumstances, 'fit the physics' desiderata so well—again, one of very many instances of Wigner's 'unreasonable effectiveness' (or Dirac's 'mathematical quality in Nature')![6]

What Veneziano (and Suzuki[7]) found was a four-particle scattering amplitude[8] that provided a model of what was expected (given features of the S-matrix, describing collision processes) to be present in an adequate theory of strong (hadronic) interactions. We will look closely at this model in Chap. 3, seeing amongst other things how the Beta function manages to perform this particular feat of representation. Firstly, in this chapter, we give a brief introductory overview of some basic historical and methodological niceties involved in the study of string theory, along with a rough picture of the way the history of string theory will be 'sliced up' in this book. We also present some very elementary physics and mathematics of strings, including a rough guide to the contemporary conception. Finally, to whet the reader's appetite, we give a brief *preliminary* snapshot of string theory's history.

In Part I, Chap. 2 will then focus on the phenomenological 'model building' period of particle physics in the 1960s from which string theory emerged, and many of whose basic principles modern day string theory still embodies. Chapter 4 examines the transitionary period in which the Veneziano dual model (Chap. 3) was reinterpreted as

---

(Footnote 4 continued)

amusement than in the aforementioned, that "[o]ne of the things that upsets me most these days is to hear that I found the Beta-function ansatz almost by chance, perhaps while browsing through a book on special functions. ...[T]here is nothing further from the truth. Had one not found a simple solution for the (average) imaginary part, and recognised the importance of imposing crossing on the full scattering amplitude, the Beta-function would have stayed idle in maths books for some time..." (p. 185).

[5] See Merton and Barber's study of serendipity for more on this notion of 'chance' versus 'preparedness' [17, see especially p. 259].

[6] Claud Lovelace [15, pp. 198–199] describes a remarkably similar route to his discovery [14] that the Poincaré theta series, which provides a method for constructing Abelian integrals, which can in turn be used to solve hydrodynamics on non-simply connected Riemann surfaces (which Holger Nielsen had already shown to correspond to "momentum flowing across the world-sheet") provides a way of visualizing the Veneziano formula. This work presaged much later work lying at the intersection of mathematics and string theory (especially that involving automorphic forms and other sectors of number theory).

[7] Mahiko Suzuki did not publicly present nor publish his result as Veneziano did. Like Veneziano, Suzuki ceased active research in the field of dual resonance models not so long after discovering the Beta function ansatz. Joseph Polchinski noted that when taking a course on dual theory given by Suzuki in 1980 (according to Suzuki on "S-matrix theory *á la* Chew with less emphasis on nuclear democracy"—private communication), he had prefaced the course by stating: "This is the last time this will ever be taught at Berkeley" (Interview with the author; transcript available at: http://www.aip.org/history/ohilist/33729.html). I describe Suzuki's route to the discovery of the formula in Chap. 3.

[8] That is, a formula describing the probabilities for a process with both an input and output of two particles.

(or, more precisely, *derived from*) a theory of strings (a quantized theory of massless, relativistic strings), albeit still describing hadrons at this stage. Chapter 5 looks at early forms of supersymmetric string theory, in which a cluster of difficult problems (absence of fermions and problems with tachyons) is disposed of—this also includes a discussion of the so-called 'zero-slope limit' in which the reduction of dual models to standard field theories was explored.

Part II will then take up the story (in Chap. 6) from the dual string theory's elevated position to its fall (as a result of the experimentally better qualified quantum field theory underlying hadrons: quantum chromodynamics [QCD][9]). Chapter 7 covers the more slowly-moving, relatively quiet transformative period in which string theory was converted into a theory of gravitation (and other gauge interactions) to (in Chap. 8) its seemingly phoenix-like rise in the form of (anomaly-free) superstring theory, where it was beginning to be widely recognised that the theory potentially offered up a genuinely plausible 'theory of everything.'

Part III begins, in Chap. 9, by studying the period in which superstring theory established its stronghold in the research landscape, with a series of discoveries (heterotic strings, Calabi-Yau compactification, and more) pointing to the possibility of realistic low-energy physics (forming, together with the anomaly cancellation results, the 'first superstring revolution'). Finally, Chap. 10 culminates in what is labeled the 'second superstring revolution,' ignited by the (re)discovery of D-branes and new duality symmetries revealing previously hidden non-perturbative aspects and surprising interconnectedness in the theoretical edifice of superstring theory (in fact, pointing *beyond* a simple picture of strings). This final chapter will also bring the story, more or less, up to the present—including brief discussions of black holes in string theory, the AdS/CFT duality, and some of the controversies string theorists find themselves embroiled in (e.g. the 'string landscape' and the anthropic principle). With Bohr's maxim in mind,[10] we end with some speculations on what the future might bring.

1.2 The Four Ages of Strings

String theory was conceived from the union of data from strong interaction scattering experiments and, we will see, the principles of S-matrix theory (together with the 'resonance-Regge pole' duality principle). Ironically, then (given the current critiques based on its dire empirical status[11]), string theory began life as a largely phenomenological endeavour: strongly data-driven. Clearly then, in order to

---

[9] Though as we shall see, this 'experimental superiority' claim is not so simple. QCD, and the theory of quarks, performed exceptionally well when applied to high-energy (deep) scattering experiments, but not so well in low-energy situations, due to its still ill-understood property of colour confinement. In these situations, string theory holds up remarkably well and, in a sense to be explained, was integrated into QCD to deal with such regimes.

[10] Namely: "Prediction is hard. Especially about the future".

[11] See, for example [1, 24, 30].

understand where string theory comes from, and why it looks the way it does today, we need to first cover some preliminary ground involving its ancestors in hadronic physics and S-matrix theory (which we attempt in the next chapter).

However, we cannot properly understand the emergence of string theory (*qua* 'theory of everything') without saying something about its curious interaction with pure mathematics, especially the theory of finite simple groups and lattices—this interaction was something that began very early on in the life of string theory and, indeed, it doesn't take long for empirical doubts about dual theory and string theory—e.g. claims about pursuing mathematical problems in string theory for their own sake—to creep into the history, despite its initially data-driven origins.[12] We will see that much of the strength of string theory flows from its immensely powerful mathematical structure, which can often be found to lie at the root of string theorists' trust in their approach. For this reason we devote a considerable amount of space to it. Besides, one cannot properly appreciate string theory's *physical* claims without also understanding the mathematical structures that support them.

The evolution of string theory can be broken up ('periodised') into four broad phases,[13] according to which the beginning of each new phase (aside from the first, which amounts to the 'origination' of the theory) heralds the resolution of some severe problem with the theory (most often some *mathematical* inconsistency, but also inconsistency with known data resulting in empirical inadequacy):

- Phase 1 [1968–1973]:
  - Phase 1A (Exploring Dual Models) [1968–1969]: the 4-point dual resonance model for hadrons is discovered by Veneziano, quickly generalised to N-point amplitudes, factorised, made unitary (with the n-loop case considered), and represented in terms of a infinite set of oscillators, which are then given an initial 'unphysical' string interpretation. The Paton-Chan procedure enables isospin factors to be added. The problem of spurious states ('ghosts') is revealed in the operator formalism, and quickly resolved via gauge fixing. This introduces a tachyon.

---

[12] The history of this fruitful interaction deserves a book of its own, and I only discuss those portions of the history that are directly relevant to the development of string theory, and must omit very many interesting examples.

[13] I make no claim for uniqueness or canonicity with this periodization. It is somewhat arbitrary, though it tries as much as possible to home in on genuine 'critical points' in the history of string theory and is underwritten, as far as is possible, by citation analysis highlighting these critical historical points by their impact on the research literature. The citation analysis reveals an explosion (or implosion) in publication numbers and citations at the outset of each new period. Of course, restricting the analysis to the published literature does not reveal the fine structure that moves a discipline, and for this reason I also utilise a range of 'external' sources, including interviews and archives. Note also that I do not always stick to a strict chronological ordering, but often cluster according to thematic links.

Virasoro discovers a second dual model, which faces the same issues as the (generalised) Veneziano model which are (broadly) dealt with in the same way. A split between an abstract operator approach and a more geometrical (string-picture) approach appears. A Feynman diagram approach is introduced, highlighting topological features of dual models.

- Phase 1B (Embryonic String Theory) [1970–1973]: The problems with dual models are brought into clearer focus, and tackled with great speed. The critical dimension $d = 26$ is discovered to be required for consistency of the theory. An action is constructed, based on minimization of string worldsheet area. New fermionic (spin) dual models are constructed, along with the fermion-emission vertex. Adding fermions is seen to have radical implications for the structure of the theory: the critical dimension shifts from 26 to 10, and (part of) the tachyon problem is eradicated. The critical dimension is also given a physical explanation in terms of zero-point energy. The 'no-ghost theorem' is proved. An internally self-consistent quantum theory of a free, massless, relativistic string is constructed and shown to reproduce the physics of dual models. (Externally, this phase also leads to *hadronic* string theory's development radically slowing down at the hands of several remaining internal difficulties coupled with the rise of quantum chromodynamics. However, the dual models' zero slope (low energy) limits of Yang-Mills gauge fields and Einstein gravity were discovered during this phase. There are also attempts to recover QCD-type phenomena from string models.)

• Phase 2 (Theoretical Exaptation) [1974–1983]: the dual resonance framework is rescaled so that it describes quantum gravity rather than hadrons. The hadronic string work continues, and integrates with gauge theory. A geometrical string picture based on functional-integral techniques (incorporating gauge elimination techniques) is devised. An interacting string picture is introduced. Spacetime supersymmetric string theory is introduced (using the GSO projection) and seen to eliminate the remaining tachyon (giving a consistent theory). It is realised that some string theories are afflicted with a chiral anomaly. During this period, additional important, though low key, work continues, including further work on the hadronic string, on the problem of dimensional reduction of the extra dimensions, and on new formulations of superstring theory and superstring field theory (by Polyakov and by Green and Schwarz). At an external level, supergravity theories are introduced and undergo a dramatic rise. They both inherit from and donate many new elements to string theory (including physicists who switch alliegance).

- Phase 3 (Superstring Phenomenology) [1984–1994]: string theory is discovered to be anomaly free. It is taken seriously as a theory of everything on account of the way the anomaly is resolved (requiring, in one instance, a phenomenologically promising gauge group). Deeper physical results are found with the introduction of the 'heterotic' string concept, embodying the gauge group. Much effort is spent on finding compact spaces (primarily Calabi-Yau manifolds) that improve string theory's phenomenological prospects. However, uniqueness is lost as a result of the compactification procedure. This is partially resolved via newly discovered equivalences, linking theories on different manifolds, which are, in part, made possible by newly devised conformal field theory techniques. Some early non-perturbative explorations involving D-branes and dualities improve this linkage, but also point to a very different kind of theory *beyond* string theory.
- Phase 4 (Beyond Strings) [1995–present]: string theory is understood to contain objects, Dp-branes, of a variety of dimensionalities (of which strings are a single example, for $p = 1$). More dualities are introduced, leading to a conjecture that the different string theories (and a further 11-dimensional theory) are simply limits of a deeper theory: M-theory. The new ideas are put to work on the study of black holes, including a successful reproduction of the Bekenstein-Hawking entropy (using a D-brane/black hole correspondence). The black hole information paradox is also explored in this new framework, and given a possible resolution that preserves information in Hawking radiation. This is linked to the Maldacena conjecture, relating a four-dimensional gauge theory and a 10-dimensional superstring (gravity) theory. This opens up new possibilities for empirical work for superstring theory and also leads to new possibilities for defining string theory. A problem with the stability of the compact spaces emerges and is fixed using D-brane ideas once again. However, the solution means that string theory proliferates into a vast 'landscape' of vacua. The anthropic principle is introduced as a way of recovering the vacuum state corresponding to our world. This introduces acrimonious debates concerning the scientific status of string theory, which are still ongoing.

These phases will form the structure of the rest of the book, with the proviso that there will sometimes be thematic grouping, as mentioned above. A few preparatory words on each of these phases is in order, since these phases, as presented, obviously hide an awful lot of 'fine structure'. For example:

- Phase 2, the rescaling of the theory[14] to render it able to cope with a distinct problem is a kind of 'theoretical exaptation' in which a theory can be seen to struggle for survival by modifying its function (see Chap. 7). In this case, the strings, previously held to describe hadrons, come to perform an entirely different role, describing all interactions, including gravity, and matter. I prefer to say that this in fact results in a new theory, despite the fact that one can tell a story according to which hadronic strings are a direct ancestor (related through the 'evolutionary parameter' α').
- Phase 3 is rather remarkable since the transition is marked not by any empirical discovery but by the resolution of a mathematical problem (though one with obvious physical ramifications). This highlights quite clearly the way constraints other than empirical ones (e.g. freedom from tachyons, ghosts, anomalies, and so on) can play a crucial role in the development of scientific theories.[15]

I end the story around 1995, leaving phase 4 largely unexplored, since beyond this the development of superstring theory—already a rather unwieldy entity from a historical point of view—becomes virtually intractable (at least for my meagre powers of analysis). I leave this as a task for future historians, once the dust has had time to settle; though I do provide a brief review and assessment of some of this recent work. I should also note that I begin with an earlier phase of research (the 'zeroth' phase), showing the pre-history leading to the emergence of dual models from Regge theory and Geoffrey Chew's 'bootstrap' approach to strong interaction physics, many of whose concepts formed the basis of the dual models at the root of string theory.

1.3 The Physics and Mathematics of Strings

Strings and their motions have been a staple of mathematical physics since the eighteenth century, when the formal description of their vibrational motion was one of

---

[14] That is, of the slope parameter in the theory, α', determining the scale at which 'stringy' effects appear.

[15] Peter Galison, in one of the few early historical studies of superstring theory, refers to the increased importance of extra-empirical constraints in string theory as "a profound and contested shift in the position of theory in physics" [6, p. 372]. Helge Kragh reiterates this in his recent book on theories of everything, writing that "[p]ublications by many physicists ...show a tendency to unrestrained extrapolation of physics into domains that according to the traditional view [of scientific methodology] are inaccessible to the methods of physics" [13, p. 367]. Here I also point the reader towards Richard Dawid's recent book [3], which focuses on the issue of how these non-empirical theory assessment criteria can lead to a reasonable level of trust in string theory. See also [20], in which I argue for the legitimacy of (a certain level of) trust in string theory (despite the absence thus far of confirmation via novel experiments predicted by the theory). A more general examination of philosophers accounts of scientific methodology in the light of string theory is [11]. John Schwarz has pointed out that experiment has undergone its own kinds of shifts—e.g. to big collaborative experimental ventures, distributed in a modular fashion across local experts, necessarily involving a different approach to the gathering of evidence—that modify its methodological foundations too, so that the two situations (theory and experiment) are not so different [23, p. 201].

the chief outstanding problems of the time. There was a deep debate over how best to resolve this problem, involving (amongst others) D'Alembert, Euler, and Daniel Bernoulli initially; and later, Fourier and Dirichlet. Furthermore, the solution to this problem inspired the field of mathematical analysis in profound ways, not entirely unlike the manner in which string theory has itself inspired algebraic geometry and other areas. Let us present some of the details of the classical theory of strings as a warm-up exercise for our main topic. We then give a quick presentation of the basic ideas of modern day string theory to give a better sense of what this history is a history *of*.

The problem of modelling the motion of a classical string is a useful starting point for understanding some of the details of string theory.[16] Let us consider an elastic string with fixed endpoints A and B, separated by a distance d, with the line connecting A and B providing an x-axis (such that $0 \leq x \leq \pi$), and with a y-axis perpendicular to this. Let the string be stretched tightly, and put into a small vibrating motion by displacing it from equilibrium along the y-axis. Determining the solution $y(x, t)$ of the equation in which y is expressed as a function of x and time t, provoked some intense debate amongst eighteenth century mathematicians.[17]

D'Alembert discovered the general form of the equation in which the relationship could be expressed, which we would now write as[18]

$$\frac{\partial^2 y}{\partial x^2} = \frac{1}{c^2} \frac{\partial^2 y}{\partial t^2} \tag{1.1}$$

This equation should 'contain' all the possible shapes that the string can possess (given the boundary conditions at A and B). Daniel Bernoulli devised the most general solution, $y(x, t)$, for a vibrating string with fixed endpoints (with vertical displacement y). Ignoring the string length, l, and setting $c = 1$, we can write this simply as the trigonometric solution:

$$y(x, t) = sin\ x\ cos\ t \tag{1.2}$$

This provides the first harmonic (or fundamental mode) of the string.[19] The second harmonic would then simply be

---

[16] A stretched string can of course possess multiple modes of vibration (the *harmonics*) which is analogous to the various modes of vibration corresponding to different particles (infinitely many of them) in the case of quantized superstrings.

[17] See [7] for a nice account of this episode, along with its impact on the development of analysis. The same episode served as a background for the development of the notion of a mathematical function.

[18] Here the constant c is the speed at which waves may propagate across the string. It has a value equal to the square root of the string tension divided by mass per unit length.

[19] The fundamental mode of the string was in fact discovered by Brooke Taylor (of Taylor series fame) and laid out, using Newton's method of fluxions, in his *Methodus Incrementorum* of 1715. However, Taylor made no attempt to account for the overtones in which a string can vibrate in many ways simultaneously—a notion that required the concept of superposition, not yet discovered.

Fig. 1.1 Daniel Bernoulli's depiction of the various possible vibrations of a string, along with their superposition (reproduced in [10, p. 126])

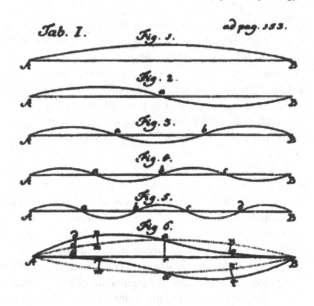

$$y(x, t) = \sin 2x \cos 2t \tag{1.3}$$

The general solution would be an infinite sum of such harmonics

$$y(x, t) = \sum_{n=1}^{\infty} a_n \sin nx \cos nt \tag{1.4}$$

All of these harmonics can then be superposed, as depicted in the diagram reproduced in Fig. 1.1. As George Mackey has noted, Bernoulli's understanding of linearity, allowing the construction of such superpositions, was not accepted at the time, delaying what could have been the beginning of the application of harmonic analysis to the solution of linear partial differential equations by over half a century [16, p. 563].

Many of these same concepts naturally flow into string theory (at least in its earliest phases). The strings of this theory are also able to vibrate in an infinite number of normal modes and frequencies and, as is well known, the particles of 'low-energy physics' are understood to be quantized modes of oscillation (an infinite tower of them, each with different values of mass and spin) of the strings. However, string theory is also a relativistic quantum theory, and the dual incorporation of quantum and relativistic aspects significantly alters the meaning of the above classical string concepts.[20] Relating to the quantum aspects, string theory can be consistently formulated only in 26 or 10 spacetime dimensions (the 'critical dimensions' for bosonic and supersymmetric [i.e. with both bosonic and fermionic degrees of freedom] strings respectively).

The string length is of vital importance in physical applications of string theory, setting the scale at which stringy aspects are non-negligible. More precisely, the tension parameter T in the string sets the scale (where the tension has dimensions: $[T] = [M^2] = [L^{-2}]$)—this manages also to impose a cutoff to control ultraviolet divergences. Note that the tension also determines the gap between the frequencies of the string's normal modes.[21] In the early string model of hadrons the string has a tension such that as pairs of quarks increase their distance from one another (stretching the string) the string stores potential energy—this energy (per unit length of string) is held constant over stretching. One interesting feature of this is that it requires infinite energy to separate two quarks. This was an early phenomenological success of the string model of hadrons since quarks always come in combinations with zero net colour charge, a property known as *confinement*.[22]

A more recent understanding of string theory is that it is a quantum theory of gravity, and other interactions, which can be understood, in one sense, in terms of the theory involving the three fundamental constants of nature: $\hbar$ (Planck's constant, governing quantum behaviour), c (the velocity of light in vacuo, governing relativistic effects), and G_N (Newton's constant, governing gravitational effects). As Planck himself knew, these constants can be uniquely combined so as to generate fundamental length, mass (energy), and time scales. However, there is no sense in which general relativity is *quantized* in string theory. Rather a theory of spinning strings is quantized and this theory has limits (roughly in which the string size goes to zero)

---

[20] We might mention Dirac's [5] earlier attempt to come up with a higher-dimensional theory (modelling it as a charged conducting surface in fact) of the electron according to which Rabi's question 'Who ordered that?' (of the muon) could be answered by treating it as the first excited state of the electron (associated with the size and shape oscillations of what is visualizable as a kind of bubble). Crucially, such a system had a finite self-energy. The idea was to have the surface tension counterbalanced by electrostatic repulsion, so as to avoid the problem of instability inherent in three-dimensional rotating objects—I should add, of course, that strings, being one-dimensional objects, do not need a special counterbalancing force since the centrifugal force alone can prevent the collapse. David Fairlie recalls attending a seminar by Dirac on this subject, accompanied by David Olive [18]—both Fairle and Olive are important figures in string theory's history, and we shall meet them again.

[21] Initially, given that string theory was applied to hadron physics, $\sqrt{T}$ was set to the mass scale of hadrons (which translates to distances of 10^{-13} cm). In the refashioning of string theory as a quantum gravity theory (amongst other things) John Schwarz and Jöel Scherk set $\sqrt{T} = 10^{19}$ MeV $= M_{\text{Planck}}$ (in terms of distances, this is 10^{-33} cm: a 20 order of magnitude adjustment!), since the Planck mass sets the scale of quantum gravitational effects.

[22] Ultimately, of course, quantum chromodynamics (with its asymptotic freedom) would win out—see 't Hooft [26] for a nice discussion of the 'rehabilitation' of field theory, leading to the construction and acceptance of the quark theory and its basis in QFT (this discussion also includes the role of string theory).

Fig. 1.2 A worldtube, worldsheet, and worldline, as generated by the time evolution of a
closed string, open string, and
point-particle respectively.
Time goes up the page, and
space across. Of course, it is
the added spatial distribution
that accounts for the special
qualities of string theory

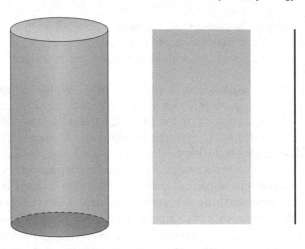

in which general relativity (and Yang-Mills theories) emerge.[23] In other words, the
classical gravitational field is part of an effective description of physics at lower
energies, that owes its existence to the string theory operating at higher energies.[24]

Of course, the existence of a length to the fundamental objects allows them to
span space (parametrized by a new coordinate σ), and not only time. This results
in a world*sheet* or world*tube* rather than a worldline (as in the point-particle case:
Fig. 1.2), and is responsible for many of the 'miracles' (mathematical and physical)
that string theory is able to perform. The fundamental idea of string theory is, then,
simple enough: instead of a local quantum field theory of point-like particles, the
theory employs one-dimensional objects, that can be open or closed—though, by
the basic splitting and joining mechanism, open string interactions (corresponding
to Yang-Mills fields) also imply the existence of closed strings (corresponding to
gravitation: hence, *all* string theories contain gravitation). Whereas in the quantum
field theory of point-like excitations we have worldlines, meeting at single distinguished points (the *vertices* at which interactions happen, with a strength determined
by the coupling constant of the relevant theory), in string theory, given the worldsheet
description (for open strings) and worldtubes (for closed strings), the interactions do
not happen at some single point: the question of whether or not there are interactions

---

[23] More precisely, certain modes (in this case spin-1 and spin-2) survive the limit-taking procedure
and become massless particles in the low-energy limiting theory. These particles turn out to have
exactly the properties of gravitons (mediating the gravitational force) and gauge bosons (such as
the photon). That there exist massless (and therefore infinite range) particles is clearly not good for
describing hadrons (which are short-range forces). This was one of the many factors responsible
for the demise of string theory as a fundamental theory of hadrons.

[24] Or, as Edward Witten succinctly puts it, string theory is "a quantum theory that looks like
Einstein's General Relativity at long distances" [30, p. 1577].

INTERNATIONAL CONFERENCE ON SYMMETRIES AND QUARK MODELS

WAYNE STATE UNIVERSITY, DETROIT, MICHIGAN, U.S.A.

June 18-20, 1969

Fig. 1.3 Yoichiro Nambu who was the first to suggest that the dual models could be understood via a theory of strings, here photographed at the conference at which the string idea received its first public airing. *Image source* R. Chand (Ed.) [2]

is answered by global properties (invariants) of the worldsheets and worldtubes—specifically, the genus g (or number of handles) of the surfaces.[25]

Like standard quantum field theory, string theory too is often presented in a perturbative fashion, with scattering amplitudes given by a series of string worldsheet diagrams, rising in powers of the coupling constant g_s of the theory. This much is in direct analogy with the point-particle case. However, where they differ is that in the case of string theory there is just a single diagram (defined by its genus) at each order of the perturbation series, thanks to the invariance properties of the theory that allow one to deform apparently distinct diagrams into one another.[26]

---

[25] The fact that the interactions are 'smeared' (or 'soft,' as is often said) renders string theory unable to cope with the hard-scattering events leading to the ultimate hadronic crowning of the quark theory, with its point-like particles. However, the same softness also impacts positively on the properties of the amplitudes, making their high-energy (short-distance) behaviour far better than for point-like theories, which must model interactions as occurring at spacetime points. This improved behaviour (which controls the usual problematic divergences) was a major motivating factor in the continued pursuit of string theory beyond hadrons. (I should, however, point out that recent work on the so-called gauge/string duality points to string theory's ability to cope with hard scattering—see, e.g., [19].)

[26] This is in exactly the same sense that a doughnut is 'the same' (topologically speaking) as a coffee cup: if all that matters to you is the holes, then they are identical. Though it was initially implicit, the conformal invariance of the theory was understood early on and heavily constrained its development.

A rather abstract, but condensed formulation of string theory can be given in terms of mappings of string worldsheets into spacetime—quantum theory and the invariances of the classical theory impose conditions on the nature of the spacetime. The initial step is to consider a map Φ from a Riemann surface Σ into the 'target space' X (on which are defined a metric G and additional background fields B^i, though we can ignore the latter for simplicity):

$$\Phi : \Sigma \longrightarrow X \tag{1.5}$$

This represents a string worldsheet sitting in spacetime.[27] For a spinning open string, for example, the map's image in X (taking as inputs (σ, τ)-parameters) would look like a helical corkscrew structure. The action, from which the equations of motion are derived, is then a function of this map, given the metric G:

$$S(\Phi, G) \tag{1.6}$$

Just as the classical action for a free point particle is essentially just the length of the worldline and is stationary for particles moving in straight lines, so the string action is essentially just the area of the worldsheet and is stationary for strings moving in a way that minimizes the area of their worldsheets—where one uses the metric on X to *induce* a metric on the embedded string worldsheet, enabling area measurement. One can then derive the dynamics of string propagation from this action. The Φ-field describes the dynamics of a two-dimensional field theory of the worldsheet relative to the fixed background metric on spacetime.

The quantum theory is then given by the path-integral[28]:

$$\mathscr{P}(X) = \sum_g \int_{moduli_g} \int \mathscr{D}\Phi\, e^{iS(\Phi, G, B^i)} \tag{1.7}$$

There are many consistency conditions that must be met by such string models. The most important of these concern the restriction of the number of Lorentz dimensions of X in order to resolve the conformal anomaly (i.e. the breakdown of classical

---

[27] This worldsheet has a metric $h_{\alpha\beta}$ defined on it in the version of string theory due to Alexander Polyakov. In the original Nambu-Gotō version the worldsheet was metric-free, with distance measurements made possible by the metric it inherits from the embedding into spacetime.

[28] In order that only *physical quantities* are included in the sum, it is performed over 'moduli space': i.e. the space of *inequivalent* 2D Riemann surfaces. The orbit space of metrics modulo conformal and diffeomorphisms symmetries is known as Teichmüller space. Moduli space is more tightly circumscribed, involving also 'large' diffeomorphisms (those not connected to the identity). When one further quotients Teichmüller space by the modular group of transformations, one has moduli space (of a Riemann surface), over which the path-integral in Eq. 1.7 is performed. As we will see, mirror symmetry (and other string dualities) have the effect of producing unexpected identifications of points in the moduli space of a string theory, further reducing it. If interactions are included, then the sum will include holed surfaces.

conformal symmetry in the transition to a quantum theory): 26 (in the bosonic case) or 10 (in the bosons + fermions case).[29]

Superstring theory is consistent, then, only if spacetime has 10 dimensions. To construct a realistic theory therefore demands that the vacuum state (i.e. the vacuum solution of the classical string equations of motion, supplying the background for the superstrings) is given by a product space of the form $\mathcal{M} \times \mathcal{K}$, where $\mathcal{M}$ is a non-compact four dimensional Minkowskian (or possibly more general) spacetime and $\mathcal{K}$ is a compact six dimensional real manifold. One gets the physics 'out' of this structure via topological invariants and gauge fields living on $\mathcal{K}$, choosing the specific form of the compact manifold to match the observed and expected low-energy phenomena in $\mathcal{M}$ as closely as possible. For example, if one wants $\mathcal{N} = 1$ supersymmetry in the non-compact dimensions, $\mathcal{M}$, then one requires a very special space for the compact dimensions $\mathcal{K}$, namely a Calabi-Yau manifold.[30]

Somewhat problematically, there are five quantum-mechanically consistent superstring theories (in 10 dimensions: here we ignore the purely bosonic case, lacking leptons and quarks), differing in the kinds of strings they can possess: Type I, $SO(32)$-Heterotic, $E_8 \otimes E_8$-Heterotic, Type IIA and Type IIB. The Type I theory and the heterotic theories differ from the Type II theories in the number of supersymmetries, and therefore in the number of conserved charges. One is able to compute physical quantities from these theories using perturbation expansions in the string coupling constant around some background. More recent developments in string theory have attempted to link these various theories together (using duality symmetries, expressing physical equivalence) in a bid to restore uniqueness. It is now believed by most string theorists that there exists a single overarching theory (M-theory) containing each of these superstring theories (and also 11 dimensional supergravity) as approximations, valid for particular parameter values. Such a theory would clearly look very different from these superstring theories. In the subsequent chapters we will explore how this structure came into being.

1.4 Historical Snapshot

Though all four interactions were known in string theory's 'prenatal' (<1968) stage, they were pursued, on the whole, as independent lines of inquiry. Physicists were partitioned into groups according to the particular interaction they worked upon: the

---

[29] Though Polyakov studied 'non-critical' string theories which departed from these constraints, there are other constraints that must be enforced to retain the conformal symmetry (which involves a so-called 'Liouville mode' on the worldsheet)—see Sect. 8.2.

[30] This is defined to be a compact n-dimensional complex Ricci flat manifold with Kähler metric, with trivial first Chern class. Ricci flatness means that the metric is a solution of the vacuum Einstein equations for general relativity. The first Chern class $c_1(\mathcal{X})$ of a metric-manifold is represented by the two-form $(1/2\pi)\rho$ (with ρ the Ricci tensor $R_{i\bar{j}}dz^i \wedge d\bar{z}^j$). Calabi proved Yau's conjecture that when $c_1(\mathcal{X}) = 0$, there exists a unique Ricci-flat Kähler metric for any choice of compact Kähler space. If one has a Ricci flat metric then one also gets the desired single supersymmetry since Ricci flatness is a sufficient condition for an $SU(3)$ holonomy group. See [9] for a good, friendly introduction to all things Calabi-Yau.

tools and concepts that worked for one group and interaction did not necessarily transfer to other groups and interactions. Recall there was as yet no unification via the concept of gauge fields and Yang-Mills theory.[31]

The state of strong interaction physics prior to Veneziano's discovery can be characterised by a series of phenomenological discoveries, with theoretical understanding lagging behind the production of data. In particular, experimental plots of particle masses (squared) against spins were found to fall into certain apparently universal patterns, known as linear Regge trajectories, that could not be derived from standard quantum field theoretical models. No other model was known at the time that could explain these features. Much effort was channeled into a departure from quantum field theory: S matrix theory.[32]

In 1968 the Veneziano model made significant inroads on this problem and was able to reproduce the patterns in the data. It was in fact an *infinite spectrum* model, describing the scattering of particles of ever-increasing spins. But there was no physical picture attached to the model, and it was initially investigated as an abstract structure using a powerful operator formalism that Veneziano also helped to develop. It was quickly realised that this amplitude might have something to do with oscillators (as originally discovered in the operator formalism). Shortly after that, the connection with strings was made, quite independently, by Yoichiro Nambu (see Fig. 1.3), Holger Nielsen, and Leonard Susskind.[33] The dual model spectrum seemed to be indicating, as Susskind put it, "the degrees of freedom of the internal state of a hadron are equivalent to those of a violin string or organ pipe" [25, p. 1182]. The Veneziano model could then be understood as encoding the behaviour of a one-dimensional quantum system (which itself 'encoded' an entire Regge trajectory of particles), describing a reaction involving a pair of incoming (open) strings scattering into two more outgoing (open) strings. At this stage, the notion that these string models might provide a unified picture of physics was nowhere to be found. The string models were entirely focused on understanding the strong interaction, and the strings themselves viewed more as an heuristic tool than potential building blocks of the real world.

The notion of what string theory *is* undergoes several quite radical transformations during its brief lifetime, as the notions changed of what the strings are (open or closed), what they represent (hadrons, gravity, photons, …), what scale they operate

---

[31] That requires some qualification: the Glashow-Weinberg-Salam theory of weak and electromagnetic unification had been developed in 1967, but stagnated for some time before the conditions were ripe enough to recognize the importance of what they had achieved.

[32] There is a certain irony in how things have developed from S-matrix theory since its primary virtue was that it meant one was dealing entirely in observable quantities (namely, scattering amplitudes). Yet string theory grew out of S-matrix theory. Of course, most of the complaints with string theory, since its earliest days, have been levelled at its *detachment* from measurable quantities— let's call it 'the tyranny of (experimental) distance' (involving what Nambu is said to have called "postmodern physics": physics without experiments!). However, S-matrix theory, though crucial for the emergence of string theory as we know it today, was a single rung on a ladder of many, and though certain philosophical residues (such as the distaste for arbitrariness in physics) from the S-matrix programme stuck to string theory, it soon became a very different structure.

[33] Initially, a host of different terms were employed to describe the one-dimensional structures: from rubber bands to threads to sticks.

at $(10^{-13}$ cm vs. 10^{-33} cm), what kind of space they propagate in (4D, 5D, 10D, 11D, 26D), whether this space has internal, compact dimensions (and if so how the compactification is achieved), what kinds of symmetries they have (worldsheet vs. space-time supersymmetry), and whether there are 'higher-order' and 'lower-order' strings (0-branes, 1-branes, 2-branes, etc.), and so on.... Although there is clear continuity of structure linking these changes, in some cases it is better to think of the resulting altered structures as *different theories*, with different intended target-systems. Doing so, somewhat paradoxically, makes for a more rational (less weird) historical progression.

The weirdest part of string theory's history (that Kaku was referring to in the opening quotation) is the switch that occurred when it changed from being a theory of strong interactions to a theory incorporating gravitational interactions and Yang-Mills fields. This is a clear case in which it makes sense to think of the resulting theory as a genuinely *new* theory, couched in a near-identical framework.[34] There was no switch; rather, a distinct theory was constructed. As I aim to demonstrate, when viewed in the right light, this was not an irrational 'act of desperation,' to save string theory at any price, but one that began to take place while string theory (and dual models) for strong interactions were still enjoying a period of popularity. It was, however, very lucky in retrospect that this work *had* begun prior to QCD's breakthroughs, since otherwise superstring theory, as we know it today, might never have had the opportunity to get off the ground.

1.5 Summary

In this opening chapter we considered the orthodox story of string theory's genesis, and indicated how it might be refined—a task we carry out in subsequent chapters, of course. We indicated the broad outlines (or 'periodisation') of the history as presented in this book. We also included a brief guide to the physics and mathematics of vibrating strings and string theory, showing their similarities and differences.

References

1. Cartwright, N., & Frigg, R. (2007). String theory under scrutiny. *Physics World, 20*(9), 15–15.
2. Chand, R. (Ed.). (1970). *Symmetries and quark models*. Singapore: World Scientific.
3. Dawid, R. (2013). *String theory and the scientific method*. Cambridge: Cambridge University Press.
4. Dirac, P. A. M. (1939). *The relation between mathematics and physics* (Vol. 59, pp. 122–129). Proceedings of the Royal Society of Edinburgh.

---

[34] Of course, though the frameworks match initially (apart from the change of scale), the fact that the new theory is a theory of gravitation and other interactions suggests a host of new possibilities for developing the framework that would simply not arise in the older theory.

5. Dirac, P. A. M. (1962). *An extensible model of the electron* (Vol. 268, pp. 57–67). Proceedings of the Royal Society of London A19.

6. Galison, P. (1995). Theory bound and unbound: Superstrings and experiments. In F. Weinert (Ed.), *Laws of nature: Essays on the philosophical, scientific, and historical dimensions* (pp. 369–408). Berlin: Walter de Gruyter.

7. Grattan-Guinness, I. (1970). *The development of the foundations of mathematical analysis from Euler to Riemann*. Cambridge: MIT Press.

8. Gross, D. J. (1988). Physics and mathematics at the frontier. *Proceedings of the National Academy of Sciences of the United States of America, 85*(22), 8371–8375.

9. Hübsch, T. (1994). *Calabi-Yau manifold: A bestiary for physicists*. Singapore: World Scientific.

10. Jahnke, H. N. (2003). Algebraic analysis in the 18th century. In H. N. Jahnke (Ed.), *A history of analysis* (pp. 105–136). New York: American Mathematical Society.

11. Johansson, L.-G., & Matsubara, K. (2011). String theory and general methodology: A mutual evaluation. *Studies in History and Philosophy of Modern Physics, 42*(3), 199–210.

12. Kaku, M. (1999). *Introduction to superstrings and m-theory*. Berlin: Springer.

13. Kragh, H. (2011). *Higher speculations: Grand theories and failed revolutions in physics and cosmology*. Oxford: Oxford University Press.

14. Lovelace, C. (1970). M-loop generalized veneziano formula. *Physics Letters, B32*, 703–708.

15. Lovelace, C. (2012). Dual amplitudes in higher dimensions: A personal view. In A. Capelli et al. (Eds.), *The birth of string theory* (pp. 198–201). Cambridge: Cambridge University Press.

16. Mackey, G. (1992). *The scope and history of commutative and noncommutative harmonic analyis*. Berlin: American Mathematical Society.

17. Merton, R. K., & Barber, E. (2004). *The travels and adventures of serendipity: A study in sociological semantics and the sociology of science*. Princeton: Princeton University Press.

18. Olive, D. I. (2012). From dual fermion to superstring. In A. Cappelli et al. (Eds.), *The birth of string theory* (pp. 346–360). Cambridge: Cambridge University Press.

19. Polchinski, J., & Strassler, M. J. (2002). Hard Scattering and gauge/string duality. *Physical Review Letters, 88*, 031601.

20. Rickles, D. (2013). Mirror symmetry and other miracles in superstring theory. *Foundations of Physics, 43*, 54–80.

21. Rosner, J. (1969). Graphical form of duality. *Physical Review Letters, 22*(13), 689–692.

22. Schwarz, J. H. (1987). The future of string theory. In L. Brink et al. (Eds.), *Unification of fundamental interactions* (pp. 197–201). *Physica Scripta*, The Royal Swedish Academy of Sciences. Singapore: World Scientific.

23. Smolin, L. (2006). *The trouble with physics: The rise of string theory, the fall of a science, and what comes next*. Boston: Houghton Mifflin Harcourt.

24. Susskind, L. (1969). Structure of hadrons implied by duality. *Physical Review D, 1*(4), 1182–1186.

25. 't Hooft, G. (1999). When was asymptotic freedom discovered? Or the rehabilitation of quantum field theory. *Nuclear Physics, B74*(1–3), 413–425.

26. Veneziano, G. (1968). Construction of a crossing-symmetric, regge-behaved amplitude for linearly rising trajectories. *Nuovo Cimento A, 57*, 190–197.

27. Veneziano, G. (1998). Physics and Mathematics: A happily evolving marriage? *Publications Mathmatiques de l'IHÉS, S88*, 183–189.

28. Wigner, E. (1960). The unreasonable effectiveness of mathematics in the natural sciences. *Communications on Pure and Applied Mathematics, 13*(1), 1–14.

29. Witten, E. (2001). Black holes and quark confinement. *Current Science, 81*(12), 1576–1581.

30. Woit, P. (2007). *Not even wrong: The failure of string theory and the search for unity in physical law*. New York: Basic Books.

Part I
The (Very) Early Years: 1959–1973

Chapter 2
Particle Physics in the Sixties

> *As you can see, the new mistress is full of mystery but correspondingly full of promise. The old mistress is clawing and scratching to maintain her status, but her day is past.*
> Geoffrey Chew, Rouse Ball Lecture. Cambridge 1963

David Gross has described the early 1960s as a period of "experimental supremacy" [28, p. 9099]. The theoretical situation was almost entirely phenomenologically-oriented, with a profusion of new particle data being generated by experiments at Brookhaven, CERN, DESY, SLAC, and elsewhere. Theory was in a rather sorry state. Most of the work was concerned with model building to try and get some kind of foothold on the diversity of new phenomena coming out of the latest generation of particle accelerators. There was genuine uncertainty about the correct framework for describing elementary particles, and even doubts as to whether there *were* such things as elementary particles.[1]

An Erratum to this chapter is available at 10.1007/978-3-642-45128-7_11

---

[1] In fact, the beginnings of an erosion of confidence in quantum field theory (the orthodox framework for describing elementary particles) can be traced back to at least the 1950s, when the likes of Heisenberg, Landau, Pauli, and Klein were debating whether field theoretic infinities could be dealt with by invoking some natural (possibly gravitational) cutoff—of course, at this time non-Abelian gauge theories (and asymptotic freedom) were not known. (There was also a positivistic distaste with the notion of unobservable 'bare' masses and coupling constants.) One might also note that a new spirit flowed through the rest of physics at this time; not simply because it was a time of great social upheaval (being post-WWII = "the physicist's war"), but also because many of the 'old guard' of physics had passed away. In the immediate aftermath of WWII, there was extreme confidence in the available theoretical frameworks, and little concern with foundational issues. By the late-1950s and into the 1960s, this confidence was beginning to wane, as Chew's remarks in the above quotation make clear—the "new mistress" is S-matrix theory, while the "old mistress" is quantum field theory (amusingly, Marvin Goldberger had used the terminology of "old, but rather friendly, mistress" to describe quantum field theory in his Solvay talk from 1961—clearly Chew's remarks are a reference to this).

D. Rickles, *A Brief History of String Theory*, The Frontiers Collection, DOI: 10.1007/978-3-642-45128-7_2, © Springer-Verlag Berlin Heidelberg 2014

One of the central problems was triggered by the *strong* interactions, involving hadrons,[2] describing the properties of nuclei. True to their name the strongly interacting particles have large coupling constants determining how strongly they interact with one another, so the standard field theoretical tool of expanding quantities in powers of these constants fails to give sensible results.[3] Steven Weinberg notes that the "uselessness of the field theory of strong interactions led in the early 1950s to a widespread disenchantment with quantum field theory" [55, p. 17]. It didn't end with the strong interactions: the weak interaction too was then described by the non-renormalizable Fermi theory. This situation led to a move to bypass quantum field theory and instead deal directly with the fundamental constraints—and other general properties characteristic of the strong interaction—on the S-matrix that are expected of a good relativistic quantum theory (i.e. the scattering probability amplitude).[4] As Pierre Ramond writes, "[i]n the absence of a theory, reliance on general principles alone was called for" [46, p. 503].

String theory did not spontaneously emerge from a theoretical vacuum; it emerged precisely from the conditions supplied by this profound foundational disenchantment.[5] With hindsight, the earliest developments in string theory—i.e. the dual

---

[2] The name 'hadron' was introduced by Lev Okun in a plenary talk on "The Theory of Weak Interaction" at CERN in 1962, invoking the Greek word for large or massive (in contrast to lepton: small, light).

[3] Recall that the combination of relativity and quantum mechanics implies that particles (quanta of the field) can be created and destroyed at a rate depending on the energy of the system. Therefore, any such combination of relativity and quantum will involve many-body physics. This is compounded as the energy is increased. If the coupling constant is less than 1 then one can treat the increasing number of particles as negligible 'corrections' to the lowest order terms—note that the simpler, non-relativistic field theoretic case (the 'potential-scattering' problem) does not involve varying particle number. If the coupling constant is greater than 1, then going to higher order in the perturbation series (and adding more and more particles) means that the corrections will not be negligible so that the first few terms will *not* give a good approximation to the whole series.

[4] Not everyone was enchanted by this new S-matrix philosophy. As Leonard Susskind remembers it, the "general opinion among leaders of the field was that hadronic length and time scales were so small that in principle it made no sense to probe into the guts of a hadronic process—the particles and the reactions were unopenable black boxes. Quantum field theory was out; Unitarity and Analyticity were in. Personally, I so disliked that idea that when I got my first academic job I spent most of my time with my close friend, Yakir Aharonov, on the foundations of quantum mechanics and relativity." [51, p. 262]. As mentioned in the preceding chapter, Susskind would go on to make important contributions to the earliest phase of string theory research, including the discovery that if you break open the black box that is the Veneziano amplitude, you find within it vibrating strings. In his popular book *The Cosmic Landscape* Susskind compares this black box ideology to the behaviourst psychology of B. F. Skinner [50, pp. 202–203].

[5] Indeed, Stanley Deser remarked that the reason he got into general relativity and quantum gravity, after a background in particle physics, was precisely because "quantum field theory appeared to be degenerating while gravitational physics looked like a new frontier" (interview with the author, 2011—available via the AIP oral history archives [Call number OH 34507]). This suggests that there was something like a 'crisis' in Kuhn's sense. It was, of course, resolved to the satisfaction of many physicists (in quantum chromodynamics [QCD]) by a complex series of discoveries, culminating in a solid understanding of scaling and renormalization, dimensional regularization, non-Abelian gauge

resonance models alluded to in the previous chapter—can be viewed as perfectly rational and progressive steps given the state of physics just prior to it. In this first part we describe this state of affairs, and introduce the mathematical and physical concepts, formalism, and terminology necessary in order to make sense of early (and large portions of later) string theory.[6] However, many of the old concepts still make an appearance in modern string theory despite being not so well-known. This part might therefore also serve as a useful primer for those wishing to learn some string theory by providing some of the original physical intuitions and motivations.

2.1 Hadrontology

Recall that hadrons come in two families: *baryons* (particles composed of three quarks, such as the protons and neutrons making up an atomic nucleus) and *mesons* (force-mediating particles composed of two quarks, (a quark and an anti-quark), such as the pions and kaons found in cosmic rays)—particles that do not interact via the strong force are called *leptons*. The interactions amongst the components of nuclei were originally thought to be mediated entirely by π-mesons (a name contracted to 'pions'). However, the early models did not consider hadrons as internally structured entities composed of point-like constituents that interact through hard collisions, but as extended objects with 'soft' interactions.[7] From our present position we would say that the models were latching on to *low-energy*, long-range aspects of hadron physics in which the pions were the tip of an iceberg. There were many more mesons lurking below the surface. Unifying the profusion of mesons and baryons posed one of the most serious challenges of mid-twentieth century physics.

(Footnote 5 continued)
theories, and asymptotic freedom—recall that QCD is based on quark theory, where the 'chromo' refers to the extra degree of freedom postulated by Oscar Greenberg (in addition to space, spin, and flavour), labeled 'colour' by Gell-Mann. I don't discuss these discoveries in any detail in this book. For a good recent historical discussion, see [2] (see also: [29, 44]). However, QCD, while an excellent description of the high-energy behaviour of hadrons, still cannot explain certain low energy features that the earliest dual models (leading to string theory) had at least some limited success with.

[6] Naturally, many important concepts (from the point of understanding the development of string theory) have fallen out of fashion as the theories and models to which they belonged have been superseded.

[7] Of course, in QCD the strong interaction is governed by the exchange of gluons (massless, spin-1 bosons) which are coupled to any objects with *strong* charge or 'colour' (i.e. quarks). This has many similarities to QED, albeit with a coupling $\alpha_{strong} = g_s^2/4\pi \approx 1$, instead of the much weaker $\alpha_{EM} = e^2/4\pi \approx 1/137$. However, in the early days of hadron physics quarks were seen as convenient fictions used as a mere book-keeping tool for the various properties of hadrons—the nomenclature had some resistance: Victor Weisskopf, for example, wanted to call them 'trions,' while George Zweig wanted to call them 'aces'! The gluons are themselves coloured which implies that they self-interact. This results in a characteristic property of quarks, namely that they are confined within hadrons, unable to be observed in their singular form. The gluons attract the field lines of the colour field together, forming a 'tube'. Accounting for this tube-like behaviour was considered to be an empirical success of the early string models of hadrons, as we see below.

The challenge was further intensified as technological advances made possible proton accelerators[8] and bubble chambers capable of registering events involving hadrons by photographing bubbles formed by charged particles as they dart through a superheated liquid, thereby superseding earlier cosmic rays observations.[9]

Of course, quantum mechanics renders events probabilistic. This infects the natural observables in particle physics too. One of the observable quantities is the *scattering cross-section* (which basically offers a measure of the scattering angle made by colliding beams, or a beam and a static target). This tells you the likelihood of a collision given that two particles are moving towards one another. The magnitude of the cross-section is directly proportional to this likelihood. The cross-section itself is a function of the energy of the incoming beams, and if one examines the behaviour of the cross-section as a function of this energy, one can find peaks such that one can ask whether they correspond to particles or not.

Analysing the data from these scattering experiments pointed to the production of very many more new particles, or 'hadronic resonances' (very short lived, 'fleeting' strongly interacting particles[10] corresponding to sharp peaks in the total cross section, as a function of the energy)—of course, the strong interaction's being strong implies that such particle production will be plentiful. As described in the Particle Data Group's documents, resonant cross sections are described using the Breit-Wigner formula:

$$\sigma(E) = \frac{2J+1}{(2S_1+1)(2S_2+1)} \frac{4\pi}{k^2} \left[\frac{\Gamma^2/4}{(E-E_0)^2 + \Gamma^2/4} \right] B_{in} B_{out} \qquad (2.1)$$

where E is the energy in the centre of mass frame, J is the resonance spin, $(2S_1+1)$ and $(2S_2+1)$ are the polarisation states of a pair of incident particles, k is the initial momentum in the centre of mass frame, E_0 is the resonance energy (again in the centre of mass frame), Γ describes the resonance width (with $\frac{1}{\Gamma}$ giving the mean resonance lifetime), and the $B_{in} B_{out}$ pair describe the resonance branching fractions for the incoming and outgoing channels, where $B_{channel}$ would be computed as $\frac{\Gamma_{channel}}{\Gamma_{all}}$

---

[8] Primarily the Proton Synchrotron [PS], turned on in 1959, becoming the highest energy accelerator at that time, attaining a beam energy of 28 GeV. By comparison, the Cosmotron at Brookhaven reached energies of just 3 GeV, though at the time of its first operation it was six times more powerful than other accelerators. For a good, technical review of these experiments see [31].

[9] The tracks of these particles are bent using strong magnetic fields. The quantum numbers of the particles can then be computed from the curvature of paths, thus enabling (under the assumption of energy-momentum conservation) the identification of various particle types.

[10] Such resonance particles are too short-lived and localised to leave a directly observable trace. Resonances possess lifetimes of the order of 10^{-24} s. They would simply not travel far enough to leave a track before decaying. Given that particles travel at the speed of light c, solving for the distance traveled gives just $\approx 10^{-15}$ m. They are simply not stable enough to warrant the title 'particle,' which implies some degree of robust and continued existence. Of course, bubble chambers cannot allow one to see such particles, but one can infer their existence by observing decay products via various channels (see Figs. 2.1 and 2.2). (However, Chew [8, pp. 81–82] argued that, since both were to be represented by S-matrix poles, particles and resonances should not be distinguished in any significant way.)

Fig. 2.1 Particle tracks showing the annihilation event of
an anti-proton within a liquid
hydrogen bubble chamber
(using the PS coupled to the
80 cm Saclay chamber used
by CERN—image taken in
1961). Decay products are a
negative kaon, a neutral kaon,
and a positive pion. *Image
source* CERN, 1971

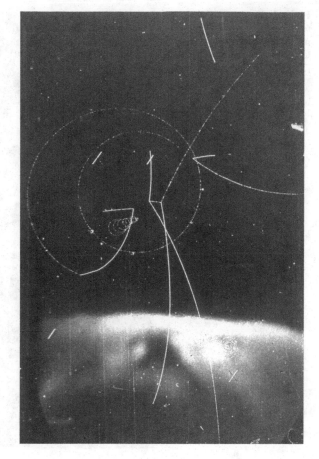

(that is, one counts the total number of decays through some channel relative to the
total number of particles produced)—see http://pdg.lbl.gov/2013/reviews/rpp2012-
rev-cross-section-formulae.pdf for more details.

The search for patterns in this jumble of data led to the discovery of a new
symmetry principle and a deeper quark structure underlying the dynamics of hadrons.
This work can be viewed in terms of a *drive to systematise*.[11] A central concern was
whether these new particles (or, indeed *any* of the particles) were 'fundamental'
(i.e. elementary)—with the sheer number of different particle types naturally casting

---

[11] As we will see below, it was consideration of *hard* scattering processes that led to quantum field
theory once again providing the framework in which to couch fundamental interactions. What such
processes revealed was a hard, point-like interior structure of hadrons, much as the classic gold foil
experiments of Rutherford had revealed a point-like atomic nucleus.

Fig. 2.2 The associated
reactions of the previous
photograph. *Image source*
CERN annual report of 1961,
[3, p. 93]

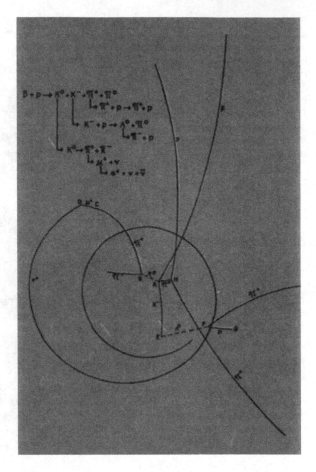

doubt on the idea that they were *all* elementary. If so, then the others might be
constructed as bound states of some small number of elementary particles.[12]

---

[12] Of course, the quark model postulated a deeper layer of elements of which the new particles
were really bound states. Although I won't discuss it, mention should be made here of the 'current
algebra' approach to strong interactions, of Murray Gell-Mann (see, e.g., [24]). In this programme,
although the underlying theory of quarks and their interactions wasn't determined, certain high-level
algebraic aspects of the free theory were, and these were believed to be stable under the transition to
the interacting theory. The current algebra is an $SU(n) \otimes SU(n)$ algebra (with n the number of what
would now be called 'flavours'), generated by the equal-time commutation relations between the
vector current $V_\mu^a(x)$ and the axial vector current $A_\mu^a(x)$. One of the crucial approximation methods
employed in the construction of dual models (that of infinitely many narrow hadronic resonances)
was developed in the context of current algebra. (See [2] for a conceptually-oriented discussion
of current algebra, including an extended argument to the effect that this amounts to a 'structural
realist' position in which the structural (broad algebraic) aspects constituted a pivotal element of
the development of the theory.)

One of the most hotly pursued approaches, S-matrix theory, involved focusing squarely on just those properties of the scattering process—or more precisely of the probability amplitude for such a scattering event—that *had* to be obeyed by a physically reasonable relativistic quantum field theory. The combination of these general principles with (minimal) empirical evidence drawn from observations of hadrons was believed to offer a way of (eventually) getting a predictive physics of strong interactions.[13] In its most radical form, espoused by Berkeley physicist Geoffrey Chew, the question of which hadrons were elementary and which were bound states was simply not appropriate; instead, one should treat them democratically, as on all fours.[14]

The S-matrix was originally developed by John Wheeler, as a way of condensing the complex bundle of information that is needed to describe a collision process, encapsulating the experimentally accessible information about any scattering experiment one could think of. Heisenberg actually named the object that achieves this condensation and imbued it with far more significance than Wheeler ever did.[15] Wheeler saw it as a mere tool "to deal with processes to be analysed by a more fundamental treatment" [56]. This might, as in the case of quantum electrodynamics [QED] be provided by a quantum field theory, which delivers up an S-matrix as an infinite expansion in the coupling constant (as we saw, in the case of QED this is the fine-structure constant $\alpha_{EM} = \frac{e^2}{4\pi}$).[16] Alternatively, one can sidestep talk of fields entirely, and focus on the scattering probability amplitude itself, which after all should contain all physically observable information (including the cross sections mentioned above, which can be written in terms of the matrix elements).

In this latter sense the S-matrix has an affinity with Bohr's positivistic strategy of ignoring what happens between energy transition processes involving electrons

---

[13] James Cushing's *Theory Construction and Selection in Modern Physics* [12] is a masterly account of the historical development of this new way of doing particle physics. In it he argues that the S-matrix methodology, of employing general mathematical principles to constrain the physics (at least, of the strong interaction), was perfectly viable and bore much fruit, despite the confirmation of QCD that knocked S-matrix theory off its pedestal. I agree with this general sentiment, and string theory can be found amongst such fruit.

[14] A concept Gell-Mann had labeled 'nuclear democracy'—surely a term coloured by the political and social climate of Berkeley in the 1960s. For a discussion of the context surrounding Chew's 'democratic' physics, see [32]. To this idea was appended the notion of 'bootstrapping' strongly interacting particle physics, in the sense that hadrons are bound states of other hadrons, that are themselves held together by hadron exchange forces—a purely endogenous mechanism.

[15] Holger Nielsen notes that he gave a talk on string theory while Heisenberg was visiting the Niels Bohr Institute at a conference given in his honour, but, as he puts it, "I do not think though that I managed to make Heisenberg extremely enthusiastic about strings" [40, p. 272]. Interestingly, David Olive also spoke on multi-Veneziano theory (that is, the generalised Veneziano model) and its relationship to quarks and duality diagrams, on the occasion of Heisenberg's 70th birthday, in Munich, June 1971. He notes that Heisenberg's reaction was a protest denying that the quark model was physics [37, p. 348].

[16] This connection was at the core of Freeman Dyson's equivalence proof of Feynman's and the Schwinger-Tomonaga formulations of QED [16], which employed the S-matrix to knit them together—the method of proof was to derive from both approaches the same set of rules by which the matrix element of the S-matrix operator between two given states could be written down.

orbiting atoms. In this case what is ignored (as unphysical or meaningless since unobservable, since too short-lived) are the unmeasurable processes occurring between initial and final states of a collision process.[17] Rather than describing what happens at the precise spacetime point (the vertex) at which the two or more particles meet (in which case there is no measurement to ascertain what is happening), one focuses on the measurable 'free' (non-interacting) situation when the systems were not and are no longer able to causally interact (mathematically speaking, at infinity, in the asymptotic limits), and therefore the particles have straight trajectories at constant velocities. In effect one draws a black box around the innards of the process and focuses on the particles entering and leaving the box and the probabilities of their doing so. This is somewhat paradoxical since the interaction between particles is described by an expression involving the particles' being far apart!

The S-matrix catalogues these possible relations between inputs and outputs along with their various probabilities. Measurable quantities such as scattering cross-sections can be written in terms of the matrix elements of the (unitary) S-matrix operator S. Recall that in quantum mechanics the state of a system is represented by a wave function $\psi(p)$, a square-integrable function of the system's momentum p (a 3-vector). For n particles it is a function of all the particles' momenta $p_1, ..., p_n$ (each a 3-vector).[18] The S-matrix is then an operation that transforms an initial state (a free wavefunction) of such incident particles to a final state (another free wavefunction), which, under the action of the unitary operator S, will have the general form of a superposition of all possible final states. The amplitude for finding one of these final states (say $|p'_1, p'_2\rangle$) in a measurement (for which the initial state is $|p_1, p_2\rangle$), is given by $\langle p'_1, p'_2|S|p_1, p_2\rangle$:

---

[17] In this sense, Heisenberg's way with the S-matrix corresponds to a repetition of the ideas that led to his matrix mechanics in the context of high-energy particle physics. Once scattering matrix elements have been fixed, then all cross-sections and observables have thereby been determined. Heisenberg's view was that one needn't ask for more (e.g., equations of motion are not required— on which, see Dirac [14]). The rough chronology that follows is that renormalisation techniques are developed, leading to quantum electrodynamics (with its phenomenal precision), leading to the demise of S-matrix theory. It was the subsequent fall from grace of quantum field theory at the hands of mesons that led to the resurrection of S-matrix theory, as we will see (see Fig. 2.5 for a visual impression of this "resurrection"). The trouble was that the finite, short range nature of the forces behind mesons seemed to imply that the particles were massive (in the context of Yukawa's exchange theory). Chen Ning Yang and Robert Mills had argued otherwise, of course, in order to preserve gauge invariance (now generalised to non-Abelian cases), but this view (famously discredited by Pauli) had to wait for an understanding of confinement and the concept of asymptotic freedom to emerge. Fortunately, by that time S-matrix theory had enough time to spawn string theory—'t Hooft gives a good description of this progression (including the impact of dual models and hadronic string theory) in [54] (see also [27, 28]).

[18] In the case of quantum mechanics this will be with respect to a Lebesgue measure, $d\mu(p) = \Pi d^3 p_i$. In the context of a relativistic quantum theory the measure must be Lorentz-invariant, so one has a mass term: $d\mu(p) = \Pi(m^2 + p_i^2)^{-\frac{1}{2}} d^3 p_i$ (with m the particle mass).

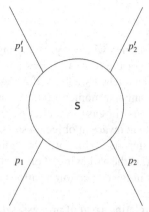

As in many episodes in the history of physics, what was essentially a mathematical result, here from complex analysis, led in 1959 to a breakthrough in physical theory. Analytic continuation allows one to extend the domain of definition of a complex function. A (complex) function is said to be *analytic* (or holomorphic, in mathematical terms) if it is differentiable at every point in some region. It was already known, thanks to the work of Gell-Mann, Chew, and others, that the S-matrix was an analytic function of its variables (representing physical quantities: the momenta of 'incoming' and 'outgoing' particles). This allowed the properties of the S-matrix to be probed almost independently of field theoretical notions in a quasi-axiomatic fashion (with very little by way of direct experimental input). The S-matrix theory (also known as the 'theory of dispersion relations',[19] though the links between dispersion relations and Heisenberg's theory took more time to emerge) then sought to derive the S-matrix by imposing various natural consistency conditions on it: Lorentz invariance, crossing,[20] unitarity, and analyticity (see the box below).

---

[19] The term 'dispersion' harks back to Kramers and Kronig's work in optics and the theory of *material* dispersion involving the absorption and transmission (in the form of a spectrum of different colours, or rainbow) of white light through a prism (or, more generally, some dispersive medium). In this case, a dispersion relation connects the frequency v, wavelength, λ, and velocity of the light, v: $v = v(\lambda)$. The spatial dispersion of light into different colours occurs because the different wavelengths possess different (effective) velocities when traveling through the prism. A good guide to dispersion relations is [41]. It was Murray Gell-Mann (at the 1956 Rochester conference [23]) who had initially suggested that dispersion relations might be useful in computing observables for the case of strong interaction physics. In simple terms, the idea is to utilise S-matrix dispersion relations to tie up experimental facts about hadron scattering with information about the behaviour of the resonances (independently of any underlying field theory). More technically, this would be achieved by expressing an analytic S-matrix in terms of its singularities, using Cauchy-Riemann equations. Chew developed this (initially in collaboration with Goldgerber, Low, and Nambu: [4]) into the general idea that strong forces correspond to singularities of an analytic S-matrix.

[20] In more orthodox terms, crossed processes are represented by the same amplitude and correspond to continuing energies from positive to negative values (whence the particle-antiparticle switch)—this corresponds, of course, to CPT symmetry. This idea of crossing also harks back to Murray Gell-Mann, this time to a paper coauthored with Marvin Goldberger [22]. Of course, if analyticity is satisfied, then the operation of *analytic continuation* can amplify knowledge of the function in

- **Lorentz invariance** is satisfied when physical quantities are unchanged by Lorentz transformations (of the form $x'^{\mu} = \Lambda^{\mu}_{\nu} x^{\nu}$ for all 4-vectors $x^{\nu} = (x^0, \mathbf{x}) = (t, \mathbf{x})$ and Lorentz tensors Λ^{μ}_{ν}). (Of course, this also implies that energy, momentum, and angular momentum are conserved.)
- **Analyticity** is satisfied just in case a scattering amplitude A is an analytic function of the Lorentz invariant objects used to represent the physical process in which one is interested. This formal condition is the mathematical counterpart of *causality* (i.e. the outlawing of effects preceding causes). (This condition has its origins in the dispersion relations of classical optics—see footnote 19.)
- **Crossing** is a symmetry relating a pair of processes differing by the exchange of one of the input and output particles (mapping particle to anti-particle and vice versa); for example, $a + b \rightarrow c + d$ and $a + \bar{c} \rightarrow \bar{b} + d$ (where $\bar{b}$ and $\bar{c}$ are b and c's anti-particles).
- **Unitarity** is simply the condition that the scattering matrix S is unitary: $S^{\dagger} S = 1$. Or, in other words, probability (that is, the squared modulus of the amplitude) must be conserved over time. (This also includes the condition of coherent superposition for reaction amplitudes.)

As indicated above, one of the central objects of the physics of elementary particle physics is the scattering (or transition) amplitude A. This is a function that churns out probabilities for the outcomes of collision experiments performed on pairs of particles[21]—note, this is not the same as the matrix of such described above. It takes properties of the particles as its argument. For example, the function might depend on the energy E of the collision event and the scattering angle θ representing a particle's deflection $f(E, \theta)$ thus encodes the nature of this interaction. The general representation involves the incoming energy and the momentum that is transferred in the collision, s and t respectively, defined as follows:

- t is the square of the difference between the initial and final momenta of the particles involved in some process (also known as "the momentum transfer"):

(Footnote 20 continued)
some region of its domain to other regions—as Cushing puts it, "an analytic function is determined globally once it has been precisely specified in the neighbourhood of any point" [13, p. 38].

[21] More generally, it is more appropriate to think about *channels* of particles. One can think of a channel, loosely, as a providing a possible 'route' from which the final state emerges. There might be many such possible routes, in which case one has a multichannel collision process, otherwise one has a single channel process. Such channels are indexed by the kinds of particles they involve and their relative properties. In scattering theory one is interested in inter-channel transitions; i.e. the transition from some process generated through an input channel and decaying through an output channel. Given a set of available channels, unitarity in this case is simply the property that every intermediate state must decay through *some* channel, so that $\sum_{out} |S_{\langle in, out \rangle}|^2 = 1$.

$$t = (p_a - p_c)^2 = (p_b - p_d)^2 \qquad (2.2)$$

- s is the square of the sum of the momenta of the initial states on the one hand and the final states on the other:

$$s = (p_a + p_b)^2 = (p_c + p_d)^2 \qquad (2.3)$$

We denote the incoming momenta of the particles, p_a and p_b, with outgoing momenta $-p_c$ and $-p_d$. In this process there is a conservation of total momentum (4-momentum); i.e. $p_a + p_b = p_c + p_d$ (also, $p_i^2 = m_i^2$, with m_i being the ith particle's mass).[22] The scattering amplitude is, then, a function of certain conserved (invariant) quantities ('channel invariants'). Suppose we have some process involving a pair of incoming particles going into some pair of outgoing particles (of the same mass m, for simplicity): $a + b \rightarrow c + d$. This will involve a 4-point amplitude $A(s,t)$. The amplitude is then written as $A(s,t) \sim \beta(t)(s/s_0)^{\alpha(t)}$ (where β is a residue function). The squared modulus of this object delivers the observable scattering cross-section discussed above.

The Mandelstam variables define reaction channels as follows (see Fig. 2.4):

- The reaction $a + b \rightarrow c + d$ occurs in the s-channel, with the physical (real) region defined by values $s \geq (m_a + m_b)^2$.
- The 'crossed' reaction $a + \overline{c} \rightarrow \overline{b} + d$ occurs in the t-channel (as noted in the box above), with the physical (real) region defined by values $t \geq (m_a + m_c)^2$.[23]

Recall that Feynman diagrams were originally intended to provide a mathematical representation of the various contributions to the S-matrix in the context of perturbative (Lagrangian) field theories. However, in the late 1950s Landau [34] had instigated the examination of the links between Feynman graphs and singularities of the S-matrix, thus liberating the former from weakly-coupled quantum field theories to which they were previously thought to be hitched. The singularity conditions that Landau found pointed to a correspondence between tree graphs[24] and poles (and loop diagrams and branch points). Thus was born the idea that general conditions

---

[22] The variables s and t are known as Mandelstam variables, with a third, $u = (p_a - p_d)^2 = (p_b - p_c)^2$, completing the set of Lorentz invariant scalars. These variables are not all independent because of the presence of the constraint $s + t + u = \sum_{i=1}^{i=4} m_i^2$, so any two variables can be used to construct the scattering amplitudes, therefore we can dispense with u for convenience.

[23] The u-channel would be obtained from the t-channel by switching particles c and d: $u = (p_a - p_d)^2 = (p_b - p_c)^2$. In the u-channel is the reaction: $a + \overline{d} \rightarrow \overline{b} + d$, where the physical region is $u \geq (m_a + m_d)^2$.

[24] In other words, a tree graph in the sense of Landau is understood to represent, directly, physical hadrons via the lines. Landau's singularity conditions are satisfied by a classical process sharing the topological (network) structure of the graph. Coleman and Norton later provided a proof of this graph-process correspondence. As they put it: "a Feynman amplitude has singularities on the physical boundary if and only if the relevant Feynman diagram can be interpreted as a picture of an energy- and momentum-conserving process occurring in space-time, with all internal particles real, on the mass shell, and moving forward in time" [11, p. 438].

Fig. 2.3 Graph showing the number of papers published on S-matrix theory (or the S-matrix) following Heisenberg's paper in 1943, with a significant growth occurring in the 1960s. In his survey of models for high-energy processes, John Jackson found that, between 1968 and the first half of 1969, "various aspects of S-matrix theory, with its ideas of analyticity, crossing and unitarity, accounted for 35% of the theoretical publications" [30, p. 13]; cf. [52, p. 285]. *Image source* Thompson-Reuters, *Web of Science*

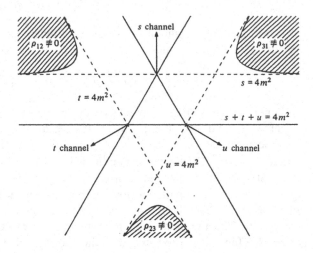

imposed on the structure of the scattering amplitude might be enough to determine the physical behaviour of particles.

These considerations led to a variety of features that could be aimed at in model building. It was from this search that the Veneziano model was born. Before we discuss that model, we first need to say something about some important intervening work, of Tullio Regge, Stanley Mandelstam, and Geoffrey Chew, that will help us make better sense of the foregoing.

2.2 Chew's Boots and Their Reggean Roots

In 1959 Tullio Regge [47] suggested that one think of solutions to the Schrödinger equation for the potential scattering problem in terms of the complex plane, using complex angular momentum variables (which, of course, take on discrete values). This ignited a surge of research in linking 'Regge theory' to the world of hadrons and high energy (special relativistic) physics.[25]

A singularity of a complex function (i.e. a point where the value of the function is zero or infinity for some argument) is known as a *pole* (a tree graph in graphical terms,

---

[25] This expansion into the complex plane has a significant impact on the mathematics employed. For example, integration takes on a different appearance since, whereas given the real numbers one follows a single path to integrate between two points, in complex analysis one can take many different paths in the plane, leading to planar diagrams and contour integration. Note, however, that all were taken with the complex expansion. 't Hooft mentions that his PhD supervisor, Martinus Veltman, was of the opinion "Angular momentum aren't complex. They're real. Why do you have to go to a complex thing? What does it mean?" (interview with the author, 10 February 2010). See [17] for a good general overview of Regge theory, including its place within Veneziano's dual resonance model.

with loops corresponding to branch points). Regge focused on the potential scattering problem, where the amplitudes become simple poles in angular momentum (i.e. at certain special values of the momenta). The locations of these poles is determined by the energy of the system and the poles themselves were taken to correspond to the propagation of intermediate particles. As one tunes the energy parameter, one gets a graph (a Regge trajectory) describing the properties of resonances and scattering amplitudes (for which the transfer of momentum is large). In the relativistic case one must introduce another class of singularity in angular momentum, in particular at $j = -1$. Stanley Mandelstam tamed these singularities by introducing a second Riemannian hyperplane[26] of the complex j-plane and performing branch cuts in the j-plane, known as "Regge cuts".[27]

A Regge pole is then the name given to a singularity that arises when one treats angular momentum J as a complex variable.[28] Physically a Regge pole corresponds to a kind of particle that 'lives' in the complex angular momentum plane, whose spin is linearly related to its mass. Tuning the energy of such a particle to a value which would spit out an integer or half-integer value for the spin would produce a particle that one ought to be able to detect. Confirmation of this relationship was indeed found in early hadron spectroscopy which generated Regge plots showing (for mass squared plotted against spin) a linearly rising family of particles on what became known as a 'Regge trajectory' (see Fig. 2.5).[29] In this way specific types of particles could be classified by these trajectories, each trajectory containing a family of resonances differing with respect to spin (but sharing all other quantum numbers).

There was a curious feature about some of the spin values,[30] as represented in the plots of Regge trajectories, namely that they were seemingly unbounded from above. Particles with large spins are more like finite-sized objects possessing angular momentum (from real rotation[31]). In the case of baryons, one can find experimentally observed examples of spin $J = 10$! According to Regge theory, the high energy

---

[26] A Riemann surface provides a domain for a many-valued complex function.

[27] To put some 'physical' flesh on these concepts, it is safe in this context to think of simple poles as particle *exchanges* at a vertex, while a cut is a singularity corresponding to pair *production* (of particles). Technically, of course, a branch cut is a kind of formal 'barrier' that one imposes on a domain in order to keep a complex function single valued.

[28] The singularity is of the form $\frac{1}{J-\alpha}$ (where α, the Regge slope, is a function of the collision energy of the process in which the particle is involved).

[29] The slope α' of the Regge trajectories was one of the concepts that would enter string theory in a rather direct way. It was suggested later that the slope has the air of a universal constant of nature, and one that might be connected to the extended, non-point-like character of hadrons, leading to a fundamental length scale set by hadron constituents, $\lambda \approx \sqrt{\alpha'}$ of the order 10^{-14} cm [36]. As Daniel Freedman and Jiunn-Ming Wang showed in 1966, in addition to the 'leading trajectory,' one would also have 'daughter trajectories' lying parallel (with spins separated by one unit), underneath the leading trajectory, and separated by a spacing of integer multiples of a half.

[30] The spin values of the resonances themselves can be inferred from the angular distribution of the decay products in the various reactions.

[31] Quantum field theories face severe problems with conservation of probability (i.e. unitarity) for particles of spins greater than 1, in which case the amplitudes diverge at high energies. One of Regge theory's key successes was the ability to deal with the exchange of particles of very high

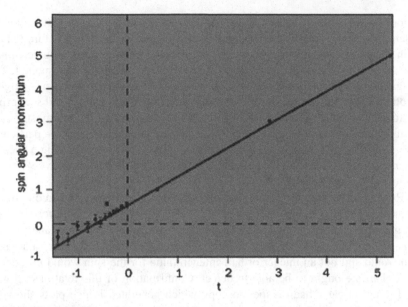

Fig. 2.4 The Mandelstam diagram providing a representation of $A(s, t, u)$ in terms of double spectral functions, ρ_{12}, ρ_{31}, and ρ_{23}, which are zero except in the shaded region (corresponding to values above the intermediate-state threshold). *Image source* [12, p. 120]

behaviour of scattering amplitudes is dominated by the leading singularity in the angular momentum Argand plane. Crucially, if such a singularity is a pole at $J = \alpha(t)$ (in other words, a Regge pole) then the scattering amplitude has the asymptotic behaviour: $\Gamma(1 - \alpha(t))(1 + e^{-i\pi\alpha(t)})s^{\alpha(t)}$ (where $s \to \infty$ and $t < 0$).

The bootstrap approach grew out of these developments of Regge and Mandelstam.[32] In dispersion theory one tries to generate physics from a few basic

---

(Footnote 31 continued)
spins by conceptualizing the process in terms of 'Reggeon' and 'multi-Reggeon' exchange (where Reggeons are composite objects associated with $\alpha(t)$).

[32] This story begins in 1958, with Mandelstam's paper marking the beginning of the so-called 'double dispersion representation' (in both energy and momentum transfer): [37]. Such double dispersion relations were later renamed the 'Mandelstam representation'. Mandelstam was explicitly taking up the suggestion made by Gell-Mann in [23], that one might "actually replace the more usual equations of field theory and ... calculate all observable quantities in terms of a finite number of coupling constants" [37, p. 1344]. Elliot Leader has written that "Tullio Regge's great imaginative leap, the introduction of complex angular momentum in non-relativistic quantum mechanics, might have ended in oblivion, weighed down by its overpowering mathematical sophistry and rigour, had not S. Mandelstam, seizing upon its crucial element and casting off the mathematical shroud, demonstrated a direct and striking consequence in the behaviour of high-energy elementary particle collision processes [35, p. 213]. Mandelstam's insight was the realization that unphysical regions of the scattering plane (involving very large values of the cosine of the scattering angle θ), for a scattering event like $A + B \to A + B$, is mathematically related to the physical reaction $A + \overline{A} \to B + \overline{B}$.

axioms, such as Lorentz invariance, unitarity, and causality discussed above. These are used as (high-level physical) constraints on the space of possible theories as input data from the world is fed in. The dispersion theory approach and the old S-matrix approach were merged together in Chew's 'bootstrap' approach to physics.[33]

A crucial component of Chew's approach was the 'pole-particle' correspondence. According to this principle, there is a one-to-one correspondence holding between the poles of an (analytic) S-matrix and resonances, so that the position of a pole in the complex energy plane gives the mass of the resonance while the residue gives the couplings. When the pole is complex, the imaginary part gives its lifetime. The idea was that the axioms of the dispersion approach would uniquely pin down the correct S-matrix, and thereby deliver physical predictions. The focus would be on the analytic properties of the S-matrix. The theory had some degree of success at a phenomenological level.

Presently, of course, our best description of nature at very small subatomic scales is couched in the framework of quantum field theory [QFT]—a framework Chew believed unhealthily imported concepts from classical electromagnetism. It is thought that there are six fundamental leptons and six fundamental quarks. These are bound together by forces that are understood as involving quantum fields. The unified theory of the weak and electromagnetic interactions, the electroweak force, is understood via the exchange of four kinds of particle: the photon, the W^+, the W^-, and the Z^0. The strong force is mediated via the exchange of eight types of massless gluon. The standard model also involves Higgs particles, H^0, whose associated field is responsible for the generation of the masses of observed particles.[34] In quantum field theory the dynamics is delivered through a Lagrangian, from which one derives

---

[33] As Chew describes the origination of the bootstrap idea, it was in discussion with Mandelstam before the 1959 Kiev Conference when they discovered that "a spin 1 $\pi\pi$ resonance could be generated by a force due to Yukawa-like exchange of this same resonance" [9, p. 605]—a resonance that was later to be named the ρ-meson. The bootstrap, more generally, refers to the notion that one can build up a pole in some variable via an infinite sum of singularities in some other variable—that is, a pole generates singularities in the crossed-channel, and these singularities generate the original pole. A pole thus generated can then be viewed as a bound state of other particles: "ρ as a force generates ρ as a particle" [9, p. 606]. Or, in more general terms, hadrons are to be viewed as bound states of other hadrons (see [5] for the more general bootstrap theory).

[34] Gravitation is not incorporated in this scheme, and is modelled only classically. The particle physics approach to quantum gravity was being pursued at around the same time that the standard model was being formed. Indeed, the tools and methods used to construct the standard model were very much bound up with work in quantum gravity. The electroweak, the strong force, and the gravitational force were, after all, described by non-Abelian theories. The properties powerfully represented by the standard model form a target that any future theory that hopes to probe still higher energies ('beyond the standard model') will have to hit. This includes string theory.

equations of motion. Essentially what Chew proposed was to eliminate equations of motion in favour of general principles. In the case of strong interactions, at least, Chew believed that a Lagrangian model simply wasn't capable of delivering up a satisfactory S-matrix.

At the root of Chew's proposal was the belief that field theory could simply not cope with the demands imposed by strong interaction physics. He wrote that "no aspect of strong interactions has been clarified by the field concept" [6, p. 1]. Though there was a family of hadrons, no family members appeared to be fundamental, and a field for each and every hadron would result in filling space with an absurd number of fields. For this reason, Chew suggested that all hadrons should be treated on an equal footing: neither more nor less fundamental than any other. The notion of fundamentality dropped out in favour of nuclear democracy, with the particles understood as in some sense composed out of each other as in footnote 33, with the forces and particles bundled together as a package deal. Chew expresses it as follows:

> The forces producing a certain reaction are due to the intermediate states that occur in the two "crossed" reactions belonging to the same diagram. The range of a given part of the force is determined by the mass of the intermediate state producing it, and the strength of the force by the matrix elements connecting that state to the initial and final states of the crossed reaction. By considering all three channels [i.e., orientations of the Feynman diagram] on this basis we have a self-determining situation. One channel provides forces for the other two—which in turn generate the first [6, p. 32].

A further development that played a crucial role was made by Chew's postdoc student at Berkeley, Stanley Mandelstam. He had discovered a way to resolve a problem in understanding the strong interaction in terms of particle exchange (à la Yukawa[35]). The problem was that the hadrons were short range, and therefore massive—Yukawa had calculated a characteristic mass of 100 MeV, corresponding to a sub-nuclear range of the strong force of 10^{-13} cm. The old cosmic ray observations delivered a candidate for such a particle in the form of the pion. Yet, by the late 1950s, particles were also being discovered with spins greater than 1, increasing linearly. This would imply that the exchange forces would also grow in such a way, without limit. Referring back to the discussion above, this would further imply that the scattering cross-section describing the size of the area over which the particles interact would also grow indefinitely. This is in direct conflict with the idea that exchanging massive particles demands *smaller* areas: the more massive the particles are, the less capable they are of covering large distances.

The solution was to treat the entire series of particles (with increasing spins) laid out along a Regge trajectory as the subjects of exchange (named a "pomeron" by Vladimir Gribov, after Pomeranchuk)—that is, rather than the individual points lying

---

[35] Yukawa had attempted to construct a quantum field theory along the lines suggested by quantum electrodynamics in 1935. His approach proposed a connection between the mass of a particle and its interaction range.

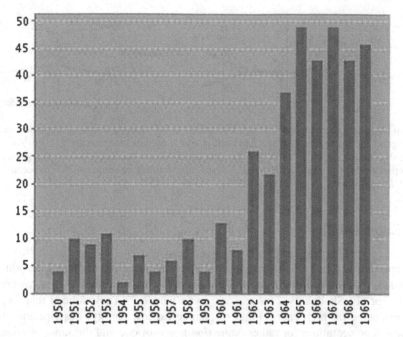

Fig. 2.5 A Regge trajectory function $\alpha(t)$ representing a rotational sequence of states (of mesons) of ever higher spins. The relationship with resonances (and bound states) comes about from the fact that when $\alpha(t)$ is a positive integer for some value of the argument t, then a bound state or resonance exists at that t-value, with spin read off the horizontal. For example, in this picture we have at $t = 3$ the resonance $\alpha(3) = 3$ and at $t = -1$ the bound state $\alpha(-1) = 0$. The various states given in this way generate a family: the Regge trajectory. A horizontal trajectory, $\alpha(t) = const.$, would represent particles of constant spin (elementary particles), while a non-zero slope represents particles of varying spin (composite particles). *Image source* [15, p. 1]

within the trajectories.[36] Applying this procedure keeps the cross-sections finite—a calculation that was performed by Chew and Steven Frautschi [5].[37]

---

[36] The Pomeron was later understood to be the trajectory given by $2 + \frac{\alpha'}{2} J^2$ (the Pomeron sector) corresponding to the massless states of gravitons and dilatons (associated with closed strings). Its defining quality is that it is, in some sense, 'without qualities,' carrying no quantum numbers (or equivalently, it has 'vacuum quantum numbers': that is, no charge, spin, baryon number, etc.). This latter basic idea of the Pomeron was introduced in Chew and Frautschi's "Principle of Equivalence for all Strongly Interacting Particles Within the S-Matrix Framework" [5]. They were to be distinguished from Reggeons (later interpreted in terms of open strings). It was subsequently found that the states of the Pomeron sit on a Regge trajectory with twice the intercept and half slope of the Reggeon trajectory. As we see, the vacuum quantum numbers are later explained by the fact that closed string worldtubes have no boundaries on which to 'attach' quantum numbers using the then standard 'Paton-Chan method.

[37] This chapter also introduced the representation of Regge trajectories (as in Fig. 2.5, now known as a 'Chew-Frautschi plot'). The original Chew-Frautschi plot consisted of a line draw between just two points (the only two then known experimentally)—cf. [5, pp. 57–58]. As Frautschi noted in an interview, "Originally, we had just drawn a straight line between two points, because two points

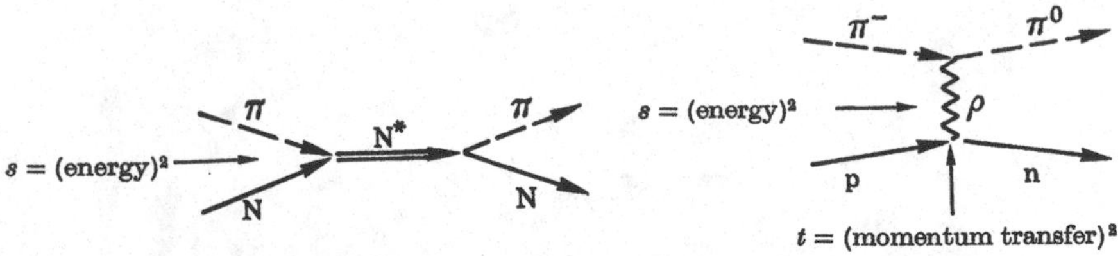

$s = (\text{energy})^2$

$s = (\text{energy})^2$

$t = (\text{momentum transfer})^2$

Fig. 2.6 Graphical representations of two descriptions of the hadronic scattering amplitude: In the left diagram one has resonance production (with π and N colliding to generate N^*, which decays after a short time back into π and N); on the right hand side one has Regge pole exchange (i.e. an interaction in which π^- and p exchange a ρ-meson, transforming quantum numbers to become π^0 and n). *Image source* C. Schmid [49, p. 257]

2.3 Enter Duality

An important step in the bootstrap approach was the principle of duality introduced by Dolen, Horn, and Schmid in 1967, at Caltech (they referred to it as "average duality" or "FESR duality", for reasons given below).[38] They noticed that Regge pole exchange (at high energy) and resonance (at low energy) descriptions offer multiple representations (or rather *approximations*) of one and the same physically observable process. In other words, the physical situation (the scattering amplitude, $A(s, t)$[39]) can be described using two apparently distinct notions (see Fig. 2.6):

- A large number of resonances (poles) exchanged in the s-channel.
- Regge asymptotics: $A(s, t)_{s \to \infty} \sim \alpha(s)^{\alpha(t)-1}$, involving the exchange of Regge poles in the t-channel.

That these are in some sense 'equivalent' in terms of the physical description was elevated to a duality *principle*[40]:

(Footnote 37 continued)
were all we had for the data. And then as more data occurred, the straight line continued through the next particle discovered and through the Yukawa exchanges in a different kinematic region. So the straight lines we'd originally drawn for our Regge particles turned out to be a pervasive feature, and eventually that came to be regarded as very strong evidence for strings. " [21, p. 19].

[38] Indeed, James Cushing referred to the combined S-matrix theory + duality framework as "the ultimate bootstrap" [12, p. 190]. However, duality really is just an implementation of the bootstrap principle of generating a pole (particle) by summing over (infinitely many) singularities in some other amplitude variable. In the case of duality one has a physical (that is, observational) equivalence between a description without forces (but with resonance production: i.e. fermions, though without spin degrees of freedom) and one with forces (mediated by an exchange particle: i.e. bosons).

[39] A simple expression of the duality is through the symmetry of the amplitude under the interchange of energy s and momentum transfer t: $A(s, t) = A(t, s)$. One can think in terms of $s - t$ duality or resonance-Regge pole duality—for this reason it is sometimes called '$s - t$ duality'.

[40] As Pierre Ramond notes, this was "elevated to a principle to be added to the Chew bootstrap program, regarding resonance and Regge trajectories as aspects of the same entities" [46, p. 505].

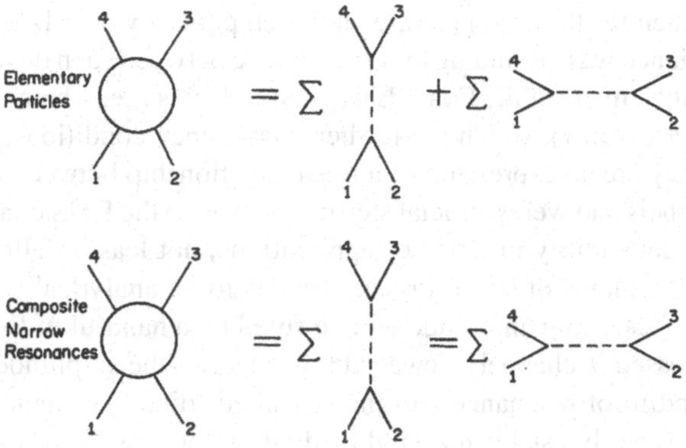

Fig. 2.7 Different methods of computing amplitudes with the interference model (*top*), associated with a picture of elementary particles, and the dual resonance model (*bottom*), associated with composite entities. In the former one sums over the contributions from both channels, while the latter identifies them in accordance with the principle of duality. *Image source* [39, p. 265]

> **DHS Duality** Direct s-channel resonance particles are generated by an exchange of particles in the t-channel.

This has the effect that the representative Feynman diagrams for such processes are identified to avoid surplus states, known as "double counting". For this reason, the two contributions to the amplitude are not to be summed together: summing over one channel is sufficient to cover the behaviour encapsulated in the other. This was matched by the experimental data. So-called "interference models" would demand that the two descriptions (both s- and t-channel contributions) be added together like ordinary Feynman tree diagrams, which would be empirically inadequate of course (see Fig. 2.7). As with any duality there is an associated *epistemic gain*: if we know about the resonances at low energies, we know about the Regge poles at high energies.[41]

One can make some physical sense of the existence of such a duality by thinking about the 'black box' nature of the scattering methodology, as discussed previously. Since one makes measurements only of the free states (the asymptotic wave-functions), one cannot discern the internal structure between these measurements, and so given that both the s-channel (resonance) and t-channel (interaction via exchange) situations have the same asymptotic behaviour, they correspond to 'the same physics'. However, the precise mathematical reason would have to wait first for the formulation of a dual

---

[41] I borrow the term "epistemic gain" from Ralf Krömer to refer to the fact that there are circumstances in which "dual objects are epistemically more accessible than the original ones" [33, p. 4]. The most significant case of this is seen in the final chapter when we look at S-duality, relating strongly coupled to weakly coupled limits of certain theories.

amplitude, and then for the string picture, at which point it would become clear that conformal invariance was grounding the equivalence between such dual descriptions.

Mention must be made of the Finite Energy Sum Rules (i.e. where the energy has been truncated or cut in s), which are further consistency conditions, flowing from analyticity.[42] They are an expression of a linear relationship between the particle in the s- and t-channels and were a crucial step on the way to the DHS duality principle. They have enormous utility in terms of applications, not least in allowing the low and high energy domains of scattering amplitudes to be analytically connected: at high energies the scattering amplitude will be ruled by a handful of Regge poles (in the so-called 'crossed' t-channel) viewed at low energies the amplitude will be ruled instead by a handful of resonances (in the so-called 'direct' s-channel), as above. Thus, the FESR already establish a kind of duality between these two regimes so that t-channel (Regge) values can be determined from s-channel resonances. More formally, one begins with the (imaginary part of the) low energy amplitude charac-terised by resonances (which sits on the left hand side of the FESR equation) and builds up the Regge terms by analytic continuation (cf. [49, p. 246]). Schematically one has (borrowing from [43, p. 204]):

$$\langle \mathrm{Im} f (\text{Resonance}) \rangle = \langle \mathrm{Im} f (\text{Regge}) \rangle \tag{2.4}$$

The averaging refers to the fact that one is integrating over Regge and resonance terms (Fig. 2.8).[43] The FESR are formally expressed as follows:

$$\mathbf{FESR}: \quad \int_o^N \mathrm{Im}\, A^{(-)}(v, t)dv = \sum_i \beta_i(t) \frac{N^{\alpha_i(t)+1}}{\alpha_i(t) + 1} \tag{2.5}$$

Hence, DHS duality is sometimes also called FESR-duality.

Though this duality in some ways embodies Chew's Nuclear Democracy (since, in the case of $\pi\pi$ scattering, both channels contain the same particles) it also paved the way for a departure from this picture. Using diagrammatic representations of the duality, Harari and Rosner reinterpreted the duality in terms of the flow of hadron constituents (quarks and anti-quarks[44]) and the exchange of such.

---

[42] According to Mahiko Suzuki, who shared an office with Horn and Schmid and collaborated with them briefly, it was Horn that coined the name "finite energy sum rule". Richard Dolen entered the collaboration (as Suzuki departed) because of his computational and data handling skills (private communication).

[43] By contrast in the competing interference model scheme, mentioned above, one would have the sum rule: f(Resonance) + f(Regge) (see [1]).

[44] At this stage the quarks were, in general, not invested with any physical reality, but were merely viewed as a kind of book-keeping method. George Zweig was entertaining the idea that quarks were real, but Gell-Mann's view that they were purely formal prevailed. Of course, he would later receive his Nobel prize, in 1969, for the discovery that hadrons are bound states of quarks. In fact, it should be pointed out that this does *not* appear to have been Gell-Mann's actual position, and his usage of the term "mathematical" to describe certain quarks was non-standard (cf. [53, p. 634]): he simply meant 'unliberated' or "permanently confined" and chose "mathematical"

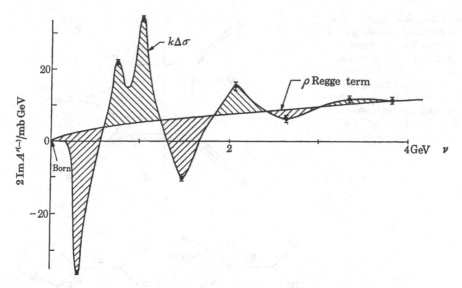

Fig. 2.8 A plot indicating duality for the πN-amplitude, $A'^{(-)}$. One can see that $2ImA'^{(-)}$ has large fluctuations at low energy values, but latches on to the ρ Regge term on the average. *Image source* C. Schmid [49, p. 260]

Although the link wasn't explicitly made at the time, these diagrams, in eliminating the links and vertices from standard Feynman graphs, already contain the germ of what would become string scattering diagrams according to which only the *topological* characteristics are relevant in the scattering process—one can easily see that the exchange and resonance diagrams are deformable and so topologically equivalent. This equivalence was given a graphical representation in the work of Haim Harari (see Fig. 2.9).

Harari was then working at the Weizmann Institute. At around the same time, at Tel-Aviv University, Jonathan Rosner also came up with the idea of duality diagrams.[45] Rosner's version can be seen in Fig. 2.10.

Since it makes an appearance in the following pair of chapters, we should also say something about the Pomeron (that is, the Pomeranchuk pole) in this context. The duality principle links Regge poles to resonances, but the Pomeron, with vacuum quantum numbers, falls outside of this scheme. It satisfies duality in a sense, but it turns out to be dual to the non-resonating background terms.

(Footnote 44 continued)

to avoid what he called "the philosopher problem"! He was worried that philosophers would grumble about the possibility of unobservable entities—and, indeed, we saw earlier that Heisenberg objected on just such grounds. David Fairlie goes further, arguing that the positivistic commandment against talking about "unobservable features of particle interactions, but only about properties of asymptotic states...inhibited the invention of the concept of quarks" [19, p. 283].

[45] Rosner notes in his paper that he became aware of Harari's work once the bulk of his own work was completed [20, p. 691]. This feature of multiple near-simultaneous discoveries is especially rife in the history of string theory—it surely points to an underlying common set of heuristics.

Fig. 2.9 Haim Harari's duality diagram for a multi-particle
process. The *top* diagram
amounts to an equivalence
class of the diagrams beneath
it, in the sense that any of the
five ordinary Feynman graphs
provides complete information about the amplitude.
Image source [9, p. 563]

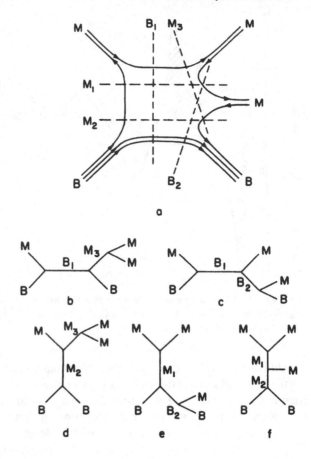

Another problematic issue was simple one pion exchange. The problem with
this case, vis-a-vis duality, is that the amplitudes for such exchange processes are
real-valued, whereas, as we have seen, duality involves only the imaginary parts of
amplitudes. Though this problem was discussed (see, e.g., the remark of Harari
following Chan's talk at a symposium on *Duality-Reggeons and Resonances in
Elementary Particle Processes*, [11, p. 399], it doesn't seem to have been satis-
factorily resolved until John Schwarz and André Neveu's dual pion model in 1971.

As we will see in the next chapter, Veneziano's achievement was to display a
solution to FESR by the Euler Beta function (thus giving an implementation of a
dual version of the bootstrap). The solution is an amplitude that displays precisely
the Regge behaviour (that is, Regge asymptotics) and satisfies all of the principles laid
out by the S-matrix philosophy (Lorentz invariance, analyticity, crossing, duality),
apart from unitarity, on account of the particular approximation scheme employed (on
which more later). The hope was that using the bootstrap principle, this framework
could then eventually be employed to predict specific physical properties of hadrons,
such as masses.

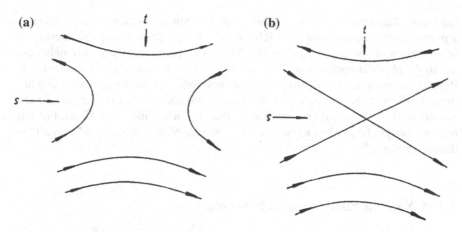

Fig. 2.10 Jonathan Rosner's graphical representation of duality. A graph will exhibit duality in only those channels in which it is *planar* (*no crossed quark lines*). Mesons are quark/anti-quark pairs, $q\overline{q}$ and baryons are triplets of quarks qqq. Here, **a** is planar in the s and t channels, with an imaginary part at high s—this represents baryons in s and mesons in t. Duality implies that intermediate baryon states build an imaginary part at high s. Graph **b** is planar only in u and t, with no imaginary part at high s. *Image source* [20, p. 689]

The ability of dual models to encompass so many, then ill-understood, features of hadronic physics led to their very quick take up. Quite simply, there was no alternative capable of doing what dual models did. Hence, though it was not then able to make novel testable predictions, even at this stage, the fact that it resolved so many thorny problems with hadrons, and explained so many features in a unified manner meant that it was still considered to be serious physics—though, it has to be said, not all were enamoured, precisely on the grounds that it failed to make experimental predictions.

Before we shift to consider the Veneziano model, a further important step towards the dual models, and away from Chew-style bootstrap models, was the introduction of the narrow-resonance (or zero-width) approximation alluded to above, which initially ignored the instability of hadrons, treating all of them instead as stable particles, with scattering and decays then progressively added as perturbations.[46] Stanley Mandelstam [38, p. 1539], wishing to model the rising Regge trajectories within the double dispersion relations approach, introduced the "simplifying assumption" that the scattering amplitude is dominated by narrow resonances (where the amplitude is understood to be approximated by a finite number of Regge poles). In this scheme, Mandelstam was able to implement crossing symmetry using the FESR. To achieve

---

[46] The resonance width gives us an indication of the uncertainty about the particle's mass. The terminology of 'narrow-resonance' is something of an oxymoron of course, since if a resonance is wide then the particle will be short-lived (a resonance particle!).

the rising, Mandelstam uses two subtraction constants,[47] which in turn generates a pair of new parameters into the scheme: the Regge slope a and the intercept b (now written, α and $\alpha(0)$ respectively). These two parameters are absolutely central to the physical implications of the early attempts to construct dual symmetric, Regge behaved models, and still play a vital role today. Mandelstam makes an additional (well-motivated) assumption that the trajectories built from these parameters, namely $\alpha(s) = as + b$, do not rise "more than linearly with s" (p. 1542). For this reason, it might be prudent to call $\alpha(s)$ the 'Regge-Mandelstam slope' rather than the Regge-slope.[48]

2.4 A Note on Early Research Networks

For reasons that should by now be clear, those working on the S-matrix programme and the bootstrap approach to strong interaction physics play a 'statistically significant' role in string theory's early life, the latter being an outgrowth of the former via the dual resonance model (as we will see in the subsequent pair of chapters). An important subset of the current string theory researcher network can be traced back quite easily to a small group of physicists from this period in the 1960s, all working in and around the S-matrix programme (or dispersion relations) and Regge theory. This is quite natural, of course, since the dual resonance models can be viewed as a culmination of the bootstrap approach (recall Cushing's remark about superstring theory constituting "the ultimate bootstrap" [12]). The lines of influence are presented below.[49]

---

[47] Chu, Epstein, and Kaus [10] argued that Mandelstam's scheme for computing the subtraction constants depends too sensitively on both the cutoff used in the FESR and on the specific value of momentum transfer s at which the FESR are evaluated.

[48] Note that Veneziano's paper, "Construction Of A Crossing-Symmetric Regge-Behaved Amplitude For Linearly Rising Trajectories," was (by far) the highest cited paper to have been influenced by Mandelstam's. Note also, that the most highly cited paper to have in turn been influenced by Veneziano's paper was Neveu and Schwarz's paper introducing the dual pion model: "Factorizable Dual Model Of Pions". Continuing, the paper on "Vacuum Configurations for Superstrings" of Candelas, Horowitz, Strominger, and Witten is the highest cited citer of the Neveu-Schwarz paper—again, by a fairly large margin. (Citation analysis performed with Thomson-Reuters, *Web of Science*.) This gives some indication of the level of continuity between the earliest work on duality and modern superstring theory.

[49] Though this is a very selective network, of course, and misses many other important contributors, many of those associated with what have been labelled 'revolutionary' developments in string theory are located on this graph. (Note that neither circle size nor overlap has any representational relevance in this diagram.)

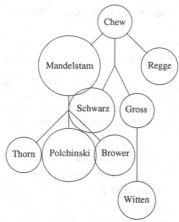

In particular, we can see a clear clustering around Geoffrey Chew and Berkeley. In a key move, Chew invited Stanley Mandelstam over to Berkeley, as a postdoc, who brought over the skills of complex analysis. It seems that Chew liked to be in close proximity to his students, and held weekly group meetings with them to discuss what they were working on. This close proximity clearly led to Chew's idiosyncratic positions being transmitted throughout the group.[50] Note that prior to joining Berkeley, Chew was based at the University of Illinois, Urbana-Champaign, together with Francis Low. Nearby, at the University of Chicago, were Nambu and Goldberger. Richard Brower, whom we will encounter later, had been at Berkeley, interacting closely with Chew and Mandelstam (his supervisor).

Note that John Schwarz was working on sum rules while at Princeton University in 1967. Schwarz's advisor was Geoff Chew. While at Berkeley, heavily influenced by Chew, he would have been steered away from work on elementary quarks[51] and quantum fields. Of course, this can't provide any explanation of why Schwarz and a few others from Chew's workshop *continued* to avoid quantum field theory. After all, David Gross (one of the few responsible for laying the finishing touches to QCD) was also a student of Chew's at roughly the same time as Schwarz and, indeed, the

---

[50] See p. 10 of Frautschi, Steven C. Interview by Shirley K. Cohen. Pasadena, California, June 17 and 20, 2003. Oral History Project, California Institute of Technology Archives. Retrieved [24th July, 2013]: http://resolver.caltech.edu/CaltechOH:OH_Frautschi_S. Frautschi, also a post-doc under Chew, shared an office with Mandelstam in 1960. Interestingly, Frautschi mentions (pp. 18–19) his later work on the so-called "statistical bootstrap" (employing some of Rolf Hagerdorn's ideas) reproduced facts of the Regge phenomenology (such as equal-spacing between successive spin states and exponential growth in particle species with mass increases) *without* invoking string theory, or being aware that what he was doing had any connection to the derivation of equal-spacing from the oscillations of a string system. By this stage, 1971–1972, Frautschi was at Cornell, and that he wasn't aware of the work that had by then been carried out using string models perhaps indicates that work on dual models and string models did not travel so widely and easily outside of the primary groups.

[51] Chew had referred disparagingly to quarks, in 1965, as "strongly interacting aristocrats" [7, p. 95].

two shared an office during their three final years (1963–6), writing a joint paper in 1965.[52]

Gross pinpoints the moment he became disillusioned with his supervisor's approach following a remark from Francis Low, at the 1966 Rochester meeting:

> I believe that when you find that the particles that are there in S-matrix theory, with crossing matrices and all of the formalism, satisfy all these conditions, all you are doing is showing that the S-matrix is consistent with the way the world is; that is, the particles have put themselves there in such a way that it works out, but you have not necessarily explained that they are there [28, p. 9101].

Gross did briefly return to the bootstrap approach with Veneziano's discovery of the beta function formula, but quickly became disillusioned once again, this time by its inability to explain scaling. As a result, Gross quickly brought himself up to speed on quantum field theory (especially renormalization group techniques) to try to find an explanation of scaling within field theory. As we see in Chap. 9, he would return to a descendent of the bootstrap programme much later, in 1985, when he helped construct the heterotic string theory.

Though things obviously become near-exponentially complicated once we move outwards from the origins of the bootstrap approach and dual models, we can trace paths of several important string researchers from Mandelstam too, including Joseph Polchinski and Charles Thorn.

There were two quite distinct styles of physics associated with the West Coast (roughly: Berkeley, Caltech) and the East Coast (roughly: Chicago, Princeton, Harvard). In particular, the East Coast seems to have been less dominated by 'physics gurus' (if I might be permitted to use that term).[53] However, this is to ignore the European influence: there is clearly a strong European component, though this will really come to dominate the theory of strong interactions in the period around Veneziano's presentation of his dual model.

This is, of course, very USA-centric, and much is missed. However, the influence spread across the Atlantic, especially to Cambridge University.[54] Mention should certainly be made too of the Japanese school. One of the initials of DHS duality (Richard Dolen) was based at Kyoto University for a time (at the Research Institute for Theoretical Physics). In his letters to Murray Gell-Mann (from 1966: in the

---

[52] Schwarz was much aided by Murray Gell-Mann's advocacy during the quieter years of superstring theory. In his closing talk at the 2nd Nobel Symposium on Particle Physics, Gell-Mann pointed out that Sergio Fubini joked that he (Gell-Mann) had "created at Caltech, during the lean years ... a nature reserve for an endangered species—the superstring theorist" [25, p. 202].

[53] At a session on dual models at CERN in 1974, Harry Lipkin put forth the following as a 'motto' of the session: "Dual theory should be presented in such a way that it becomes understandable to non-dualists. At least as understandable as East Coast theories are for West Coast physicists and vice versa" (http://www.slac.stanford.edu/econf/C720906/papers/v1p415.pdf). John Polkinghorne speaks of "Californian free-wheeling (bootstrappers)" and "New England Sobriety (field theory)" [45, p. 138]. Peter Woit's book [57, p. 150] includes a discussion of the East-West divide.

[54] Michael Green speaks of Cambridge as being "under the spell of the bootstrap ideas" [26, p. 528], with the standard graduate text being Eden, Landshoff, Olive, and Polkinghorne's *The Analytic S-Matrix* [18].

Gell-Mann archives of Caltech [Box 6, Folder 20]) he explicitly mentions interactions with several local physicists that went on to do important work on dual models and string theory—including Keiji Kikkawa, who later visited Rochester in 1967.[55]

Though it involves jumping ahead a little, much of the early detailed dual model work (including string models) took place at CERN. As has often been pointed out, this had much to do with the strong leadership and dual-model advocacy of Daniele Amati.[56] One could find David Olive (who would later take a post as a staff member, rather than a regular visitor, turning his back on a tenured position at Cambridge University), Peter Goddard, Ian Drummond, David Fairlie, and very many more centrally involved in the construction of string theory from the early dual resonance models.[57] Olive captures the hub-like dual model scene at CERN in the early 1970s as follows:

> Amati had gathered together from around Europe a galaxy of young enthusiasts for this new subject as research fellows and visitors. This was possible as centres of activity had sprung up around Europe, in Copenhagen, Paris, Cambridge, Durham, Torino and elsewhere. I already knew Peter Goddard from Cambridge University who was in his second year as Fellow, Lars Brink from Chalmers in Gothenburg was just starting, as was Jöel Scherk from Orsay, in Paris, all as Fellows, and destined to be collaborators and, particularly, close friends. Also present as Fellows were Paolo Di Vecchia (who arrived in January 1972), Holger Nielsen, Paul Frampton, Eugène Cremmer, Claudio Rebbi and others. Many visitors came from Italy, Stefano Sciuto, Nando Gliozzi, Luca Caneschi and so on. Visiting from the United States for the academic year were Charles Thorn and Richard Brower. Summer visitors included John Schwarz, and later Pierre Ramond, Joel Shapiro, Korkut Bardakçi, Lou Clavelli and Stanley Mandelstam, all from the United States [42, p. 349].

The early phase involving dual models was a particularly interconnected one, then, and also one featuring very many collaborative efforts. Stefano Sciuti, who had earlier been a part of Sergio Fubini's group in Turin, explicitly refers to the willingness to "join forces, cooperating rather than competing" as "fruit of the spirit of 1968" ([48], p. 216).

2.5 Summary

We have shown how the difficulties faced by quantum field theory in advancing beyond QED led to various models, one of which was Regge theory, with the addition of the dual resonance idea. This model achieved significant empirical successes, had

---

[55] Kikkawa later joined CUNY (with another dual model/string theorist Bunji Sakita) in 1970.

[56] I might add to this brief review of networks the fact that Amati took a sabbatical year in Orsay, while Andrè Neveu and Jöel Scherk were doing their PhDs there, spreading the gospel of dual resonance models to two of its future central proponents. Note that Neveu and Scherk later joined Schwarz in Princeton (in 1969) on NATO fellowships. However, since French higher degrees were not called PhDs, Neveu and Scherk were mistakenly classified as graduate students and assigned to Schwarz as such.

[57] David Fairlie himself oversaw a significant dual model group at Durham University in the UK, supervising several PhD theses on the subject in the 1970s.

several powerful theoretical virtues, and was therefore pursued with some excitement. We traced the story from Regge's introduction of complex angular momentum into quantum mechanics, to its extension into the relativistic domain. This combined with 'bootstrap' physics according to which the properties of elementary particles, such as coupling constants, could be predicted from a few basic principles coupled with just a small amount of empirical input. This journey culminated in the finite energy sum rules of Dolen, Horn, and Schmid, which were elevated to the status of a duality *principle*. The primary researcher network guiding research in this period was fairly narrowly confined, and can be charted quite precisely, with Geoff Chew as a key hub leading an anti-QFT school, as far as strong interactions were concerned. The bulk of later developments which place Regge-resonance duality at the heart of hadron physics (and the true beginnings of string theory) take place across the Atlantic, at CERN. We turn to these in the next chapter in which we discuss the Veneziano (dual resonance) model and its many extensions and generalisations.

References

1. Barger, V. & Cline, D. (1968). *Phenomenological theories of high energy scattering.* New York: Benjamin.
2. Cao, T. Y. (2010). *From current algebra to quantum chromodynamics: A case for structural realism.* Cambridge: Cambridge University Press.
3. TC Division. (1961). Track Chambers. CERN Annual Reports E (pp. 91–99): http://library. web.cern.ch/library/content/ar/yellowrep/varia/annual_reports/1961_E_p91.pdf.
4. Chew, G. M. L., & Goldberger, F. E. (1957). Application of dispersion relations to low energy meson-nucleon scattering. *Physical Review, 106,* 1337–1344.
5. Chew, G., & Frautschi, S. (1961). Principle of equivalence for all strongly interacting particles within the S-Matrix framework. *Physical Review Letters, 7,* 394–397.
6. Chew, G. (1962). *S-Matrix theory of strong interactions.* New York: W.A. Benjamin.
7. Chew, G. (1966). *The analytic S-matrix: A basis for nuclear democracy.* New York: W.A. Benjamin.
8. Chew, G. (1968). Aspects of the resonance-particle-pole relationship which may be useful in the planning and analysis of experiments. In G. Puppi (Ed.), *Old and new problems in elementary particles* (pp. 80–95). Amsterdam: Elsevier.
9. Chew, G. (1989). Particles as S-Matrix poles: Hadron democracy. In L. Hoddeson et al. (Eds.), *Pions to Quarks* (pp. 600–607). Cambridge: Cambridge University Press.
10. Chu, S.-Y., Epstein, G., & Kaus, P. (1969). Crossing-symmetric rising regge trajectories. *Physical Review, 175*(5), 2098–2105.
11. Coleman, S., & Norton, R. (1965). Singularities in the physical region. *Il Nuovo Cimento, 38*(1), 438–442.
12. Cushing, J. T. (1990). *Theory construction and selection in modern physics.* Cambridge: Cambridge University Press.
13. Cushing, J. T. (1985). Is there just one possible world? Contingency vs. the bootstrap. *Studies in the History and Philosophy of Science, 16*(1), 31–48.
14. Dirac, P. A. M. (1970). Can equations of motion be used in high-energy physics? *Physics Today, 23*(4), 29–31.
15. Donnachie, S. (1999). Probing the pomeron. *CERN Courier,* Mar 29: http://cerncourier.com/cws/article/cern/27985/2.
16. Dyson, F. (1948). The radiation theories of Tomonaga, Schwinger, and Feynman. *Physical Review, 75,* 486–502.

17. Eden, R. (1971). Regge poles and elementary particles. *Reports on Progress in Physics, 34*, 995–1053.
18. Eden, R. J., Landshoff, P. V., Olive, D. I., & Polkinghorne, J. C. (1966). *The analytic S-Matrix.* Cambridge: Cambridge University Press.
19. Fairlie, D. (2012). The analogue model for string amplitudes. In A. Capelli et al. (Eds.), *The birth of string theory* (pp. 283–293). Cambridge: Cambridge University Press.
20. Frautschi, S. (1995). Statistical studies of hadrons. In J. Letessier et al. (Eds.), *Hot hadronic matter: Theory and experiment* (pp. 57–62). Plenum Press.
21. Frautschi, S. C. (2003). Interview by Shirley K. Cohen. Pasadena, California, June 17 and 20, 2003. Oral History Project, California Institute of Technology Archives:http://resolver.caltech.edu/CaltechOH:OH_Frautschi_S.
22. Gell-Mann, M. G., & Goldberger, M. (1954). The scattering of low energy photons by particles of spin 1/2. *Physical Review, 96*, 1433–1438.
23. Gell-Mann, M. G. (1956). Dispersion relations in pion-pion and photon-nucleon scattering. In J. Ballam, et al. (Eds.), *High energy nuclear physics, In: Proceedings of the sixth annual Rochester conference.* (pp. 30–6). New York: Interscience Publishers.
24. Gell-Mann, M. G. (1964). The symmetry group of vector and axial vector currents. *Physics, 1*, 63–75.
25. Gell-Mann, M. G. (1987). Superstring theory. *Physica Scripta, T15*, 202–209.
26. Green, M. B. (2012) From String to Superstrings: A Personal Perspective. In A. Capelli et al. (eds.), *The Birth of String Theory* (pp. 527–543). Cambridge: Cambridge University Press.
27. Gross, D. (1992). Gauge theory—past, present, and future. *Chinese Journal of Physics, 30*(7), 955–971.
28. Gross, D. (2005). The discovery of asymptotic freedom and the emergence of QCD. *Proceedings of the National Academy of Science, 102*(26), 9099–9108.
29. Hoddeson, L., Brown, L., Riordan, M., & Dresden, M. (Eds.). (1997). *The rise of the standard model: Particle physics in the 1960s and 1970s.* Cambridge: Cambridge University Press.
30. Jackson, J. D. (1969). Models for high-energy processes. *Reviews of Modern Physics, 42*(1), 12–67.
31. Jacob, M. (Ed.). (1981). *CERN: 25 years of physics, physics reports reprint book series*, (Vol. 4). Amsterdam: North Holland.
32. Kaiser, D. (2002). Nuclear democracy: Political engagement, pedagogical reform, and particle physics in postwar America. *Isis, 93*, 229–268.
33. Krömer, R. (2001). The duality of space and function, and category-theoretic dualities. Unpublished manuscript: http://www.univ-nancy2.fr/poincare/documents/CLMPS2011ABSTRACTS/14thCLMPS2011_C1_Kroemer.pdf.
34. Landau, L. D. (1959). On analytic properties of vertex parts in quantum field theory. *Nuclear Physics, 13*, 181–192.
35. Leader, E. (1978). Why has Regge pole theory survived? *Nature, 271*, 213–216.
36. Lusanna, L. (1974). Extended hadrons and Regge slope. *Lettere Al Nuovo Cimento, 11*(3), 213–217.
37. Mandelstam, S. (1958). Determination of the pion-nucleon scattering amplitude from dispersion relations and unitarity general theory. *Physical Review, 112*(4), 1344–1360.
38. Mandelstam, S. (1968). Dynamics based on rising Regge trajectories. *Physical Review, 166*, 1539–1552.
39. Mandelstam, S. (1974). Dual-resonance models. *Physics Reports, 13*(6), 259–353.
40. Nielsen, H. (2012). String from Veneziano model. In A. Capelli et al. (Eds.). *The birth of string theory* (pp. 266–274). Cambridge: Cambridge University Press.
41. Nussenzveig, H. M. (Ed.). (1972). *Causality and dispersion relations.* North Holland: Elsevier.
42. Olive, D. I. (2012). From dual fermion to superstring. In A. Cappelli et al. (Eds.), *The birth of string theory* (pp. 346–360). Cambridge: Cambridge University Press.
43. Phillips, R. J. N., & Ringland, G. A. (1972). Regge phenomenology. In E. Burhop (Ed.). *High energy physics.* Massachusetts: Academic Press.

44. Pickering, A. (1984). Constructing quarks: A sociological history of particle physics. Chicago: University of Chicago Press.
45. Polkinghorne, J. (1989). *Rochester Roundabout*. London: Longman.
46. Ramond, P. (1987). The early years of string theory: The dual resonance model. In R. Slansky & G. B. West (Eds.). *Proceedings of Theoretical Advanced Study Institute Lectures in Elementary Particle Physics* (pp. 501–571). Singapore: World Scientific.
47. Regge, T. (1959). Introduction to complex angular momenta. *Il Nuovo Cimento, 14*(5), 951–976.
48. Sciuto, S. (2012). The '3-Reggeon Vertex'. In A. Capelli et al. (Eds.), *The birth of string theory* (pp. 214–217). Cambridge: Cambridge University Press.
49. Schmid, C. (1970). What is duality? *Proceedings of the Royal Society of London, Series A, Mathematical and Physical Sciences, 318*(1534), 257–278.
50. Susskind, L. (2006). *The cosmic landscape*. USA: Back Bay Books.
51. Susskind, L. (2012). The first string theory: Personal recollections. In A. Capelli et al. (Eds.), *The birth of string theory* (pp. 262–265). Cambridge: Cambridge University Press.
52. Taylor, J. R. (2000). *Scattering theory: The quantum theory of nonrelativistic collisions*. New York: Dover.
53. Teller, P. (1997). The philosopher problem. In L. Hoddeson, L. Brown, M. Riordan, & M. Dresden (Eds.), *The rise of the standard model: Particle physics in the 1960s and 1970s* (pp. 634–636). Cambridge: Cambridge University Press.
54. 't Hooft, G. (1999). When was asymptotic freedom discovered? Or the rehabilitation of quantum field theory. *Nuclear Physics, B74*(1–3), 413–425.
55. Weinberg, S. (1977). The search for unity: Notes for a history of quantum field theory. *Daedalus, 106*(4), 17–35.
56. Wheeler, J. (1994). Interview of John Wheeler by Kenneth Ford on March 28, 1994, Niels Bohr Library and Archives, American Institute of Physics, College Park, MD USA, www.aip.org/history/ohilist/5908_12.html.
57. Woit, P. (2007). *Not even wrong: The failure of string theory and the search for unity in physical law*. New York: Basic Books.

Chapter 3
The Veneziano Model

*All through 1969 people were adding legs to the Veneziano
amplitude, or chopping it in half.*

Claud Lovelace

3.1 Duality and the Beta Function Amplitude

In this chapter we look at the birth of dual models (or, in full, dual resonance models)
and the beginnings of dual *theory* (of which the hadronic string theory is an example,
providing an interpretation of its oscillator formalism). Given that these were found
to admit an interpretation as a string system, they are usually believed to constitute
the simultaneous birth of string theory. This is somewhat inaccurate since the very
earliest work on dual models, as with the work on duality that preceded it, had no
explicit connection whatsoever to string models. They were an attempt to incorpo-
rate the FESR duality together with the other S-matrix principles in a single model
describing hadrons. There was, at best, some indirect evidence of non-locality from
the high spins that (after the fact) might have been seen as a result of coming from
an underlying string system. There was also the Regge behaviour that could also
be reinterpreted in string theoretic terms after the fact. Indeed, one might as well
mark the notion of the Regge trajectory (with its peculiar regularities) as the birth
of strings if one is allowing *post hoc* string interpretations. I would urge that we
should understand the birth of string theory (hadronic string theory, or 'old' string
theory, that is) as taking place with the (multiple independent) discoveries of the
idea that a possible system *responsible* for 'generating' the Veneziano formula is a
family of harmonic oscillators, and of a very specific type like the string of a guitar
(this we discuss in the next chapter). To claim otherwise is clearly to project the later
interpretation onto the earlier work. Having said that, Veneziano's formula clearly

D. Rickles, *A Brief History of String Theory*, The Frontiers Collection,
DOI: 10.1007/978-3-642-45128-7_3, © Springer-Verlag Berlin Heidelberg 2014

paved the way for such string interpretations, which were of course interpretations of the structure it revealed.

The crucial step connecting duality to the other desirable properties of the scattering amplitude (such as Regge behaviour and crossing) was, then, taken by Gabriele Veneziano.[1] He observed that the Euler Beta function was able to model these features in a nice condensed (and, indeed, closed) form. Veneziano's model performed two feats: (1) it captured the empirical linear Regge trajectories relating M^2 and J; (2) it incorporated the mathematical properties of scattering amplitudes expected of strongly interacting systems (including the DHS duality identifying t and s processes) but, on account of the narrow-resonance approximation used, it did *not* satisfy unitarity. It was therefore (*almost*) a complete solution of the bootstrap—a model satisfying the conditions imposed on the S-matrix without employing quantum field theory. This was the first example of a dual resonance model. Veneziano presented his idea in July 1968, to a seminar group at which Sergio Fubini was in attendance. Encouraged by Fubini's response, Veneziano published soon after ([46, p. 214]—see Fig. 3.1).[2]

Veneziano was able to construct a representation of the 4-meson process $\pi\pi \rightarrow \pi\omega$, written in closed form, using products of Gamma functions:

$$A(s, t, u) = \frac{\Gamma[1 - \alpha(s)]\Gamma[1 - \alpha(t)]}{\Gamma[-\alpha(s) - \alpha(t)]} + \frac{\Gamma[1 - \alpha(s)]\Gamma[1 - \alpha(u)]}{\Gamma[-\alpha(s) - \alpha(u)]}$$
$$+ \frac{\Gamma[1 - \alpha(t)]\Gamma[1 - \alpha(u)]}{\Gamma[-\alpha(t) - \alpha(u)]} \tag{3.1}$$

Each summand will have a singularity (i.e. a pole) at negative integer values $(0, -1, -2, ...)$ of the argument.[3] These singularities point to locations of particles on Regge trajectories (i.e. poles in s or t). Hence, the Γ-function singularities reproduce the spectrum of particles lying on linearly rising Regge trajectories of ever

---

[1] As Christoph Schmid points out, achieving this was not an obvious possibility at the time: "[I]et me remind you that many people published 'proofs' that duality was impossible ... until Veneziano (1968) published his beautiful model. Since one example is stronger than a thousand 'proofs' to the contrary, people had to accept the fact that duality was possible" [42, p. 125]. The discovery also seems to have opened the floodgates, for some, as regards the possibility of saying something profound about hadronic scattering amplitudes (*behind* the various approximations). As David Fairlie writes, "to everyone's complete surprise Gabriele Veneziano came up with his famous compact form for a dual scattering amplitude, which encompassed contributions from many towers of resonances, and I felt that this was for me!" [20, p. 283]. It is, of course, often the case that the impact of some result is all the more impressive when its prior probability is very low. We see the same 'high impact' phenomenon following Michael Green and John Schwarz's anomaly cancellation result (for specific string theories) which had also been assigned a vanishingly small prior probability by the community of physicists working on it before their discovery.

[2] As David Olive recalls, Veneziano presented his discovery "in the ballroom of the Hofburg ... during the Vienna Conference on High Energy Physics (28 August–5 September 1968)" [38, p. 346].

[3] The Euler beta function is related to the Gamma function a follows: $B(a, b) = \frac{\Gamma(a)\Gamma(b)}{\Gamma(a+b)}$. Hence, we can also write Eq. 3.1 simply as $A(s, t, u) = A(s, t) + A(s, u) + A(t, u)$. This expression encodes the three possible permutations of the four scattered particles' labels (that are neither cyclic nor

IL NUOVO CIMENTO   VOL. LVII A, N. 1   1° Settembre 1968

Construction of a Crossing-Simmetric, Regge-Behaved Amplitude for Linearly Rising Trajectories.

G. VENEZIANO (*)

CERN · Geneva

(ricevuto il 29 Luglio 1968)

Crossing has been the first ingredient used to make Regge theory a predictive concept in high-energy physics. However, a complete and satisfactory way of imposing crossing and crossed-channel unitarity is still lacking. We can look at the recent investigations on the properties of Reggeization at $t=0$ as giving a first encouraging set of results along this line of thinking [1]. A technically different approach, based on superconvergence, has been also recently investigated [2], and the possibility of a self-consistent determination of the physical parameters, through the use of sum rules, has been stressed.

In this note we propose a quite simple expression for the relativistic scattering amplitude, that obeys the requirements of Regge asymptotics and crossing symmetry in the case of linearly rising trajectories. Its explicit form is suggested by the work of ref. [3] and contains only a few free parameters [**].

Our expression contains automatically Regge poles in families of parallel trajectories (at all t) with residue in definite ratios. It furthermore satisfies the conditions of superconvergence [4] and exhibits in a nice fashion the duality between Regge poles and resonances in the scattering amplitude.

_____

(*) On leave of absence from the Weizmann Institute of Science, Rehovoth. Address after 1 September 1968: Department of Physics, M.I.T., Cambridge, Mass.

(1) For a general review of these problems see L. BERTOCCHI: *Proc. of the Heidelberg International Conference on Elementary Particles* (Amsterdam, 1967).

(2) Such an approach was proposed independently by M. ADEMOLLO, H. R. RUBINSTEIN, G. VENEZIANO and M. A. VIRASORO: *Phys. Rev. Lett.*, **19**, 1402 (1967) and *Phys. Lett.*, **27**, B 99 (1968), and by S. MANDELSTAM: *Phys. Rev.*, **166**, 1539 (1968). Further developments and a number of references to related works can be found in ref. (2).

(3) M. ADEMOLLO, H. R. RUBINSTEIN, G. VENEZIANO and M. A. VIRASORO: Weizmann Institute preprint (1968), submitted to *Phys. Rev.*

(**) We shall mostly work here in the approximation of real, linear trajectories and consequently of narrow resonances. We briefly discuss the effects of a nonzero imaginary part in the trajectory function which, in any case, we demand to have a linearly rising real part.

(4) For superconvergence we mean both the original sum rules proposed by V. DE ALFARO, S. FUBINI, G. FURLAN and C. ROSSETTI: *Phys. Lett.*, **21**, 576 (1966), and the more recent generalized superconvergence (finite-energy) sum rules (see ref. (2) for detailed references). A unified treatment of all superconvergence sum rules has been given by S. FUBINI: *Nuovo Cimento*, **52** A, 224 (1967).

Fig. 3.1 The first page of the chapter that is often seen as marking the origin of string theory (Photo credit: Springer, [49])

higher rotations. Notice also that $A(s, t) = A(t, s)$, so that the formula satisfies DHS duality. This was no mean feat and formed the basis of a mini-industry of research on dual models, leading later to dual theory, and from there to both aspects of string theory and (via the string picture) QCD.

It is simpler to see this duality by writing the amplitude just as a function of s and t, giving (where, again, $\alpha(s)$ is the Regge-Mandelstam trajectory[4]):

$$A(s, t) = \frac{\Gamma[-\alpha(s)]\Gamma[-\alpha(t)]}{\Gamma[-\alpha(s) - \alpha(t)]} = B(-\alpha(s), -\alpha(t)) \tag{3.2}$$

Here, $B(-\alpha(s), -\alpha(t))$ is the Euler Beta function and can be represented as an integral:

$$B(-\alpha(s), -\alpha(t)) = \int_0^1 dx \, x^{-\alpha(s)-1} (1 - x)^{-\alpha(t)-1} \tag{3.3}$$

The formula describes (for the specified linear trajectories $\alpha(s)$) an infinite set of (zero width) poles.[5] Once this model was out, the immediate challenge was to add unitarity; generalise it from four-particle to multi-particle amplitudes; add spin and isospin; and understand it from a more physical point of view (that is, understand *what* it is a model *of*). The first step crucial along these latter lines was the development of the harmonic oscillator representation of the generalised amplitude [24], which allowed for a demonstration of factorization (on which, see Sect. 3.4).

Though Veneziano's amplitude satisfied duality, it was valid only in an extreme approximation, namely the 'narrow resonance approximation' (in fact, with *infinitesimally* narrow resonances), with infinitely many poles—so: an infinite set of particles is found to be sufficient for both resonances and Regge poles (exchange particles). Recall that the width is related to stability, so that in the limit of zero-width a particle would never decay.[6] Further, the 'zero-width' approximation used implies that an infinite family of 1-hadron states make up the intermediate states *and* implies that an infinite family of such states will be exchanged, giving the strong forces. Duality requires that we don't add these together, but treat them as different approximations to the same physical process. The narrow resonance approximation is a very useful

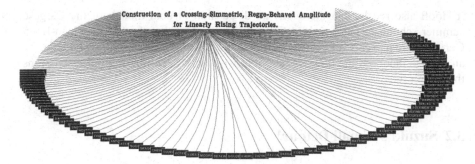

Fig. 3.2 The dramatic impact of Veneziano's paper. This network map shows the number of authors referring to the paper within a year of its publication, with each line representing a paper discussing the Veneziano model (Data generated using Thompson-Reuters, *Web of Knowledge*.)

tool for studying duality since one can get around the business (real though it is) of hadronic instability.

It was natural, therefore, to extend Veneziano's work to get around this short-coming. The race to generalise, extend , and otherwise make sense of the Veneziano model was very dramatic, as can be seen in Fig. 3.2, which shows the number of papers citing Veneziano's paper within one year of its publication. There were a great many instances of independent simultaneous discoveries of the same results, making it particularly hard to pin down priority claims—and as a result I shall generally avoid this, except where priority is obvious.

It is worth pointing out that Veneziano's paper did not take with all strong interaction physicists; it was primarily deemed to be of importance to those already working on Regge theory and duality. For example, Gerardus 't Hooft's experience on first encountering it was as follows[7]:

> [W]e didn't understand it, I think. The understanding came much later, that this really was a string theory. So, Veneziano's paper was understood, but not as a string theory. It was some sort of abstract notion and ... it sounded like something very complicated, very difficult ... in particular how to show these theories of dispersion relations, are they unitary? And, all we knew was that the problem was complicated. People didn't give unique answers. Some people said, "yes," some people said, "no," some people said, "maybe." I mean, you had expressions which have pole singularities in the propagators. They have the right structure. They could obey dispersion relations. It's nearly right. So, in general the theory is nearly right, but there are still some things missing. And ... without the string interpretation these theories look very complicated. ... Later we realized ... that the particles are string-like. It became much more transparent. But then, I always thought there were fundamental difficulties with it, even in my days at CERN. Now and then I tried these theories, and often the theories were the strong interactions. But, later with QCD I realized these theories could be a good approximation ... to QCD. But, even as approximation[s] ... I failed to make sense of them. (Interview with the author, 10th February, 2010—AIP, p. 31; transcription courtesy of the American Institute of Physics)

---

[7] Note that 't Hooft's comments have some overlap with my reasons for being cautious about marking the birth of string theory with the construction of the Veneziano's model.

't Hooft also recalls that there was no discussion of dual models at the Cargèse summer school on physics, in 1970, despite the fact that the likes of Jean-Loup Gervais were present—of course, 't Hooft's mind was focused firmly on field theory and gauge theory at the time, so he might well have been filtering out discussions that didn't fit.

3.2 Suzuki's 'Small Detour'

> For me, the beta function amplitude was a small detour in my long career.
> Mahiko Suzuki

Before we proceed to the near-industrial scale refinement job that followed the publication (and spread) of Veneziano's discovery, I would first like to consider another figure that is often referred to as a kind of 'co-discoverer' of the link between the beta function and a dual amplitude: namely, Mahiko Suzuki, referred to in the first chapter. Suzuki's story paints an interesting (and more complete) picture of the research landscape around the time of Veneziano's own discovery.[8]

Suzuki received his PhD from the University of Tokyo in 1965, under the supervision of Hironari Miyazawa, who, curiously enough, himself proposed a fledgling version of supersymmetry for hadrons in 1966 (relating mesons and baryons)—though his version involved internal rather than space-time transformations. Miyazawa arranged for Suzuki to skip his final year of graduate school and join Gell-Mann's Caltech group, as a Fulbright scholar, from 1965 to 1967—upgraded to a Richard Chase Tolman Research Fellowship in 2 years, thanks to Gell-Mann. Among those he shared an office with were David Horn and Christoph Schmid (the H and S from DHS-duality), and also Roger Dashen and Stephen Adler. Indeed, one of his early collaborations was on the FESR with Horn and Schmid. Though he left this collaboration early on, he readily admits that the FESR work was vital background for his own discovery of the beta function amplitude. Suzuki spent a year at the IAS in Princeton following his Caltech fellowship, and coincidentally Gell-Mann took his sabbatical there that year, cunningly co-arranged with Low, Goldberger, and Kroll, who also took their sabbaticals at the IAS that year. Also present as postdocs were Daniel Freedman and Jiunn-Ming Wang, who had just come up with their result about parallel Regge trajectories (see footnote 29). Suzuki came up with the idea that the beta function must be the scattering amplitude that incorporates duality during this visit, during the end of term break.

---

[8] Of course, it is the dissemination that (quite rightly) holds the weight in matters of scientific discovery, so I don't mean to reduce Veneziano's place in the history of dual theory and string theory with this discussion. My aim is to flesh out the background to the discovery and to present a piece of the history that has hitherto remained under wraps—my sincere thanks to Professor Suzuki for sharing his story with me.

Suzuki seems to have followed a similar path to Veneziano, namely going to the zero-width limit to achieve a simplification of the scattering problem and to make duality easier to satisfy in a transparent way. He describes his next steps as follows:

> I needed a gamma function of a Regge trajectory to incorporate a family of the Regge trajectory and its daughters in the intermediate state channel, and another gamma function for the Regge family in the force channel. After taking product (not sum) of these two gamma functions, I need the third function to make the high-energy (Regge) asymptotic behaviour in agreement with experiment. In the spring of 1968, I tried to realize this ideal limit of the hadron amplitudes. Once the problem was simplified so much, I had only to look for a right product of gamma functions (Private communication).

Thus the stage was set in such a way as to allow for a methodical search for an appropriate mathematical expression that supplied the required product. In this sense one can see very clearly that luck plays no role, and I expect that the same methodical procedure lay behind Veneziano's discovery. The book of formulas that Suzuki found the correct expression in was the 3-volume set *Sugaku Koshiki*,[9] by Moriguchi, Udagawa and Hitotsumatsu (Iwanami Shoten, 1956). The necessary function was on p. 2 of volume 3, in an entry entitled: "The asymptotic behavior of a ratio of gamma functions". Though the asymptotic behavior of the *Beta* function was not covered, Suzuki had no problem transforming it into that of the beta function.[10]

On this discovery, that ignited so much subsequent work after the publication of Veneziano's paper, he writes that "it was a small (I thought so, then) exciting discovery for me" (private communication). Hence, this reveals an interesting parallel discovery of the beta function-dual amplitude connection, replete with what looked like the right scientific context (with Horn, Schmid, and Gell-Mann all in Suzuki's loop). Suzuki prepared a paper containing the result, and planned to submit it to *Physics Letters B*, once he had arrived at CERN, where he was due to stay as a visiting scientist for a few months after his Fulbright had expired. He didn't feel any pressure since, as he puts it, he "did not anticipate anybody could possibly come up with this esoteric amplitude". On his arrival at CERN Suzuki handed over his manuscript (handwritten) to the secretary (a Madame Fabergé) for typesetting ready for mailing to the journal. However, Suzuki was told that the paper had first to be approved by a senior physicist. Suzuki went to Leon van Hove's office and explained his beta function amplitude idea, after which van Hove agreed to read the manuscript, whereupon he placed the paper in a drawer. It turned out that Schmid was a postdoc at CERN at the same time, and so he was there to greet Suzuki on his first day. On explaining his work to Schmid (following the customary academic greeting of 'what are you working on?'), Schmid pointed out that a young postdoc by the name of Veneziano had written a preprint of work that sounded similar. Suzuki rushed off

---

[9] Meaning "mathematical formulas". These volumes traveled with Suzuki when he left Japan for life in Pasadena. They still grace his shelves in Berkeley. Following an expression of interest in them from David Horn, Suzuki had a set mailed over as a present to Horn in 1966.

[10] Though it seems he initially stuck to writing it as the ratio of gamma functions, as expressed in his book of formulas. It was in fact Ling-Lie Wang (now Chau, currently at UC Davis) that mentioned that this ratio was simply the beta function, while chatting about their current research topics in the IAS library reading room.

to get a copy of the preprint and discovered that it was virtually identical, save for the choice of properties of the scattered particles, $\pi \pi \to \omega \pi$ in Veneziano's case and $\pi^- p \to \pi^0 n$ in Suzuki's case (where he had omitted spins as an inessential complication).[11] Suzuki realised he'd been "scooped" and retracted his paper from van Hove's office.

Thus goes the story of a parallel discovery of the beta function-dual amplitude. Had the timing been a little different, Suzuki might have been able to give a seminar on the result at the IAS, or at least discussed it further with people like Horn, Schmid, and Gell-Mann. Of course there are many such instances of multiple, parallel discoveries in science—not least the discovery of 'the Higgs mechanism'! Interestingly, Suzuki suggests that he and Veneziano were not alone in their search for an appropriate function. Chatting to Nambu during the late summer of 1968, while attending the biennial international high energy conference in Vienna, Suzuki discovered that Nambutoo was "casually combing for fun some mathematical books to search a function that satisfies the nuclear democracy or the duality" (private communication). Though Nambu didn't come to the beta function in his search, it highlights the fact that there was scientific convergence and, once again, discredits the notion that there was any kind of randomness involved in the discovery of the beta function amplitude.

It seems that Murray Gell-Mann had been aware of Suzuki's parallel discovery for he explicitly credits him as "co-inventor" in a reference for a position at Berkeley in 1969 (see Fig. 3.3). Suzuki recalls George Trilling, then Chairman of physics at Berkeley, introducing him as "co-discoverer of the Veneziano amplitude" during his department colloquium in the late summer of 1969.

3.3 Unitarity, Generalisations, and Extensions

The Veneziano model, though impressive, did not involve *all* the desirable properties of a good S-matrix bootstrap: it violated unitarity (i.e. the preservation of the probabilities summing to one [= unity], at each instant of time) and involved only four particles. Both a tree-level N-point amplitude and a treatment of loop amplitudes were needed to patch these problems and achieve a complete theory. Adding unitarity to the model would produce a representation of the world of an infinite number of resonances. This problem was solved while the model was still floating free of the string interpretation. As we see in the next section, which overlaps temporally with many of the issues of this section, ghost (negative probability) states were introduced along the way, and a framework also needed to be developed to remove these from the space of states, isolating a physical subspace.

---

[11] Not long after the publication of Veneziano's paper, both Claud Lovelace [33] and Joel Shapiro [43] independently constructed a Veneziano-like formula for the reaction involving $\pi + \pi \to \pi + \pi$ scattering.

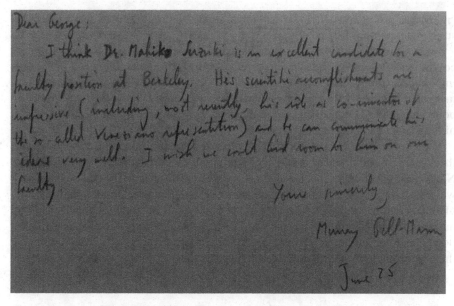

Fig. 3.3 Letter of reference for Mahiko Suzuki (addressed to George H. Trilling; dated 25th June, 1969), by Murray Gell-Mann, crediting Suzuki with independent discovery of the Beta function amplitude. *Image source* Gell-Mann papers, Caltech [Box 19, Folder 14]

Veneziano's original amplitude was for 4-particles, 2-in and 2-out, and only held in an extreme approximation of zero width resonances.[12] It was also an orbital model, lacking spinning particles, but this took a little longer to correct, as the search began for more realistic dual models.[13] First, in order to properly establish the Veneziano model, an N-point amplitude involving any number of loop contributions was required, as mentioned above. The first issue was quickly generalised to so-called 'production amplitudes,' involving more output particles than went in to the process. For example, Korkut Bardakçi and Henri Ruegg (both visiting CERN at the time) [3] first generalised the Veneziano model to five particle scattering, 2-in and 3-out. Miguel Virasoro [50] also found a 5-point amplitude. Soon after, Chan Hong-Mo [10] further extended this to a six-point function and from there to the N-point case—independent results of this kind were obtained again by Bardakçi and Ruegg [3], and also Goebel and Sakita [26] and Koba and Nielsen [32].

As Chan noted, in 1970, these many-particle generalisations of Veneziano's amplitude offer a better implementation of the bootstrap idea, since all particles can be treated as bound states of the other. It led some to think that, in Chan's words, it was

---

[12] This approximation basically means, in modern terms, that it is only carried out at the tree level, with 'external' particles, not to all orders.

[13] However, an early study of the problem of incorporating spin, in some detail, was that of Yasunori Miyata in Tokyo [36]. Later studies, as we see in a later chapter, correspond to what are now viewed as the first spinning string models of Pierre Ramond and Andrè Neveu and John Schwarz.

"more than just phenomenology" and might mark "the beginning of a new theory of strong interactions" [11, p. 379].

Within the context of these generalisations, David Fairlie and Keith Jones discovered the existence of a tachyonic ground state (i.e. for which $m^2 = -1$): "if one imposes the (unphysical) condition $\alpha(0) = 1$ demanding that the ground state is a tachyon (i.e. possesses a particle of negative mass squared), then the four- and five-point amplitudes can be expressed as integrals of a single integrand over the whole of the real line and the plane respectively" [21, p. 284].[14]

Ziro Koba[15] and Holger Nielsen established a highly influential framework for the N-particle amplitude in 1969—they too initially focused on the 5-point function. Nielsen had just finished his *Candidatus Scientiarum* degree at the end of 1968, and was able to briefly refer to the Veneziano model, after learning about it during a talk of Hector Rubinstein's that he'd seen at the Niels Bohr Institute (and considered to have been an highly influential episode).[16] The basic idea was to choose as variables points on a line in the projective plane ('Koba-Nielsen variables'). They presented their work at CERN shortly after developing the idea.

In the Koba-Nielsen framework (developed in [32]) the Beta function looks like:

$$B(\alpha(s), \alpha(t)) = \int_0^1 (1 - x)^{-\alpha' t - \alpha_t - 1} x^{\alpha' s - \alpha_s - 1} \mathrm{d}x \qquad (3.4)$$

The N-point amplitude takes the form (with subscripts labelling particles such that i refers to the ith particle, possessing momentum p_i):

$$A(s, t) = \int_{-\infty}^{\infty} \frac{\mathrm{d}^N z}{\mathrm{d}V_{abc}} \prod_1^{N-1} (z_i - z_{i+1})^{\alpha(0)-1} \prod_1^{N-1} \theta(z_i - z_{i+1}) \prod_{i<j} (z_i - z_j)^{2\alpha' p_i \cdot p_j} \qquad (3.5)$$

---

[14] Fairlie compares the problem of the dual model tachyon to that weighing on Yang-Mills theory in the early days of its existence because of the zero mass particles described by the theory, which seemed clearly inadequate in accounting for spin-1 strongly interacting particles [20, p. 284]. As he notes, though his colleagues were sceptical of resolving the problem, the ground state tachyon was indeed eliminated thanks to a clever projecting out of the physically irrelevant sector of states by Gliozzi, Scherk, and Olive, in 1977 (see Sect. 7.3).

[15] The Japanese physicist Ziro Koba died on 28th September 1973. He had been a student of Tomonaga's. Apparently, he had once shaved his head as a self-punishment for making an error in a self-energy calculation he was carrying out for Tomonaga (see Madhusree Mukerjee's article on Nambu in *Scientific American*, February 1995, p. 38). He had neglected to include certain processes involving virtual pairs created via the Coulomb self-interaction of a vacuum electron—see *Progress in Theoretical Physics* 2(2), pp. 216–217 (for the original calculation) and p. 217 for the retraction. Curiously, Koba had once shared an office with Yoichiro Nambu (who would later become the first to give the full string interpretation of the Veneziano formula) in Tokyo, just after the second world war.

[16] See http://theory.fi.infn.it/colomo/string-book/nielsen_note.txt. It seems that Hector Rubinstein was very effective in spreading the news of the Veneziano model. Leonard Susskind also credits Rubinstein with bringing the Veneziano amplitude to his attention while Rubinstein (then based in Israel) was visiting him in New York [48, p. 204].

Here $dV_{abc} = \frac{dz_a dz_b dz_c}{(z_b - z_a)(z_c - z_a)(z_a - z_c)}$ is a measure intended to formalise the conformal invariance of the formula, which is expressed in terms of invariance with respect to the modular (or Möbius: $SL(2, \mathbb{R})$) group $z' \to \frac{az+b}{cz+d}$ (with unit determinant: $ad - bc = 1$). The Möbius invariance of the amplitude ensures that duality is preserved. The conformal invariance was later interpreted via an analogy with electrostatics, by Fairlie and Nielsen (see Sect. 4.1). This approach was crucial in unpacking some of the deeper elements of the mathematical structure underpinning the dual models, especially that pertaining to string worldsheets.

In 1969, Jack Paton and Chan Hong-Mo [39] further generalised the (already generalised, as above) Veneziano model by adding isospin factors (extending the reach of Veneziano models to non-neutral mesons). Though this later became the orthodox method of attaching quantum numbers to open string end-points, it should be understood that there is no question of the factors being associated with strings at this stage. Rather, the analysis is done using external lines (corresponding to the external particles) in a standard graph picture—though they do presciently mention in closing a similarity between their solution and quark pictures of the Harari-Rosner type.[17] The method is to assign an element of $SU(3)$ (that is, the 3×3 λ matrices of $SU(3)$) to the external lines (here denoted by π_a, K, and $\overline{K}$) of a scattering diagram (where τ_a is a Pauli matrix and a_i is the isospin label):

$$\pi_a \sim \begin{pmatrix} \tau_a & 0 \\ 0 & 0 \end{pmatrix}, \quad K \sim \begin{pmatrix} 0 & K \\ 0 & 0 \end{pmatrix}, \quad \overline{K} \sim \begin{pmatrix} 0 & 0 \\ K & 0 \end{pmatrix} \qquad (3.6)$$

The isopsin factors associated with some particular ordering (say, 1, 2, 3, ..., N) are then given by the trace formula: $\mathrm{Tr}(\tau_{a_1} \tau_{a_2} ... \tau_{a_N})$. Amplitudes are multiplied by such factors in order to implement isospin.

Virasoro [51] was also able to construct a novel dual amplitude, distinct from Veneziano's, but sharing analyticity, crossing-symmetry, Regge behaviour, and the other desirable properties. This took the form:

$$A(s,t,u) = \frac{\Gamma(\frac{1}{2} - \frac{1}{2}\alpha(s)) \Gamma(\frac{1}{2} - \frac{1}{2}\alpha(t)) \Gamma(\frac{1}{2} - \frac{1}{2}\alpha(u))}{\Gamma(1 - \frac{1}{2}\alpha(t) - \frac{1}{2}\alpha(u)) \Gamma(1 - \frac{1}{2}\alpha(s) - \frac{1}{2}\alpha(u)) \Gamma(1 - \frac{1}{2}\alpha(s) - \frac{1}{2}\alpha(t))} \qquad (3.7)$$

This reduces to Veneziano's formula for intercept 2 ($\alpha(s) + \alpha(t) + \alpha(u) = 2$: to eliminate ghosts). Just as the Veneziano formula was taken up and extended and generalised in various ways, so too was Virasoro's version. Joel Shapiro constructed the N-point generalisation of Virasoro's original 4-point amplitude in 1970 [44], with the integral form:

---

[17] Rather interestingly, these factors would undergo successive transformations ("theoretical exapation" in the terminology of Chap. 7) as the understanding of dual models underwent its own transformation, first into string theory (amounting to an index known as the "Chan-Paton factor") characterising the endpoints of open strings) and then later as a result of developments leading to D-branes (amounting to an index characterising the surfaces the endpoints of open strings terminate on). (I should also point out, as a matter of historical accuracy, that Chan-Paton factors ought really to be called Paton-Chan factors, given that Paton was the lead author on the original paper.)

$$\int d^2z |z|^{k_1 \cdot k_2 / 2} |1 - z|^{k_2 \cdot k_3 / 2} \tag{3.8}$$

Integrations go over the complex plane, and in contrast to the Veneziano amplitude (in Koba-Nielsen form), the integrand possesses $SL(2, \mathbb{C})$ invariance. Shapiro also generalised the electrostatic analogy to the Virasoro case, with the external particles living on the surface of the Argand plane.[18]

The next section deals with the factorization of the amplitude, which is an essential step in the study of radiative corrections (i.e. loop amplitudes). This latter task was initiated by Kikkawa, Sakita, and Virasoro [31]. This was a fairly radical (in the dual model context), intuitive attempt to restore unitarity in which the Veneziano model is considered to be the lowest-order term (i.e. a Born term) in a perturbation series, approximating a more complete unitary theory (with unitarity emerging via the loop expansion). The idea was to work by analogy with standard Feynman-diagram technology to get a "Feynman-like" theory, generating a perturbation series which, when summed, would give a unitary amplitude of resonances with non-vanishing widths.[19] This work depended on a thorough understanding of the factorization properties of tree diagrams which was achieved using an operator formalism developed by Sergio Fubini, David Gordon, and Gabriele Veneziano [24]. Using this formalism they were able to construct loop diagrams of any order by sewing together tree diagrams. The resulting loop amplitudes, constructed later, admitted a physical interpretation just in case Virasoro's unit intercept condition held and if a condition on the dimensionality

---

[18] Note that there exists a two-to-one correspondence between the operators in the original Veneziano model and in the Virasoro-Shapiro model (an important implication of this is a doubling of masses in the latter, as compared to the former). However, both the original Veneziano model and the Virasoro-Shapiro model are consistent only in $d = 26$ (a discovery that would be made in the year following these generalizations). The Virasoro-Shapiro model was later interpreted to be a closed-string analogue of the original Veneziano model.

[19] David Kaiser refers to superstring theory as a "sign of the S-matrix program's afterlife" which came about through "the transmogrification of Gabriele Veneziano's 1968 'duality' model" [29, p. 385]. He argues that Feynman diagrams ("paper tools") were at the heart of this transmogrification, claiming that "[t]oday's superstring theories owe their existence" to such tools. I would, however, say that the duality programme (initiated by DHS duality) initially marked a rather dramatic *failure* of Feynman diagrams, pointing to a need for diagrams ('duality diagrams') with very different representational characteristics. These map in an even more indirect way onto their target processes, as the Harari-Rosner diagrams make clear—functioning as equivalence classes of Feynman diagrams and thus superseding them. It took a little longer to interpret this equivalence class as emerging from the invariance properties of string worldsheets, and strictly speaking there was a discrete jump from pre- to post-duality programme Feynman diagrams. The Kikkawa, Sakita, Virasoro paper [31] was pivotal in the restablishment of Feynman diagram (or 'Feynman-like' diagrams, as they make clear) techniques, as were Holger Nielsen and David Fairle's 'fishnet diagrams' (discussed in Chap. 4). Subsequent usage of Feynman-like diagrams in superstring theory truly superseded their original role, since one eventually finds (thanks to modular invariance) that only one diagram is needed at each order of perturbation theory, which defeats their original purpose.

of spacetime held (namely, $d = 26$). The more general n-loop case was investigated by Kaku and Yu [30], Claud Lovelace[20] [34], and Alessandrini [1] (see also: [2]).

3.4 Factorization and the Beginnings of Dual Theory

In the context of a narrow resonance approximation, unitarity must be secured via a demonstration of factorizability. Physically, this procedure allows one to split the process up into incoming particles and outgoing particles.[21] In the case of the generalised N-point Veneziano amplitude, this lets one express the amplitude as a chain product of graph nodes and lines (or, more physically, vertices and propagators). The formal procedure corresponding to this involves isolating the residues at the poles of the amplitude (as functions of s). An amplitude factorizes just in case such a residue can be written as an expression for which each term must be the product of two factors, describing the number of incoming particles and outgoing particles respectively (along with their momenta). For the Veneziano amplitude, factorizability was proven independently by both Bardakçi and Mandelstam [5] and by Fubini and Veneziano [23]. Nambu too came up with a formulation in terms of infinitely many oscillators: [37]. The resulting system was found to be an infinite family of harmonic oscillators, $\alpha_n^\mu = \sqrt{n}a_n^\mu$ and $\alpha_{-n}^\mu = \sqrt{n}a_n^{\dagger\mu}$, allowing for an expression of the dual model spectrum.[22] The oscillators satisfied the following relations:

$$[\alpha_m^\mu, \alpha_n^\nu] = m\eta^{\mu\nu}\delta_{m+n,0} \tag{3.9}$$

The realisation of [5, 23], and others that the Veneziano formula admitted a factorization in terms of an infinite set of harmonic oscillators then paved the way for a better understanding of the formula, offering a very clear path to a physical theory rather than an abstract model. The harmonic-oscillator formalism also opens up new computational and mathematical directions, of which the operator formalism for dual models was an instance.[23] The operators in this case were creation and annihilation operators for the oscillators. From this one can construct the dual model's spectrum as an infinity of states forming a Fock space, built up by the creation and annihilation

---

[20] Here, the loop corrections (known as 'M-loops') were conceptualised as integrals over a holed surface, with the number of handles (the genus) corresponding to the order in the perturbation series, much as in the modern string theoretic sense.

[21] Of course, this corresponds to the "chopping it in half" part of the Claud Lovelace quotation opening this chapter.

[22] Note that in their model of rising Regge trajectories in 1968 (still pre-Veneziano's model), Chu et al. [13] had guessed at the existence of a possible harmonic-oscillator potential as the 'force' causing the trajectories to rise.

[23] Pierre Ramond describes the creation/annihilation operator formalism as "clearly the window into the structures behind the dual models" ([41, p. 362]).

operators.[24] Once these were given, the problem of factorization of the generalised Veneziano amplitude (now written using the operator formalism) was a somewhat trivial matter to show.

A 'twisting operator,' generating twisted propagators by switching external particle lines in duality diagrams, was constructed soon after the operator formalism was devised, in September 1969, by Caneschi, Schwimmer, and Veneziano [8]. According to Stefano Sciuto, Fubini believed that all that remained to establish the dual theory on firm foundations at this point was to find the expression of the vertex for emission of a general state.[25] As he recalls, what Fubini said was:

> Now we know the spectrum, we have the propagator and we have the vertex for the emission of the lowest lying states, we only miss the vertex for the emission of a generic state: if we were able to get it, we would have a **theory**, not only a model [46, p. 215].

Fubini and Veneziano later constructed 'untwisted' vertex operators $V(z; p)$ in order to represent scattering amplitudes at vertices, showing also that the amplitude factorised:

$$V(z; p) =: e^{ip \cdot Q} \quad (Q^\mu = x^\mu + i \sum_{n \neq 0} \alpha_n^\mu / n) \tag{3.10}$$

This expression allows one to represent the creation or emission of a state at an interaction point using creation operators. And, likewise, joining or absorption of a state, in terms of annihilation operators. Not long after, it would be realised that such states admitted an interpretation in terms of strings.

The operator formalism was without a doubt a pivotal point in the history of S-matrix theory, dual theory, and string theory. In many ways it severed the umbilical cord between dual theory and S-matrix theory, allowing more orthodox tools and concepts from quantum field theory (such as Fock space, with creation and annihilation operators, control over physical and unphysical sectors, and so on) to be adapted to dual models.[26] This made the properties of dual models especially transparent. This formalism itself, as we will see in the following chapter, played a key role in pointing to a string picture because of the nature of the oscillators. Strings were by

---

[24] The oscillators were Bose fields in the first dual models. In the next chapter we look at the attempt to generalise this to include a fermionic sector, and also a combination of a bosonic and fermionic sector of states.

[25] As John Schwarz writes, "it suggested that these formulas could be viewed as more than just an approximate phenomenological description of hadronic scattering. Rather, they could be regarded as the tree approximation to a full-fledged quantum theory." [45, p. 55]. A fact that came as a surprise to Schwarz, and many others.

[26] Elena Castellani quite rightly puts great stress on the "continuous influence exercised by quantum field theory" in the development of string theory [9, p. 71]. Quantum field theory had at its disposal very many powerful tools for dealing with problems faced by dual models, not least the elimination of the ghosts from the spectrum of states, which were eliminated using a gauge-symmetry device known from QED—though, as we will see, an infinite-dimensional symmetry is required in the case of dual models (a result that would be understood in the string picture as arising from the infinitely many ghosts corresponding to string's infinite tower of vibrational modes).

no means a strange choice of system given the state of play after the construction of the operator formalism.

However, the oscillator formalism also revealed a serious problem: ghosts were seen to be exchanged at poles of the amplitude. In 1969, Fubini and Veneziano [23] showed that the dual resonance model's spurious states—the problematic time component of one of the modes of oscillation (caused by the indefinite metric of the Lorentz group)—could be viewed as unphysical degrees of freedom, and eliminated via a gauge choice, thus restricting the Hilbert space to physical states.[27] But there were *infinitely* many ghosts to fix, and Fubini and Veneziano's method was not general enough to cover all cases. At the close of 1969, Miguel Virasoro [52] devised an infinite Ward-like class of gauge (or 'subsidiary') conditions that *could* serve to cancel the infinity of ghosts, via a one-to-one correspondence (one subsidiary condition per mode of oscillation), for the (physically unrealistic) unit intercept case, $\alpha(0) = 1$, which meant that the lowest lying particle (the leading trajectory) was a tachyon with a massless first excited state.[28] The tachyon was seen as the price one had to pay for ghost elimination.

Virasoro was able to construct an infinite-dimensional (gauge) algebra from the oscillators, with generators $L_m = \frac{1}{2}\Sigma_n : \alpha_{m-n} \cdot \alpha_n$.[29] In the context of a two dimensional field theory one can label a point of the field with complex coordinates which will have the effect of singling out two classes of symmetry generator: $L[\xi]$ (responsible for generating holomorphic diffeomorphisms, or motions) and $\overline{L}[\bar{\xi}]$ (responsible for generating anti-holomorphic motions). The Virasoro algebra[30] is then the infinite-dimensional Lie algebra with basis $\{L_n | n \in \mathbb{Z}\}$, obeying the following commutation relations:

$$[L_m, L_n] = (m - n)L_{m+n} + \frac{c}{12}(m^3 - m)\delta_{m+n,0} \qquad (3.11)$$

The central (or 'anomaly') term c here is simply a c-number term that commutes with all other operators—that is, c is in the subgroup of operators that maximally commute with other elements (including non-symmetries): $[c, L_n] = 0$, $\forall n \in \mathbb{Z}$. Noether's theorem leads to c being referred to as the central *charge*, since conserved

---

[27] This is analogous to the situation that arises with timelike photons in QED, in which the spurious states also decouple.

[28] Virasoro was, of course, well aware of the overly restrictive nature of the unit intercept case, but expected that his method could be generalised to more physically realistic cases. Fubini and Veneziano [22] later did this using a projective operator language, providing ghost-cancellation up to the third excited level. Brower and Thorn [6] still later extended this to the ninth excited level.

[29] Here he was building on earlier work of Gliozzi [25], who had constructed a similar set of operators: $L_n = -\frac{1}{2}\sum_{m=-\infty}^{\infty} : a_{-m} \cdot a_{m+n}$, $n = -1, 0, 1$ (note that this is Mandelstam's condensed version of the Gliozzi operators: [35, p. 282]. However, Virasoro extended this to *all* values of n.

[30] Note that the Virasoro algebra is the central extension of the 'Witt algebra' (over $\mathbb{C}$) defined by $[L_m, L_m] = (m - n)L_{m+n}$ (generators are $\{L_n : n \in \mathbb{Z}\}$)—see [53]. It acts as $[L_m, L_m]f = \{-z^{n+1}\frac{d}{dz}, -z^{m+1}\frac{d}{dz}\}f$. This is the Lie algebra associated with diffeomorphisms of the circle.

quantities are referred to as charges.[31] As Ramond notes, before this was understood there was "an intermediate period during which it was just written as 1/3 before it was realized that there were 26 (and not four) oscillators per mode" [40, p. 538]. The elements L_-, L_0, and L_+ also generate a Möbius algebra (a real subgroup of the projective algebra), satisfying:

$$[L_0, L_\pm] = \pm L_\pm, \quad [L_+, L_-] = -2L_0 \tag{3.12}$$

Using the operator formalism of [23], these operators (defining Virasoro's gauge conditions) can be written as (see [16, p. 93]):

$$L_0 = -\frac{p_0^2}{2} - \sum_{n=1}^{\infty} n a_n^\dagger a_n \tag{3.13}$$

$$L_+ = i p_0 a_1 - \sum_{n=1}^{\infty} \sqrt{n(n+1)} a_n^\dagger a_{n+1} \tag{3.14}$$

$$L_- = -i p_0 a_1^\dagger - \sum_{n=1}^{\infty} \sqrt{n(n+1)} a_n a_{n+1}^\dagger \tag{3.15}$$

Virasoro's algebra emerges precisely when the intercept is unity and a larger set of symmetries is induced, in which case the invariance group becomes infinite. Del Giudice and Di Vecchia [15] showed, shortly after Virasoro published his algebra, that physical (non-spurious) states (on the mass shell) must satisfy $L_l | \psi, \pi \rangle = 0$ and $[L_0 + 1] | \psi, \pi \rangle = 0$ (that is, physical states are orthogonal to spurious ones).[32] In 1972, Del Giudice and Di Vecchia, together with Fubini [17], constructed the space of physical (transverse, positive-norm) states (later called "DDF states") for the unit intercept case.[33] The DDF scheme involved the crucial result that the action of a vertex operator $V(z; p)$ on a physical state would spit out another physical state—a result that Brower would later build upon in [7].

Jumping ahead a little, the condition of unit intercept can be seen to follow from the definition of the appropriate physical vertex operators for the emission of ground

---

[31] The central charge c in the algebra is credited to Joseph H. Weis. Weis died, aged just 35, in a climbing accident in the French Alps (on the Grandes Jorasses) in August, 1978—he was killed with his climbing partner, and CERN physicist, Frank Sacherer. He had taken his PhD under Mandelstam at Berkeley, before taking up a postdoctoral position at MIT. He discovered the central charge during his study of 2D QCD. Though he never published it, he seems to have communicated his discovery to several people, and one can find him credited in, e.g., [6, p. 167], [22].

[32] These physical states were also shown to be eigenstates of the twist operator mentioned above [15, p. 587].

[33] Goddard and Thorn [27], and also Brower [7], later proved in 1972, that this space is complete when the number of transverse components of the oscillators is 24 (that is to say, the physical Hilbert space of states has dimension $\mathscr{T}^{24}$). There are no ghosts present when this condition, in addition to the unit intercept condition, is met. For $D > 26$ (where D is the spacetime dimension) ghosts appear, for $D < 26$ there are no ghosts.

state strings, $v_0(k)$. We can write this (following Clavelli, [14, p. 11]) as:

$$v_0(k) = \int d\tau : e^{i\sqrt{2\alpha'}k\cdot x(0,\tau)} \tag{3.16}$$

In order to constitute a physical state, this had better commute with the Virasoro constraints, $[L_N, v_0(k)] = 0$, which gives:

$$[L_N, v_0(k)] = N(1 - \alpha'k^2) \int d\tau e^{-iN\tau} : e^{i\sqrt{2\alpha'}k\cdot x(0,\tau)} \tag{3.17}$$

One can achieve consistency here iff $1 - \alpha'k^2 = 0$. Since the intercept state is given by $\alpha(0) - \alpha'k^2 = 0$, it follows that $\alpha(0) = 1$ for the restriction to physical states. It also quickly follows from the unit intercept condition that the spin-1 state, $\alpha(0) - \alpha'k^2 = 1$, must have zero mass, $k^2 = 0$.

Finally, we should mention that the discovery of vertex operators in the context of the dual resonance model is curiously intertwined with their appearance in the theory of affine Lie algebras. At the root of the connection are the vertex operators developed in the course of the discovery of the operator formalism for the dual resonance models. The mathematical link here goes back to the fact that Regge trajectories involve an infinite number of resonances, so that symmetries based on these (the DHS duality symmetry) will involve infinite-dimensional groups. This physical situation led to the physicist's discovery of Kac-Moody algebras (later formulated as infinite-dimensional extensions of Lie algebras) within the context of strong interaction physics—see [19] for a historical discussion of the interplay between Kac-Moody algebras and physics.[34]

3.5 Summary

The Veneziano model, or rather the wider project of generalising it to multi-particle, multi-loop situations, was considered to be a genuinely possible route to a full theory of strong interaction physics. Accordingly, many physicist-hours were spent labouring on it, despite the fact that the framework was in many ways utterly detached from most areas of particle physics. The clear early problems with the Veneziano model (the lack of unitarity and the restriction to the 4-point scattering scenario) were resolved with remarkable speed and skill, well within two years, as was the problem of formulating the appropriate mathematical framework (replete with an understanding of the consistency conditions one must impose). The result was a general, elegant operator formalism for dual models that clearly pointed towards some underlying system responsible for generating the excitation spectrum.

---

[34] As Goddard and Olive note [28, p. 121], there is even an earlier precedent in the form of Tony Skyrme's construction of a fermionic field operator for the soliton in the sine-Gordon model [47].

The dual models were still facing problems on several fronts, including the lack of fermions in the spectrum. This would wait until the dual models had already undergone significant reinterpretation (including the beginnings of an interpretation as a theory of strings), though work had already begun prior to this interpretation and much of the initial work floated free of the string interpretation, based instead on the operator formulation, as we will see. The tachyon remained an issue, though it would be tamed to a certain extent when fermions were included.[35] Once the string picture begins to emerge, from 1969 onwards, we see a division into two classes of approach to dual models: a geometrical approach based on the string idea (and its associated worldsheet) and a more abstract approach based on the operator formalism. It would take some time for the string picture to fully take centre stage and develop computational prowess to rival the operator approach.

References

1. Alessandrini, V. (1971). A general approach to dual multiloop diagrams. *Il Nuovo Cimento A*, *2A*, 321–352.
2. Alessandrini, V., & Amati, D. (1971). Properties of dual multiloop amplitudes. *Il Nuovo Cimento A*, *4*(4), 793–844.
3. Bardakçi, K., & Ruegg, H. (1969). Meson resonance couplings in a 5-point Veneziano model. *Physics Letters*, *B28*, 671–675.
4. Bardakçi, K., & Ruegg, H. (1969). Reggeized resonance model for arbitrary production processes. *Physical Review*, *181*, 1884–1889.
5. Bardakçi, K., & Mandelstam, S. (1969). Analytic solution of the linear-trajectory bootstrap. *Physical Review*, *184*, 1640–1644.
6. Brower, R. C., & Thorn, C. B. (1971). Eliminating spurious states from the dual resonance model. *Nuclear Physics*, *B31*, 163–182.
7. Brower, R. C. (1972). Spectrum-generating algebra and no-ghost theorem for the dual model. *Physical Review D*, *6*, 1655–1662.
8. Caneschi, L., Schwimmer, A., & Veneziano, G. (1969). Twisted propagator in the operatorial duality formalism. *Physics Letters B*, *30*(5), 351–356.
9. Castellani, E. (2012). Early string theory as a challenging case study for philosophers. In A. Cappelli et al. (Eds.), *The birth of string theory* (pp. 63–78). Cambridge: Cambridge University Press.
10. Chan, H.-M., & Tsou, T.-S. (1969). Explicit construction of *n*-point function in generalized Veneziano model. *Physics Letters*, *B28*, 485–488.
11. Chan, H.-M. (1970). The generalized Veneziano model. *Proceedings of the Royal Society of London A.*, *318*, 379–399.
12. Chew, G. F. (1970). Hadron bootstrap: Triumph or frustration? *Physics Today*, *23*(10), 23–28.
13. Chu, S., Epstein, G., Kaus, P., Slansky, R. C., & Zacharia, F. (1968). Crossing-symmetric rising Regge trajectories. *Physical Review*, *175*(5), 2098–2105.
14. Clavelli, L. (1986). A historical overview of superstring theory. In L. Clavelli & A. Halperin (Eds.), *Lewes string theory workshop* (pp. 1–20). Singapore: World Scientific.
15. Del Giudice, E., & Di Vecchia, P. (1970). Characterization of the physical states in dual-resonance models. *Il Nuovo Cimento A10, 70*(4): 579–591.

---

[35] A complete understanding of the elimination of tachyons would take the best part of a decade to achieve, once supersymmetry was better understood.

16. Del Giudice, E., & Di Vecchia, P. (1971). Factorization and operator formalism in the generalized Virasoro model. *Il Nuovo Cimento A, 5*(1), 90–102.
17. Del Giudice, E., Di Vecchia, P., & Fubini, S. (1972). General properties of the dual resonance model. *Annals of Physics, 70*(2), 378–398.
18. Di Vecchia, P. (2008). *The birth of string theory*. In M. Gasperini & J. Maharana (Eds.), *String theory and fundamental interactions, Lecture Notes in Physics* (vol. 737, pp. 59–118). Berlin: Springer.
19. Dolen, L. (1995). The beacon of Kac-Moody symmetry for physics. *Notices of the American Mathematical Society, 42*, 1489–1495.
20. Fairlie, D. (2012). The analogue model for string amplitudes. In A. Capelli et al. (Eds.), *The birth of string theory* (pp. 283–293). Cambridge: Cambridge University Press.
21. Fairlie, D. B., & Jones, K. (1969). Integral representations for the complete four and five-point Veneziano amplitudes. *Nuclear Physics, B15*, 323–330.
22. Fubini, S., & Veneziano, G. (1970). Algebraic treatment of subsidiary conditions in dual resonance models. *Annals of Physics, 63*, 12–27.
23. Fubini, S., & Veneziano, G. (1969). Level structure of dual resonance models. *Il Nuovo Cimento, A64*, 811–840.
24. Fubini, S., Gordon, D., & Veneziano, G. (1969). A general treatment of factorization in dual resonance models. *Physics Letters, B29*, 679–682.
25. Gliozzi, F. (1969). Ward-like identities and twisting operator in dual resonance models. *Lettere al Nuovo Cimento, 2*(18), 846–850.
26. Goebel, C. J., & Sakita, B. (1969). Extension of the Veneziano form to *n*-particle amplitudes. *Physics Review Letters, 22*, 257–260.
27. Goddard, P., & Thorn, C. B. (1972). Compatibility of the dual pomeron with unitarity and the absence of ghosts in dual resonance models. *Physics Letters, 40B*, 235–238.
28. Goddard, P., & Olive, D. I. (1988). Kac-Moody and Virasoro algebras: A reprint volume for physicists. Singapore: World Scientific.
29. Kaiser, D. (2005). *Drawing theories apart: The dispersion of Feynman diagrams in postwar physics*. Chicago: University of Chicago Press.
30. Kaku, M., & Yu, L. (1970). The general multi-loop Veneziano amplitude. *Physics Letters B, 33*(2), 166–170.
31. Kikkawa, K., Sakita, B., & Virasoro, M. A. (1969). Feynman-like diagrams compatible with duality. I. Planar diagrams. *Physical Review, 184*, 1701–1713.
32. Koba, Z., & Nielsen, H. (1969). Reaction amplitude for N-Mesons: A generalization of the Veneziano-Bardakçi-Ruegg-Virasoro model. *Nuclear Physics, B*(10): 633–655.
33. Lovelace, C. (1968). A novel application of Regge trajectories. *Physics Letters B, 28*, 264–268.
34. Lovelace, C. (1970). *M*-loop generalized Veneziano formula. *Physics Letters B, 32*(8), 703–708.
35. Mandelstam, S. (1974). Dual-resonance models. *Physics Reports, 13*(6), 259–353.
36. Miyata, Y. (1970). Pseudoscalar-vector meson scattering amplitude reduced from six-point Veneziano model. *Progress of Theoretical Physics, 44*(6), 1661–1672.
37. Nambu, Y. (1970). Quark model and the factorization of the Veneziano amplitude. In R. Chand (Ed.), *Symmetries and quark models* (pp. 269–277). Singapore: World Scientific.
38. Olive, D. I. (2012). From dual fermion to superstring. In A. Cappelli et al. (Eds.), *The birth of string theory* (pp. 346–360). Cambridge: Cambridge University Press.
39. Paton, J. E., & Chan, H.-M. (1969). Generalized Veneziano model with isospin. *Nuclear Physics, B10*, 516–520.
40. Ramond, P. (1987). The early years of string theory: The dual resonance model. In R. Slansky & G. B. West (Eds.), *Theoretical advanced study institute lectures in elementary particle physics 1987: Proceedings* (pp. 501–571). Singapore: World Scientific.
41. Ramond, P. (2012). Dual model with Fermions: Memoirs of an early string theorist. In A. Capelli et al. (Eds.), *The birth of string theory* (pp. 361–372). Cambridge: Cambridge University Press.
42. Schmid, C. (1970). Duality and exchange degeneracy. In A. Zichichi (Ed.), *Subnuclear phenomena, Part 1* (pp. 108–143). New York: Academic Press.

43. Shapiro, J. A. (1968). Narrow-resonance model with regge behavior for $\pi\pi$ scattering. *Physical Review, 179*(5), 1345–1353.
44. Shapiro, J. A. (1970). Electrostatic analogue for the Virasoro model. *Physics Letters B, 33*, 361–362.
45. Schwarz, J. H. (2001). String theory origins of supersymmetry. *Nuclear Physics, B101*, 54–61.
46. Sciuto, S. (2012). The 3-Reggeon Vertex. In A. Capelli et al. (Eds.), *The birth of string theory* (pp. 214–217). Cambridge: Cambridge University Press.
47. Skyrme, T. H. R. (1961). Particle states of a quantized meson field. *Proceedings of the Royal Society of London A, A262*, 237–245.
48. Susskind, L. (2006). *The cosmic landscape*. New York: Back Bay Books.
49. Veneziano, G. (1968). Construction of a crossing-symmetric, Reggeon-behaved amplitude for linearly rising trajectories. *Il Nuovo Cimento, A57*, 190–197.
50. Virasoro, M. (1969). Generalization of Veneziano's formula for the five-point function. *Physical Review Letters, 22*, 37–39.
51. Virasoro, M. (1969). Alternative constructions of crossing-symmetric amplitudes with Regge behavior. *Physical Review, 177*(5), 2309–2311.
52. Virasoro, M. (1969). Subsidiary conditions and ghosts in dual-resonance models. *Physical Review D, 1*(10), 2933–2936.
53. Wanglai, L., & Wilson, R. L. (1998). Central extensions of some lie algebras. *Proceedings of the American Mathematical Society, 126*(9), 2569–2577.

Chapter 4
The Hadronic String

> [T]he string model originated as a model for the S-matrix, and it
> may well not have been discovered if S-matrix theory had not
> been vigorously pursued at the time.
>
> Stanley Mandelstam

The idea that the Veneziano model might have a basis in a theory of strings was recognised independently by several physicists.[1] The puzzle was to find out what 'lay behind' the Veneziano model and thereby attempt to reconstruct the formula and its predictions from a more fundamental physical picture. Indeed, the notion of providing a 'picture' of the Veneziano model (primarily the N-particle generalisation) was a key feature of this work. Key target features included the Regge trajectories (the tower of resonances described by the Veneziano model) and the DHS duality (linking apparently different kinds of particle processes).[2] Susskind, Nambu, and Nielsen all surmised that the regular presentation of the tower meant that it was being generated by some internal oscillatory motions, lying within hadrons. That is to say, the spectrum of the dual model was suggestive of the spectrum of an oscillating system.

Despite the fact that many of the modern concepts of superstrings come from this work—including the notion of a worldsheet and the idea that the above mentioned duality is a manifestation of the conformal symmetry of the worldsheet—the earliest

---

[1] As mentioned earlier, this kind of convergence is common in scientific discovery and often points to 'being on the right track'—or at least to an overall consistency in the methods of science. In the present case, this is not in the least surprising since the Veneziano formula (its spectrum of states) implied that, at the very least, it was describing an infinite family of harmonic oscillators and the harmonic oscillator is one of the best-known examples in physics.

[2] While a mystery from a point-particle point of view, DHS duality is directly understood in hadronic string terms of a perfectly natural geometrical outcome of having a physics of extended objects. Gomez and Ruiz-Altaba explain it thus: "[d]uality is... intrinsic to the string picture, because a Feynman diagram where two rubber bands merge into one and then become two rubber bands again allows for arbitrary definition of s, t, or u channels" [24, p. 54]—they suggest that a more appropriate picture, given this malleability, would be "chewing gum"!

D. Rickles, *A Brief History of String Theory*, The Frontiers Collection,
DOI: 10.1007/978-3-642-45128-7_4, © Springer-Verlag Berlin Heidelberg 2014

string interpretations can be seen in each case (at best) to 'hedge' on the issue of
realism about strings. In much the same way as quarks were used in this period, one
finds that the string concept is used heuristically to suggest new research directions
or tools to apply, or used analogically, again more for convenience than to provide
a faithful representation of actual hadronic processes. It would take further work
in establishing the *consistency* of the theory, as well as its demonstrable ability to
reproduce the physics of dual models, before strings could be taken seriously as a
realistic model of our world. Unfortunately, by the time this was achieved, the dual
bootstrap picture was being replaced by quantum field theory, Chew's "old mistress".
Initially, there was certainly not the slightest intimation that the strings would have
anything to do with physics beyond hadrons.

4.1 The Multiple Births of Strings

As we have seen, the Veneziano model as it was understood in the immediate aftermath of its construction was not thought to have anything to do with a dynamical
theory of extended objects. The Veneziano formula was exactly that: a formula. It was
an example of a mathematical object that would deliver probabilities for scattering
events. As such it did not offer any kind of mechanism, or any physical picture, for
the kinds of systems that would satisfy it and generate its spectrum. However, what
was present was the infinite set of oscillators revealed by the formalism of Fubini
and Veneziano. The string models were the fruit of the effort to restore some physical
intuition to the dual resonance amplitude: the dual model spectrum could be viewed
as issuing from the quantum mechanical behaviour of vibrating, rotating strings. The
infinite set of oscillators were modes of vibration of the string.[3]

The intuitive picture that we can draw from this is that a particle is 'composed' out
of an open spinning string with unconstrained end-points. The string has a tension
along its length, and given its rotation is also subject to a centrifugal force. The
spin is maximal when the string is straightest, corresponding to the leading Regge
trajectory. Given that the strings have a finite, fundamental length one can also see how
the slope of the Regge trajectories is determined from the string picture. 'Classical'
string aspects naturally arise in the various 'harmonics' that also contribute to the
motion of the string and correspond to the daughter trajectories lying under the
leading (fundamental) trajectory. This correspondence between string properties and
the Regge trajectories suggest that it might have been possible to guess in advance
(without the benefit of the Veneziano model and its operator formulation) that the
Regge trajectory was pointing to a unified system generated as the infinitely many
excitations of a single fundamental string.

---

[3] Note that the naming convention for 'strings' appears to have only been firmly established in 1974,
in Claudio Rebbi's survey of dual models and relativistic quantum strings for *Physics Reports* [? ,
p. 4].

This chapter focuses on the key elements involved in this crucial transition in the understanding of the Veneziano amplitude, leading to 'dual string models'. Note, however, that when we speak of the Veneziano model, we are referring in each case (Susskind, Nambu, Nielsen) to the N-point generalisations that occurred later.

Susskind: A Rebel Without a Theory

Just before finding out about the Veneziano amplitude, Leonard Susskind[4] had been investigating relativistic quantum theory in the so-called infinite momentum frame.[5] Susskind argued that physics in the infinite momentum frame was Galilean invariant, implying that standard quantum mechanics was applicable. He wanted to apply this to the problematic hadrons to investigate their internal structure with well-worn tools. He heard about the Veneziano formula in 1968—cf. [57]. There then followed an interaction between his then current research programme and aspects of this formula. In his Galilean-invariant framework m^2 is the energy term, and the same term determines the spacing of the poles of the Veneziano amplitude. Hence, given this correspondence, the energy levels were equally spaced, implying that the system generating the Veneziano formula must be some kind of quantum harmonic oscillator. The remaining task was to try to construct an oscillator model that reproduced the Veneziano formula. As Susskind points out [57, p. 263], while the Veneziano formula looked like:

$$A(s, t) = \int_0^1 x^{-s}(1 - x)^{-t}dx \tag{4.1}$$

his own (modeling a hadron oscillating in the infinite momentum frame) looked like

$$A(s, t) = \int_0^1 x^{-s}(\exp - x)^{-t}dx \tag{4.2}$$

where the two formulas are clearly related by the interchange $(1 - x) \rightarrow (\exp - x)$. When making a comparison with the more general Veneziano formulae (for more than 4 particles), he found agreement too (modulo the same interchange). Susskind viewed the correspondence as highlighting a potentially fruitful *analogy* rather than anything profound. Hence, the paper containing this idea was entitled "Harmonic Oscillator Analogy for the Veneziano Amplitude" [52]. However, soon after publishing this paper Susskind realised that certain of the properties ("higher harmonics") could

---

[4] The subtitle 'Rebel Without a Theory' refers to a remark made by Susskind in *The Cosmic Landscape* [56, p. 204], in which he discusses his distaste for S-matrix ideology.

[5] This is now more commonly referred to as the "light-cone frame", an idea introduced by Fubini and Furlan in 1965 [16] (see [10, pp. 50–52]). The idea is to boost the velocities to such a degree that the dilation effect on the system is very large, allowing one to explore its motions and internal structure more easily.

only be generated by a very specific oscillator such as a violin string rather than, say, a spring [54]:

> I was able to produce formulas that looked a lot like Veneziano's from the simple weight and spring model, but they weren't quite right. During that period I spent long hours by myself, working in the attic of my house. I hardly came out, and when I did I was irritable. I barked at my wife and ignored my kids. I couldn't put the formula out of my mind, even long enough to eat dinner. But then for no good reason, one evening in the attic I suddenly had a "eureka moment". I don't know what provoked the thought. One minute I saw spring, and the next I could visualise an elastic string, stretched between two quarks and vibrating in many different patterns of oscillation. I knew in an instant that replacing the mathematical spring with the continuous material of a vibrating string would do the trick. Actually, the word *string* is not what flashed into my mind. A *rubber band* is the way I thought of it: a rubber band cut open so that it became an elastic string with two ends. At each end I pictured a quark or, more precisely, a quark at one end and an antiquark at the other [47, p. 206].

It quickly became clear to Susskind that the Veneziano amplitude could be given an interpretation in terms of scattering elastic bands, coming in from infinity, then merging to form a single band, and then splitting before moving out to infinity again. Of course, this achieves a high degree of theoretical (and, correspondingly, onto-logical) simplification and economy. What was considered to be a series of excited hadronic states (on a trajectory) corresponding to *distinct* particles or resonances, are, in a string picture, states realised in one and the same object (much like Bernoulli's diagram of the superpositions of vibrational modes of a violin string that we saw in Chap. 1).

Susskind notes that the string model for hadrons was "not an immediate success" [56, p. 217]. He traces this to a widespread negative stance against theories that tried to 'picture' what was happening in the world; cracking open the black box of S-matrix theory and looking within the collision region. However, while I think there was an S-matrix motivated reaction against the string model, I don't think that visualisability was the problem. Rather it was a reaction against what the string model *revealed* within the black box, and that was something in direct conflict with nuclear democracy and bootstrap ideals. If there is just a single system underlying all of the different resonances, than that implies a fundamentality that was anathema to most physicists working on the dual resonance models, steeped, as they were, in Chew's philosophical ideas and the notion that duality implied nuclear democracy. Hence, while it is often said that the S-matrix theory was abandoned because of the rise of QCD, it might be said that the S-matrix programme, when carried to its completion with the implementation of duality and the Veneziano model, contained the seeds of its own destruction.[6]

---

[6] Susskind claims to pinpoint the time at which Murray Gell-Mann became interested in string theory, and as a result put the 'stamp of approval' on it (ibid). He first refers to a discussion he had with Gell-Mann at a conference in Coral Gables, Florida, in 1970, where Gell-Mann had simply laughed when he mentioned his idea of a string structure of hadrons. Then two years later, at the 'Rochester conference,' at Fermilab, Gell-Mann apparently apologised for his earlier behaviour and pointed out that he was interested in such work, and indeed spoke a little on the subject at the conference. Gell-Mann both outlined his theory of quarks and, though it was not yet known by the name, quantum chromodynamics (with Fritzsch) he gave a summary talk, which is presumably

Nambu's Tale

Yoichiro Nambu too initially spoke in rather more non-committal terms than is often supposed in the more recent literature discussing the origins of string theory. Here too it was by *analogy* that the string picture was initially suggested. There was plenty of historical precedent for the idea that the fundamental objects might not be point particles, but some non-local entities.[7] However, all of these had failed to provide a consistent picture. Hence, there was good reason to be somewhat tentative at this stage. Thus, he writes[8]:

> [T]he internal energy of a meson is analogous to that of a quantized string of finite length (or a cavity resonator for that matter) whose displacements are described by the field $\phi_\alpha(\xi)$ [a Bosonic field—DR] [40, p. 275].

This connection Nambu made on the basis of his derivation of the following expression for the quantum number representing the resonance energy N:

$$N = \frac{-1}{\pi} \int_0^{2\pi} : (\partial_\xi \phi(\xi) \cdot \partial_\xi \phi(\xi) + \pi(\xi) \cdot \pi(\xi)) : d\xi \qquad (4.3)$$

where the Bose field and its conjugate are decomposed as follows (with a and a^+ the creation and annihilation operators):

(Footnote 6 continued)
what Susskind is referring to. Gell-Mann had already taken to strings by then. He had been a visitor at CERN in 1971 while Lars Brink and David Olive were also there. Moreover, John Schwarz heard that he would not receive tenure at Princeton in early 1972 and received a position as a research associate at Caltech shortly afterwards (within in a matter of months), to work on string theory (http://resolver.caltech.edu/CaltechOH:OH_Schwarz_J, p. 22). The field of string theory had advanced a great deal within the 2 years from 1970 to 1972. Statistical analysis of citations before and after Gell-Mann's remarks do not reveal any significant differences, though it certainly cannot failed to have put string theory more on the map relative to researchers from the general particle physics community.

[7] Indeed, as Nambu mentions in his reminiscences [43, p. 278], he had been working on non-local field theories as a way of reproducing the seemingly infinite number of Regge trajectories of ever higher spins. His approach involved representing them via a master wave equation in terms of infinite-dimensional representations of groups containing the Lorentz group (in what he called an *infinite-component wave equation*). The work was abandoned since it violated too many desirable properties of quantum field theory, such as CPT, microcausality, tachyon-freedom, crossing, and so on. However, the ability to manipulate creation and annihilation operators in this work would be recycled in his work on the string model.

[8] Nambu's initial suggestion that a string picture might lay behind the Veneziano model was delivered at a somewhat low key conference on Symmetries and Quark Models, held at Wayne State University, June 18–20, 1969. However, this certainly appears to have been the first public mention of the idea that the excitation spectrum of dual models was reducible to a vibrating string. Note also that Nambu's primary concern was not with elucidating the physical content of dual models, but with a factorization method for the Veneziano amplitude.

$$\phi_\alpha(\xi) = \sum_{r=1}^{\infty} \frac{1}{\sqrt{2r}} (a_\alpha^{(r)} + a_\alpha^{\dagger(r)}) \cos r\xi \tag{4.4}$$

$$\pi_\alpha(\xi) = \sum_{r=1}^{\infty} i\sqrt{\frac{r}{2}} (a_\alpha^{(r)} - a_\alpha^{\dagger(r)}) \cos r\xi \tag{4.5}$$

The displacements of the resonator are given by $\phi_\alpha(\xi)$, though as one can see, Nambu had some uncertainty as to the precise nature of the resonator at this stage: it could refer to the oscillations of a string, or the oscillations within a hollow body. Note also that Nambu made crucial usage of Veneziano and Fubini's work on the level structure of the dual amplitude (at the time only available as an MIT preprint). Following his abandonment of his earlier infinite-dimensional representation approach, Nambu describes the initial path to this string idea as follows:

> The Veneziano model realized the linear Regge trajectories and the duality of scattering amplitudes in a simple formula. So I got fascinated by ... First of all, what physics lies behind it? It is a mysterious formula nobody really understands. You can write down the formula in various ways. I wanted to decompose the formula as an infinite sum of Breit-Wigner resonances to see how many of them are at each resonant energy, what their spins are, and if the residues are positive or that they are real physical states. When I started doing this, there was a postdoc, Paul Frampton, arriving from Oxford. So I got his assistance right away and we were more or less convinced that all Breit-Wigner poles were positive [hence, physical—DR]. We also found that the degeneracy of states goes up exponentially with energy in the asymptotic limit.
>
> [...]
>
> After that I worked further on analyzing the structure of the Veneziano amplitudes. I started from the Koba-Nielsen representation of the beta function, and in the course of this analysis, I discovered that the resonances can be interpreted as the excitations of a string.[9]

Here we see that for Nambu, the Koba-Nielsen Beta function expression was crucial, as it was for Nielsen himself, though inspiring a different approach. His route to the spectrum so suggestive of a harmonic oscillator system proceeded through manipulations of this expression, starting with its factorization. Nambu's factorization yielded:

$$(1-x)^{-\alpha't - \alpha_t - 1} = e^{\alpha'(p_1 \cdot p_2 - C)(x + x^2/2 + x^3/3 + \cdots)} \tag{4.6}$$

with C describing mass dependence and dependence on the intercept value: $C = m_1^2 + m_2^2 + \alpha_t + 1$. He was then able to link this up to an expression for vector fields decomposed in terms of creation and annihilation operators (as presented above, in Eq. 4.4), giving:

$$\langle \phi(x)\phi(y) \rangle = \sum_k \frac{1}{2E_k} \exp[ik \cdot (x - y)] \tag{4.7}$$

---

[9] Interview of Yoichiro Nambu by Babak Ashrafi on July 16, 2004, Niels Bohr Library and Archives, American Institute of Physics, College Park, MD USA http://www.aip.org/history/ohilist/30538.html. Note that, in the portions relevant to string theory at least, Nambu appears to consistently have his dates a year out in this interview. Hence, for years between 1967 and 1970, simply advance them by 1 to get the correct figures.

What is being derived here, from the quantized string picture, is an expression for an N-particle amplitude, along with the Hilbert space and operator formalism (in terms of vertex operators, initially only for open strings). Once one has a picture of one dimensional oscillators, one can imagine their evolution, which will generate a diagram of a kind that exactly satisfies the DHS duality.

The underlying idea is quite simple. Given the Regge trajectories, $J = \alpha(0) + \alpha' m^2$, one can make sense of the spin-mass relationship in terms of a rotating string, with quarks as end-points. As the string rotates it generates a centrifugal force, pushing the quarks outwards, thus stretching the string apart. The longer the string, the more energy per unit length, and therefore the more mass. This physical picture corresponds to the mathematical relationship for mesons. Nambu claims to have realised the confining implications of a string interpretation right away, but was ambivalent about it on account of his own alternative field theoretic approach:

> I hit upon this string interpretation of the Veneziano model, and immediately I knew that that it could confine the quarks, because quarks attached to the ends of the string and can not separate. On the other hand, if you work out the mathematics, it did not quite work out well. Sooner or later people found out that you needed 26 dimensions, something like that. And in the meantime, there emerged a new gauge theory of color which was very nice in explaining the possibility of quark confinement. That is the usual quantum field theory, and my string theory is not quantum field theory but a more general one which also has various problems. Some of them are theoretical so far, they were found already when I worked out this infinite component wave equation. So I was in a sort of quandary, in the following sense. I knew that strings can confine quarks. On the other hand, I had also my pet theory of integral charged colored quarks, so the quarks were free to come out. But anyway, it explained the stability of color-neutral particles, hadrons.
>
> So I was in a quandary of which theory I should really side with. And I knew that string theory had mathematical problems. So probably it would not quite work out. Many decided to abandon string theory. I think it was around 1973. There was a summer institute at Aspen, and string theorists of the day got together. And more or less around the time people realized that we had to give up the hadronic string theory.[10]

It is important to note that physicists were not forced down the string theoretic path initially. One could also, as was suggested, substantiate the Veneziano formula in terms of an infinite-component field theory. The latter had very many serious problems, but the dual string model could hardly be said to be problem-free.

Nielsen's 'Almost Physical' Model

Holger Nielsen seems not to have publicly presented his paper, "An Almost Physical Interpretation of the Integrand of the n-point Veneziano Model," though it is often

---

[10] Interview of Prof. Yoichiro Nambu by Babak Ashrafi on July 16, 2004, Niels Bohr Library and Archives, American Institute of Physics, College Park, MD USA, http://www.aip.org/history/ohilist/30538.html.

claimed that he presented it at the 1970 'Rochester' conference in Kiev.[11] However, in a note that he appended to the recent release of his original preprint containing the string idea,[12] Nielsen states that he was discussing the idea with people in the spring of 1969—especially at the Lund International Conference on Elementary Particles, held from June 25-July 1 (though his 'official' talk was on the Koba-Nielsen formalism)— and circulated an initial paper identical to the later preprint save for the absence of the word 'almost' (this was a Nordita preprint). It seems that he never published any reference to the work himself, nor did he publicly speak about the work. However, Bunji Sakita did give a review of Nielsen's work at the Kiev conference, though this also was not published in the proceedings. Veneziano did, however, publish a review paper in the proceedings, and mentioned Nielsen's approach, amounting to the first published version of Nielsen's 'fishnet' approach (cf. [45, p. 272]).[13]

Nielsen's approach to a model of hadrons as strings[14] was significantly different to Nambu's and Susskind's (which followed a more or less similar path to one another). Nielsen sought a link to the standard Feynman diagram methodology for representing amplitudes and scattering. He argued that the duality diagram (of the Harari-Rosner sort) could be seen as the limiting case of a class of infinitely complicated Feynman diagrams that formed a mesh, or "fishnet" as it was later labelled.[15] In other words, something like a string world sheet is generated as an approximation to the underlying complex tangle of propagators. However, though Nielsen clearly has in mind a surface generated by the evolution of his threads, he characterises it in terms of two-dimensional conducting disc, with the Harari-Rosner quark flow lines forming its boundary. External lines are characterised as current-carrying 'electrodes' on the conductor's boundary (the analogues of the momenta of the external particles of the Veneziano model). Hence, Nielsen's model employs an electrostatic analogy which allows the integrand of the N-particle Veneziano model to be computed, as the exponential of heat produced by a steady current on the surface of the disc—conformal invariance implies that the result one gets is independent of local

---

[11] Veneziano, Zumino, Gervais, Sakita, Volkov, Gross, Migdal, and Polyakov were all present at this conference.

[12] The preprint is [46], and the brief note describing the origins of this preprint can be found at http://theory.fi.infn.it/colomo/string-book/nielsen_note.txt (on the website for the book *The Birth of String Theory*).

[13] Soon after, Sakita, together with Virasroro, published a proof (based on functional integration methods) of Nielsen's fishnet-based claim that the N-particle Veneziano provides an approximate description of planar fishnet-Feynman diagrams in the large-N limit [50] (see below). They also extend the principle to non-planar diagrams: non-simply connected and non-orientable.

[14] Nielsen speaks variously of "one dimensional structures", "thread like structures", "chain molecules", and even "sticks"!

[15] Note that the point particle description was an integral part of the early string proposals, usually entering in the form of an infinite limit of point particles to construct the string, and, as we will see later, an infinite limit of parallel point particle worldlines to construct the worldsheet. Nielsen's fishnet diagrams were precisely of this form, namely a chain of particles linked together (with nearest neighbour interactions).

stretching and rotation of the surface provided one can map it conformally to a disc (cf. [11, p. 71] and [15, p. 286]).[16]

The basic idea underlying Nielsen's path to the string (and the surfaces traced out by such string), from the Veneziano model is quite intuitive: if one is dealing with strong interactions (featuring a large coupling constant) from a Feynman diagram point of view, then one expects higher-order diagrams to dominate (cf. [45, p. 270]). The question he posed to himself was, therefore, whether one could build up the Veneziano amplitude from such high-order Feynman diagrams. The answer was yes, given an $n \to \infty$ limit (where n is the order of the diagrams) and so long as the diagrams are planar (with no crossed lines). The surfaces of evolution of threads are, as he puts it, "very rough pictures of very complicated Feynman diagrams" such that "only Feynman diagrams having the large scale topological structure of a two-dimensional network are of importance" [46, p. 18]. This provides Nielsen with his picture of the generalised Veneziano model. Nielsen also provided a qualitative account of the splitting and joining of his threads, as follows:

> Hadronic interactions are conceived of then as processes in which threads are connected at the end points into (at first) longer threads which are then again split up into (at first) shorter threads. In fact the mapping $V^\mu : \phi \to$ "Minkowski space" described by the potential of equation (25)[17] could be conceived of as describing the time track of a thread moving around in physical three space [46, p. 13].

Though it is out of chronological sequence (coming in 1973), we should mention Nielsen's later work, with Poul Oleson, which utilised a different analogy, this time involving a (type II) superconductor.[18] The work in question sparked off a field known as *Dual Superconductor Models of Colour Confinement*. The idea here was to provide a physical *grounding* for the still then rather abstract strings, by deriving string-like structures from a local field theory, and from the standpoint of such 'non-fundamental' strings, one could reproduce the behaviour captured by the Veneziano

---

[16] It seems that this component of Nielsen's work was devised in close collaboration with David Fairlie, then at Durham University—indeed, the electrostatic analogy was due to Fairlie. Amusingly, Fairlie claims to have come up with the idea of shape independence from a Philips advert in *Scientific American*, showing Ohm's Law (see [15, pp. 286–287]). A general solution (for 'discs' or surfaces of arbitrary genus) of Nielsen and Fairlie's analogue model was later provided by Alessandrini [2], where he understands the problem to consist in solving a harmonic problem on a Riemann surface. He employs Burnside's 1891 analysis of automorphic functions (which had been introduced to study harmonic problems on surfaces with circular holes). This was understood independently by Lovelace, as mentioned in footnote 6. Given this ancestry, David Fairlie makes the following amusing counterpoint to a famous quote from Witten: "Edward Witten was fond of quoting that 'String theory is a piece of mathematics which has fallen out of the twenty first century into the twentieth'. It has seemed to me more like something dragged out of the nineteenth!" [15, p. 285].

[17] The key featue of the potentials is that they satisfy Ohm's law.

[18] The type II refers to the fact that such superconductors live in a mixture of non-superconducting and superconducting regions, which results in vortices of superconducting current surrounding the non-superconducting regions. The magnetic field enters the interior of the superconductor through such vortices, or 'Abrikosov flux tubes' as they are sometimes called. Such flux tubes have energy per unit length just like strings, growing linearly, with each tube containing one unit of quantized flux (with the number of such tubes determined by controlling the external field)—cf. [39].

model. The motivation was to make sense of the fact that on the one hand the string picture copes remarkably well with the nice features of the Veneziano model (capturing experimental results), yet many of the principles leading to the Veneziano model were drawn from field theory (e.g. crossing symmetry):

> We have good reasons to believe that both field theory (of a kind which is so far not known) and dual strings (with some yet unknown degrees of freedom) are in fact realzed in nature. It is likely that nature has decided to merge some field theory with some dual string structure [47, p. 45]

The Nielsen-Oleson model was thus supposed to offer a compromise, pointing the way from S-matrix theory and bootstrap philosophy, back to more orthodox quantum field theory. Indeed, Nielsen and Oleson hoped that by merging the two in this way, the latter might serve to tame some of the troubling issues facing the former. For example, choosing a positive definite Hamiltonian might eliminate the presence of tachyons in the dual model's spectrum. Moreover, it might serve to throw light on curious features of dual models, such as the condition for a critical dimension of $d = 10$ or $d = 26$. The strategy was to "translate" such notions from dual resonance model language to field theory language (ibid., p. 46) and see if they could be reconceptualised (e.g. as internal symmetries). It was also suggested that such a field theoretic translation might point to potential generalizations of dual models. Hence, in many ways, initially at least, this field-theoretic correspondence was used as a kind of exploratory tool.

Nielsen-Oleson vortex strings were devoid of endpoints, and hence were either closed or infinitely long. A little later, Nambu[42] extended the Nielsen-Oleson idea to the case of open strings (with endpoints), using a formalism developed by Michael Kalb and Pierre Ramond [32]. This paper introduced Dirac monopoles, leading to the conclusion that quarks are sources of magnetic charge, permanently confined by their string bonds—in other words, in order to be of finite length the Nielsen-Oleson had to end on a monopole, to 'capture' the flux.[19] Such 'string/gauge' analogies have continued to play an important role in the development of string theory—a point we shall return to in the final chapter of this book. Further, though we will return to it in the next chapter, we should pause to mention that this Nielsen-Olesen interpretation of dual strings (as Abrikosov flux lines) was highly influential in the subsequent understanding of colour confinement in QCD. Hence, the important role of the hadronic dual string in the construction of QCD should not be underestimated.

4.2 Geometrical Interpretation

The generalised Veneziano model amounts to a formula for computing N-point functions in a field theory. String theory emerged from the recognition that these N-point functions are in exact correspondence with the (expectation values of) N

---

[19] This would go on to inspire Gerard 't Hooft [59] in his work on magnetic monopoles in unified gauge theories (cf. [44, p. 381]). We discuss this further in Sect. 6.3.

vertices of a theory of strings (i.e. the two-dimensional surfaces swept out by strings). Hence, one often sees it said that the Veneziano amplitude was 'really' a theory of strings. There are, of course, two stories at work here. On the one hand, there is the more abstract algebraic description based on the operatorial formalism, and on the other there is a geometrical approach based on the worldsheet picture. Although the geometric string picture was in some sense derived from the abstract operator approach, the two led curiously separate lives afterwards.

The first published discussion of the string worldsheet (and the coining of the term) appears to have been Susskind [54].[20] Armed with the worldsheet concept, Susskind was also able to show how the duality that kick-started the new work on string theory could be explained as an implication of conformal invariance (though this seems to have been suggested by others too). This must have been roughly contemporaneous with Nambu's researches (see below) which were then still unpublished. In this section we also see how the worldsheet concept was involved in the construction of the (quadratic) string action.

I mentioned that the worldsheet concept appears to have originated with Susskind. Not long after this, as part of a collaboration with Aage Kraemmer and Holger Nielsen, he wrote down an action principle for strings[21]:

> One of the most exciting things that happened was a correspondence that started when Holger Bech Nielsen sent me a handwritten letter explaining his 'conducting-disc' analogy ... Holger understood that the conducting disc was just the world-sheet and that the relation between his work and mine was simply momentum-position duality. Nielsen came to visit me in New York and we excitedly explored the possibility that the world-sheet is a dense planar Feynman diagram which we connected with Feynman's parton ideas. I believe the paper that we wrote with Aage Kraemmer was the first to contain the quadratic world-sheet action [57, p. 264].

Susskind describes the idea of the worldsheet as follows:

> Let us suppose that a meson is composed of a quark-antiquark pair at the ends of an elastic string as described in previous Sections. As the string moves in space-time a two-dimensional strip bounded by the trajectories of the quarks is generated. In analogy with Minkowski's world-line we call such a configuration a world-sheet [54, pp. 483–484].

Susskind provides variables, θ and τ, specifying coordinates, labeling the points of the worldsheet, with the dynamical variables given by $X_\mu(\theta, \tau)$ satisfying:

$$\frac{\partial^2}{\partial \tau^2} X_\mu(\theta, \tau) - \frac{\partial^2}{\partial \theta^2} X_\mu(\theta, \tau) = 0 \qquad (4.8)$$

---

[20] Susskind notes that an earlier version of the paper was rejected by *Physical Review Letters* "on the basis of not having any new experimental prediction" [57, p. 264].

[21] Susskind [55, p. 234] credits Ed Tryon [61] with the discovery that the energy of a string (in the sense of string theory) is proportional to its length. This seems to be true, so far as I have been able to ascertain. This is clearly a forerunner of the idea that the action for a string's worldsheet is proportional to its area. In the same place (p. 235) Susskind also mentions that one of his students, Henri Noskowitz (sic.), came up with the idea of starting out with an area action from which one can derive the string equations of motion. At the time Susskind pooh poohed the suggestion.

Which he understands simply as a generalization of the equation of motion for the worldline of a point, where $X_\mu \to$ field, $\theta \to$ space and $\tau \to$ time.

It was Nambu who, in 1970, extrapolated the notion of a (quasi-geometric, surface) action for a zero-dimensional point particle to that of a one-dimensional string. This generated a dynamics analogous to the minimisation principle for particle worldlines, only in this case the minimisation principle applied to the surface area that an evolving string would trace out in spacetime (roughly, $\int d$(worldsheet *area*)). His path to the worldsheet idea was based on a conception of the string as a "limit of a chain of N mass points as $N \to \infty$" [41, p. 285]. By thinking of each point within the string evolving in the same way as in the classical theory of the motion of a free mass point particle, so that each traces out its own worldline, one can easily see that in the limit of infinitely many such particles, evolving in parallel, one will generate a two-dimensional sheet with the principle of minimisation of worldline length being modified to the area of the sheet. In Nambu's own notation, the sheet is parameterised by two (intrinsic) coordinates: ξ (such that $0 \leq \xi \leq \pi$) and τ (such that $-\infty \leq \xi \leq +\infty$). The action integral is then:

$$I = \frac{1}{4\pi} \int \int \left(\frac{\partial X_\mu}{\partial \tau} \frac{\partial X^\mu}{\partial \tau} - \frac{\partial X_\mu}{\partial \xi} \frac{\partial X^\mu}{\partial \xi} \right) d\xi d\tau \tag{4.9}$$

From which he derives:

$$(\partial^2/\partial\tau^2 - \partial^2/\partial\xi^2)X^\mu = 0 \quad (\partial X^\mu/\partial\xi = 0, \text{ when } \xi = 0, \pi) \tag{4.10}$$

In more modern terms, we would write the action as:

$$S_{Nambu} = -T_0 \int_{\tau_1}^{\tau_2} d\tau \int_{\sigma_1(\tau)}^{\sigma_2(\tau)} d\sigma \sqrt{\left(\frac{\partial X_\mu}{\partial \tau} \frac{\partial X^\mu}{\partial \sigma} \right)^2 - \left(\frac{\partial X_\mu}{\partial \tau} \frac{\partial X^\mu}{\partial \tau} \right) \left(\frac{\partial X_\mu}{\partial \sigma} \frac{\partial X^\mu}{\partial \sigma} \right)} = \int d\tau d\sigma \mathscr{L}$$
$$\tag{4.11}$$

The idea here involves parameterising the string worldsheet, via parameters σ and τ, which will provide coordinates for the worldsheet in spacetime. T is the tension of the string, and is related to the Regge slope term via $\frac{1}{T} = 2\pi\alpha'$. One then considers the relationship between a point on the worldsheet, labeled by (σ, τ), and the spacetime in which the string is embedded, giving $X^\mu(\sigma, \tau)$ ($\mu = 1, ..., d$, with d the dimension of spacetime)—the action then depends on this rather than a specific parameterisation. In other words, the action is invariant under arbitrary changes of the parameters, $\delta X^\mu(\sigma, \tau) = \xi^\alpha \partial_\alpha X^\mu(\sigma, \tau)$. The symmetry group of such reparameterisations (that leave invariant the action) is infinite dimensional.[22] The clear analogies between the particle theory action and the string theory action enables the carrying over of powerful techniques and ideas from the former to the latter case.[23] The reparameterisation invariance (with respect to σ, τ, or ξ, τ in Nambu's notation),

---

[22] The group also reduces to the symmetry groups of general relativity and Yang-Mills theory as the Regge-Mandelstam parameter is sent to zero, though that was not known at the time.

[23] Of course, there are crucial differences too. The worldlines of interacting particles are given by graphs (with nodes) and therefore are not manifolds, whereas the worldsheets of splitting and joining strings *are* smooth manifolds (cf. [31, pp. 50–51].

suggests that the observable (physical) oscillations must be those *transverse* to the worldsheet, since any motions within the sheet can be gauged away by a suitable transformation of the worldsheet variables, σ and τ. In this way, the choice of action gives a representation involving the intrinsic, physical structure (with no ghosts).[24]

Nambu's original presentation of the action that now bears his name was included in notes (on "Duality and Hadrodynamics") that he had prepared in advance for a high energy physics symposium in Copenhagen in August 1970, though had in fact missed as a result of car trouble. Nambu had been invited to speak at the 1970 Copenhagen conference by Koba and Nielsen, and was also due to speak at the Rochester conference. Nambu describes the ensuing events:

> [I]n 1970, I remember there was a Rochester conference in Kiev to which I was invited, so I wanted to attend it. Now, I'd gotten my citizenship in 1970. I got a relief from my problem —I had some sort of immigration problem, but it was solved by then so I was able to go to Kiev. At the same time I got an invitation from the Copenhagen people for a summer institute or something, so I wanted to attend that too, to give a talk on my theory. It was in the summer of 1970 before going to Europe, I wanted to deposit my family in California with my friends. So we drove out from Chicago, and unfortunately on the way we had an accident on the road in the Salt Lake Desert. Actually the whole cooling system ruptured and the engine overheated and was destroyed. So we had to stay three days in the desert to fix it, and managed to get to California. But by then I had missed a plane connection, so I gave up going there and came back to Chicago. But in the meantime I had sent my manuscript to Copenhagen, hoping that it would come out eventually at the proceedings, which it did not.[25]

In fact, the notes were not available in published form until 1995, with the release of his collected papers [41]. Goddard claims that Nambu was known to have considered the geometric action in the advance copy of his paper for the Kiev conference, and news of his idea quickly spread by word of mouth [23, p. 238].[26]

Tetsuo Gotō covered much the same ground as Nambu independently in 1971 [25], with a more detailed (and published!) account of the same action (now called the Nambu-Gotō action).[27] Gotō referred to the string systems in terms of a one-dimensional mechanical continuum ("a finite one-dimensional continuous medium"), following an earlier paper by Takehiko Takabayasi [58].[28] He was primarily concerned with providing a string model explication of the Ward-like ghost

---

[24] Lay Chang and Freydoon Mansouri [12] replicate Nambu's basic result, with a solid focus on analogies with gauge theories, constructing a diffeomorphism-invariant action from which gauge symmetries written in terms of string variables are constructed.

[25] Interview of Prof. Yoichiro Nambu by Babak Ashrafi on July 16, 2004, Niels Bohr Library and Archives, American Institute of Physics, College Park, MD USA, http://www.aip.org/history/ohilist/30538.html.

[26] This is a rare instance in (recent) history of science in which a name has been attached to a concept with neither a public presentation nor a published article.

[27] Gotō mentions Nambu's construction of the same concepts in a footnote, crediting Prof. Iisuka with informing him of Nambu's work after he had arrived at similar results independently [25, p. 1562].

[28] In fact, Gotō is well aware of the "elastic string" terminology, but he does not find elasticity in his model, only straight lines. Therefore, he suggests calling it a "linear *rod*" instead [25, p. 1568].

Fig. 4.1 Motion of Gotō's
one-dimensional medium
as represented by a
two-dimensional world
sheet (embedded in four-
dimensional Minkowski
spacetime). *Image source*
[25, p. 1561]

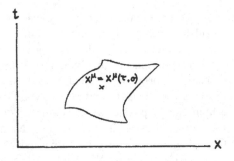

cancellation mechanism of Fubini, Veneziano, and Virasoro: "relativistic quantum mechanics of a one-dimensional object with uniform mass density is equivalent to the so-called 'string' model of hadrons with Virasoro's subsidiary conditions" (p. 1560).

The terminology of 'world sheet' was used in his discussion, and the notation was much the same as Nambu's, though with the σ (rather than Nambu's ξ), for the spatial worldsheet coordinate. Thus, his positional coordinates on the worldsheet are $X^\mu = X^\mu(\tau, \sigma)$, which he labeled "Lagrange coordinates" (see Fig. 4.1).

His action takes the (generally invariant) form (where κ_0 is a mass density):

$$ L = \int \int d\tau d\sigma \kappa_0 \sqrt{-\det g} \tag{4.12} $$

Note that neither Gotō nor Nambu (nor Susskind) considered the propagation of the strings in anything other than four-dimensional Minkowski spacetime—though Nambu did suggest including an additional fifth dimension for the oscillators, in order to avoid the problem of being forced into choosing a unit intercept, though this simply trades one unphysical feature for another. The full geometric spacetime interpretation of the quantised string, in terms of worldsheets being swept out in spacetime, was given in Goddard, Goldstone, Rebbi, and Thorn [GGRT: [22]] in which the theory is canonically quantized. When quantized, the action generates the parallel, linear Regge trajectories associated with the particles of the dual model.

One of the major consistency issues with the dual models was the fact that the Fock space of vertex operators includes ghost states (of negative norm). Negative probabilities do not make sense in quantum theory, so some method is needed for eliminating such states, giving the physical space of states.[29] This was achieved

---

[29] We might note that Landau became convinced of the inconsistency of quantum electrodynamics as a result of similar ghost states (now called 'Landau ghosts') that appear in the computation of the self-energy of an electron, though the mass of the state was so minuscule as to render the inconsistency empirically inconsequential. In that case one employs the electromagnetic gauge invariance to eliminate the ghost states (the process involves the imposition of certain subsidiary (gauge) conditions that we will see are analogous to those used in the dual model case). (Note that this is not 'ghosts' in the sense of the so-called Faddeev-Popov ghosts, in which additional *unphysical* fields are integrated over to preserve the unitarity of gauge theories.)

through no-ghost theorems, which show how the physical amplitudes for processes do not depend on the ghosts. Such a theorem would give a zero value for amplitudes involving ghosts on external lines, and would ban ghosts from appearing as intermediate states (internal lines) connecting physical states. The same situation can be found in QED, in which the time component of the photon is unphysical, with the physical states being those satisfying the Gauss constraint. Lars Brink and David Olive [5] constructed a projection operator $\tau(k)$ onto physical states (where k is a vector used to build DDF states), allowing them to calculate planar loops with only physical states propagating internally.

The recognition of dual model ghosts came almost immediately with the construction of the operator approach to the dual resonance model devised by Fubini, Gordon, and Veneziano [17]. Recall that this had been initially constructed in order to understand the dual theory's spectrum by establishing factorization of the N-particle generalization of Veneziano's original 4-point amplitude. This involved a representation of the states built up from the vacuum state $|0\rangle$ via the action of an infinite collection of harmonic oscillators, a_n^μ (where the superscript $\mu = 0, 1, 2, 3$ is the Lorentz index, for a flat $\langle -, +, +, + \rangle$ spacetime, implying that the oscillator states must transform as representations of the Lorentz group; the subscript n refers to the integral mode number characterising a bosonic model). The problem with the Landau ghost states (that is the negative-norm states) stemmed from the existence (mathematically speaking) of the timelike modes a_n^0 which automatically point to negative-norm states. A sector of the ghost states was removed by the imposition of $SO(2, 1)$ symmetry, reducing out some of the surplus states. This was discovered independently by both Fubini and Veneziano [18] (while at MIT) and Bardakçi and Mandelstam [3] (at Berkeley). Within the Koba-Nielsen (complex) formalism Möbius invariance can likewise be imposed to reduce out the problematic states. In both cases, the complete elimination would require an infinite family of conditions, one for each possible timelike component in a_n^0—cf. Goddard [23, p. 237]. Interestingly, as mentioned in the previous chapter, the infinite set of such subsidiary conditions was isolated by Virasoro [62], with the infinite set of operators (underlying the conditions) generating what is now known as the Virasoro algebra.[30] But this solution, though it indeed eliminates all time components, had an unwelcome side-effect comparable to the ghost states it was devised to cure. In order to work, the Regge slope of the leading trajectory[31] $\alpha(0)$ had to be *fixed* at a value corresponding to an intercept of 1, which in turn implied that the ground state had $M^2 < 1$ (i.e. a tachyon). Negative probability was thus replaced with negative mass, again trading in one unphysical feature for another.

The *no-ghost theorem* begins with the Nambu-Gōto action and its great virtue of focusing the attention on the physical observables by means of the reparametrization

---

[30] These operators satisfy the conditions for the algebra of 2D conformal mappings with a central charge. A fact that led to significant overlap with areas of pure mathematics.

[31] That is, lightest exchanged particle for a given spin.

invariance, forcing physical status on the transverse oscillations only.[32] Their approach was to canonically quantize the transverse degrees of freedom. The consistency conditions, of fixed intercept and spacetime dimension of 26,[33] were shown to arise once again in this context, here demanded by Lorentz invariance. Part of the machinery used was supplied by Brower, in the form of a spectrum-generating algebra, which provides a means of building up a space of physical states from a given physical state (e.g. the vacuum state) by the action of an appropriate operator (the Virasoro generators) [7].

Brower [7] and also Goddard and Thorn [21] were able to prove, independently, the "no-ghost theorem", according to which the dual theory in 26 dimensions doesn't possess negative norm vectors. Again, this number $d = 26$ was clearly playing a central role and, therefore, could not be dismissed so lightly as previously thought. The construction was based on vertex operators and propagators and the association of operator expressions to these. One can then build the S-matrix as a sum of contributions of such terms, in the standard way.

With the clarity provided by the GGRT paper, the string 'picture' was put onto a firmer footing. As Ferdinando Gliozzi puts it:

only with the GGRT paper were all the consequences of this Nambu-Goto action correctly derived and it became completely clear, even at the quantum level, that the relativistic string was not simply an analogue model used to help intuition, but that it described the underlying microscopic structure of the DRM [20, p. 448].

That is, the string picture could be seen as a genuine physical interpretation of dual-resonance models such that there exists a correspondence between dual amplitudes and the amplitudes for strings. Despite coming after the initial 'golden age' of strings, the no-ghost theorem was pivotal in the theory's development since it firmly established its full mathematical consistency. However, problems still remained: there were massless particles that were not found at the predicted energy scales, and there were spatial dimensions that were demanded by the theory but not observed. Hence, the theory was still inconsistent with physical reality. We return to these problems below.

This marked a stage of development whereby the string theory was somewhat freed from its origins in the dual resonance model on which it had, up to this point,

---

[32] Searching for a way to produce string theories in $D = 4$, Bardeen, Bars, Hanson, and Peccei [4] attempted to reintroduce the longitudinal modes into the theory, treating them as analogous to kink solutions of a nonlinear field theory.

[33] One of the additional consistency consistency checks converging on the meaningfulness of the $d = 26$ result (discussed more fully in the next section) was the realisation of the ghost-generating nature of variations of d above 26—for $d < 26$ Brower recalls running a recursive algebraic computer program to enumerate physical states to 30th level finding no ghosts for $d \leq 26$ [9, p. 317]. 26 will reduce to 10 in the case where fermions are included, as we will see in the discussion of supersymmetric strings in the next chapter. These implications, and the demonstration of the reduction to $D = 10$, are laid out in [21]. They had initially believed that this reduction might open up the possibility that "it will be four in some more realistic model" [21, p. 235]. The reasoning is clear: if adding additional structure, such as fermionic coordinates, can reduce the critical dimension so radically, then perhaps there are other structures, not yet understood, that could reduce this all the way down to $D = 4$.

been parasitic: strings were the *reason* for the dual resonance model and so could be pursued in their own right. We might further speculate that this detachment from the dual resonance model contributed to the transition from a model of strong interactions to other interactions. By 1973, then, thanks to the paper by GGRT, there was a fairly complete picture of the quantum mechanics of a relativistic string, albeit with the still peculiar restriction on the space-time dimensionality: physical states of the theory were defined as transverse modes of oscillation of a massless, relativistic string propagating in 26 dimensions.[34]

GGRT only studied the case of *free* strings, in which they can pass through one another. An important task that remained to be solved was that of incorporating inter-actions, and in such a way so as to not fall foul of the no-ghost theorem (enforcing the restriction to transverse states). Though the idea had been proposed in a qual-itative fashion by several people, Mandelstam was responsible for making precise the idea that the scattering represented by the Veneziano amplitude could be under-stood in terms of the successive splitting and joining of strings, invoking Susskind's worldsheet idea: the dynamics of a string theory is fixed once the vertex for splitting and joining of strings is found.[35] This involves the overlap integral between the two input strings and the output string (or two input strings and a final string: in between interactions, the strings move freely).

One has a many-string formulation once one has the capacity to talks of split-ting and joining. The operator formalism (Hilbert space) encodes this. For a non-interacting theory one simply has a term corresponding to the standard Nambu-Gōto Lagrangian per number of strings. Interactions are represented by an interaction vertex term which adds (splits) or subtracts (joins) strings Fig. 4.2.

Mandelstam's interacting strings model was able to recover the dual resonance model in $D = 26$. His method involves an extension of the results of GGRT to inter-acting particles.[36] String theories come in two varieties: bosonic and supersymmetric

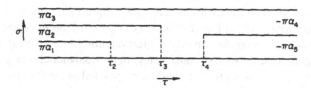

Fig. 4.2 Mandelstam's picture of string interaction. Here, three strings ($\pi\alpha_1$, $\pi\alpha_2$, and $\pi\alpha_3$) come in from $\tau = -\infty$, two of them join at τ_2, after which the resulting string joins a third at τ_3. The single string then splits into two at τ_4, which go out to $\tau = +\infty$. *Image source* [37, p. 208]

---

[34] Note that the 'light-cone frame' was introduced in GGRT to provide a formalism in which the dynamics of (bosonic) strings was given in terms of $D-2$ oscillations propagating transverse to the string's length. This was used directly to quantise the string action, and has since been used many times to make calculations easier.

[35] He later extended this to the Ramond-Neveu-Schwarz model [38].

[36] Roscoe Giles and Charles Thorne [19] developed a lattice version of Mandelstam's argument to get around certain divergences associated with using a continuum (see also [60]).

(i.e. including fermions). The bosonic theory came first and is much simpler than the supersymmetric version. Mandelstam included fermions by adding a spinor field $(S_1(x_\mu), S_2(x_\mu))$ (describing a spin-wave respectively going from right to left across the string, and from left to right across the string).[37]

Mandelstam considered the first quantized theory in which the variables are the string coordinates, giving a kind of two-dimensional field theory on the worldsheet. The second quantized version of the theory (a purely bosonic string field theory based on "multilocal relativistic strings") was developed by Kaku and Kikkawa [29, 30]. As they note:

> Notice that, though the string picture presented so far resembles a second-quantized theory because of the presence of an infinite number of harmonic oscillators, it is actually only a first-quantized theory, because we are only quantizing the coordinate $X - \mu(\sigma)$. There are an infinite number of oscillators only because they represent the normal modes of the string, i.e., because the first quantization is performed over an extended object [29, p. 1113].

This was only carried out within the non-supersymmetric case. It wasn't until Green and Schwarz's work in the early 1980s that the superstring field case (in which quantum string-fields create and annihilate complete strings) was considered: [26].

4.3 Bootstrapping Spacetime

I think it is fair to say (and many others have said it) that just after Veneziano's paper was published, the centre of the dual model universe was CERN. Many of the key pieces of the theory of hadronic strings were put into place either by permanent staff members of CERN, visitors, or those just passing through. One of those was Claud Lovelace, who discovered the famous dual model consistency condition that demanded 26 spacetime dimensions—a kind of bootstrapping of spacetime.[38] Lovelace discovered, in 1970, that only if there were 26 dimensions of spacetime would certain problematic branch cuts become simple poles (thereby avoiding a violation of unitarity), with a Regge trajectory possessing an intercept of 2 and a slope of $\frac{1}{2}\alpha'$ (thus allowing for a particle interpretation, then given in term of the Pomeron). That is, unitarity (and so consistency) of the dual model seemed to *demand* $D = 26$. The set of properties (of the pole) corresponded to the Pomeron trajectory, as mentioned. It would not long after be reinterpreted as describing closed strings, eventually associated with the graviton.[39]

---

[37] Influenced by Mandelstam's work, the (tree-level) treatment of interactions of closed with open strings was completed soon after, [1] (see also [20, p. 451].

[38] As Gomez and Ruiz-Altaba put it, "[t]he magic of the string approach to quantum gravity is that spacetime is not an ingredient put into the analysis from the start. It is philosophically astounding that spacetime is actually an output of string theory" [24, p. 83].

[39] David Olive recalls 'implanting' in Lovelace, at CERN in late June 1970, the idea that if the branch point singularity (then already identified by Lovelace) could be a simple pole instead of a branch cut then it could be interpreted as corresponding to the propagation of a new kind of particle [48, p. 348]. The branch point in question was discovered in a 1970 paper by Gross, Neveu, Scherk, and Schwarz [28]

In a paper written in the final period of Chew's original bootstrapping approach to physics,[40] he points out that, even if one could produce a unique S-matrix from his scheme, the S-matrix would depend on an underlying *a priori* space-time:

> We must not forget that, in the final analysis, the S-matrix depends on the arbitrary concept of space-time. From an ultimate bootstrap point of view, *all* concepts should be justified by self-consistency, none should be accepted on an *a priori* basis [13, p. 24].

Though it wasn't taken seriously at the time of discovery[41] (and for some time afterwards), the requirement on the space-time dimension, discovered by Lovelace, removes the arbitrariness in the way desired by Chew.[42] It might appear unphysical, but *d is* fixed by consistency conditions in dual models.

Lovelace was English by birth, born to a very wealthy family.[43] They moved to Switzerland when he was young, and he then did his undergraduate studies in Cape Town. In fact, he switched to architecture after completing his Bsc, but later switched back to physics, studying at Imperial with Abdus Salam. He describes his trajectory to the $D = 26$ result as follows:

> Gross et al. ... at Princeton, and Frye and Susskind... at Yeshiva had both found a very strange singularity in the one-loop amplitude. Like everyone else I thought that open strings were Reggeons, so this [singularity] must be the Pomeron, which would be very interesting to a phenomenologist. Unfortunately, it was tachyonic with a continuous spectrum. My notebooks show that I started redoing their calculations on 1 October 1970 at CERN. I needed a realistic model for phenomenology, so the Pomeron intercept had to be 1. By next day I had concluded that the intercept was $D/2$ in spacelike dimensions (i.e. those with oscillators). However, the ensuing calculations turned the cut into a pole by arbitrarily deleting log R factors. They go on for 88 pages until a note written in Princeton in early January says 'I think we need 24 spacelike and 2 *timelike* dimensions to get complete cut cancellation.' I suspect the correct solution came to me suddenly at night, since this note is in different ink. Thus in 26 dimensions, and assuming that two sets of oscillators decoupled, the Pomeron spectrum became discrete... There was still one tachyon, but the next particle had zero mass and spin two.[44] This matched the Shapiro-Virasoro formula [36, p. 199].

---

[40] However, the approach did morph, taking in some of the features of duality, into "Dual Topological Unitarization" [DTU], again with the motto of 'no arbitrary parameters' centre stage—see, e.g., [14].

[41] With a few exceptions, by the late 1960s and early 1970s the notion of Kaluza-Klein compactification and theories invoking extra dimensions to perform various functions (though once popular) had dropped out of fashion.

[42] At the end of the same article, Chew writes: "it is plausible that to understand zero-mass phenomena through self-consistency may require bootstrapping space-time itself" (ibid., p. 28).

[43] Lovelace died in 2012, leaving to Rutgers $1.5 million for a chair in *experimental* physics. Clavelli notes that when he arrived in Rutgers, Lovelace was "still living in a motel and driving a rental car" (http://bama.ua.edu/~lclavell/papers/Tension1.pdf).

[44] Note that Lovelace writes in terms of Reggeons (with worldsheets described in "ribbon" terms) and Pomerons (with world*tubes*, or the surface of a closed tube). These correspond to what we would now think of open and closed strings. The Pomeron described by Lovelace in this paper was later identified with the graviton (once a scale change had been implemented). In his reminiscences about this paper he writes that, given his knowledge of unified field theory and Kaluza-Klein mechanisms (not least as a result of his studies with Salam), "I was inexcusably stupid not to see in 1971 that my Pomeron was the graviton" [36, p. 199]. This is, of course, overly harsh since there was

This appearance of the spacetime condition comes from the definition of the Pomeron propagator (see footnote 44):

$$(2\pi)^{-1} \int_0^1 dR \int_0^{2\pi} d\sigma \, R^{-1-\alpha_0^P - \frac{1}{2}\alpha' p^2} \mu(R)(Re^{i\sigma})^{n\Sigma n a_n^\dagger a_n}(R)(Re^{i\sigma})^{n\Sigma n b_n^\dagger b_n}$$

(4.13)

where

$$\alpha_0^P = \frac{(D-E)}{12}$$

(4.14)

$$\mu(R) = (\frac{-\pi}{\log R})^{1-E/2}\omega^{(D-E/24-1)}(1-\omega)^F$$

(4.15)

$$\omega = e^{2\pi^2/\log R}$$

(4.16)

The Pomeron–Reggeon coupling constant f (with g being the 3-Reggeon) is defined by:

$$f^2 = (2\pi)^{-3}2^{-D/2}g^2$$

(4.17)

One has a self-consistent situation when $D = 26$, $E = 2$, and $F = 0$. Initially, Lovelace did not take the result at all seriously, in the sense of pointing to something deep about the physical world. He notes that in a seminar he gave at the Institute for Advanced Study in Princeton, in February 1971, he made the joke that he had "bootstrapped the dimension of spacetime but the result was slightly too big" (ibid.). In other words, self-consistency had forced the spacetime dimension to be 26, but at this time, of course, there was no connection to spacetime physics or gravitation. Nobody else took it seriously. This was supposed to be a theory of hadrons, pure and simple. It was only after the number $D = 26$ began to reappear, in the context of other consistency conditions such as the no-ghost theorem, that it was taken seriously as something potentially more significant.[45] Despite not thinking much of the result, Lovelace did nonetheless publish, albeit very briefly and with the qualifying remark

---

(Footnote 44 continued)
at the time no reason whatsoever to connect up dual models with gravitational physics; that was something that would require the additional investigation of the zero-slope limit of dual models.

[45] Interestingly then, the no-ghost theorem demanded that the maximum number of spacetime dimensions (or a ghost-free theory) be 26, thus providing independent confirmation of the earlier critical dimension result of Lovelace. The decoupling of negative-norm states occurs only for $d \leq 26$ (with the additional Virasoroan condition that the Regge intercept $\alpha(0) = 1$). In this way a kind of mathematical unity was achieved, in which troubles of formalism (tachyon and $d = 26$) were integrated into a single scheme, and shown to be related. Clavelli and Shapiro combined the no-ghost theorem with Lovelace's earlier work on Pomeron factorization to argue forcefully for the existence of a critical dimension in ghost-free dual models such that in this dimension the Pomeron singularity becomes a factorizable Regge pole (which can, therefore, be viewed as a real particle). In the case of the Neveu-Schwarz model (discussed in Sect. 5.2), performing the same kind of procedure Lovelace had applied in the case of Pomerons (reducing cuts to poles, and preserving unitarity), they find $D = 10$, $E = 2$, and $F = 0$. Hence, the restriction on the number of spacetime dimensions was tightly bound to the consistency of the theory. Richard Brower was able to show, using his spectrum-generating algebra that $D = 26$ provides a maximum density of states consistent

that "$D = 26$ is obviously unworldly" [35, p. 502]. He also spread the idea around various colleagues, including many dual theorists at CERN. Since CERN was at the time a hotbed of activity on dual models, the idea was able to infiltrate the research landscape.

The root of the condition is the requirement that the action principle for string theory be conformally invariant. Conformal symmetry allows one to identify any diagrams (or processes) for which all angles are preserved. The laws of string theory are insensitive to conformal transformations of the string worldsheet. An anomaly refers, in this context, to a symmetry that is obeyed at the classical level, but violated quantum mechanically. Hence, given some operation $\mathscr{O}$ for which $\{\mathscr{O}, H\} = 0$, we have $[\hat{\mathscr{O}}, \hat{H}] \neq 0$. If one has such non-conservation for gauge currents (like the conformal symmetry) then the quantum theory is not consistent: it is found to violate unitarity and possibly will be rendered non-renormalizable. String theory was found to have such an anomaly concerning conformal symmetry. That this conformal anomaly cancels in 26 dimensions forms the heart of Lovelace's result.

In 1973, Holger Nielsen and Lars Brink published a paper [6] which analysed the notion of the critical dimension more deeply, providing an explanation (deriving it from a more physical argument)—this analysis covered both the 26 dimensional case, for strings with geometrical degrees of freedom, and the 10-dimensional case, with fermionic degrees of freedom too (i.e. the Neveu-Schwarz model). As they conclude: "we have found a physical interpretation of the ground state mass squared in string models as zero point fluctuations" pointing out that their result makes it "difficult to escape the dependence on the dimension of space-time for such models" [6, p. 336]. The argument was based on the idea that the physical degrees of freedom correspond to transverse degrees of freedom. A radically abbreviated run through goes as follows. The zero point energy of the (ground state) string is given as:

$$E_{zero} = d_{eff} \sum_{n=1}^{\infty} \frac{1}{2}\omega_n = d_{eff} \sum_{n=1}^{\infty} \frac{1}{2}\frac{1}{\alpha'}\frac{n}{2E} \qquad (4.18)$$

As they note, for a string with only transverse modes, $d_{eff} = d - 2$. Next, the string is considered in the infinite-momentum frame, and the zero point energy is written as the difference between the quantum mechanical ground state and the classical version: $E_{zero} = E - |p|$, which in the infinite-momentum frame gives $2E = E + |p|$. This lets them rewrite Eq. 4.18 as:

$$E^2 - |p|^2 = \frac{d_{eff}}{4\alpha'}E(E + |p|)\int_0^{\infty} dy y f(y) - \frac{d_{eff}}{24\alpha'} + O\left(\frac{1}{E}\right) \qquad (4.19)$$

(Footnote 45 continued)
with a positive-norm space (i.e. an absence of ghosts). He also argues that "at saturation ($D = 26$)" (that is, when thus fixed) the loop theory achieves its "most elegant" form, being non-renormalizable above this value [7, p. 1661]. Note that the $D = 10$ critical dimension was originally discovered by John Schwarz in [51].

They are able to show from this that the theory has a lowest state of mass squared: $m_0^2 = -d_{eff}/24\alpha'$. Since the string has only transverse degrees of freedom, Lorentz invariance forces the spin-1 particle on the leading Regge trajectory to have a mass. This implies:

$$m_0^2 = \frac{1}{\alpha'} = -\frac{d_{eff}}{24\alpha'} = -\frac{d-2}{24\alpha'} \tag{4.20}$$

This latter expression clearly demands $d = 26$.

4.4 Summary

By 1973/4 it was known that the quantization of free open strings and closed strings reproduces the spectra of the generalized Veneziano model and Shapiro-Virasoro models respectively: the oscillators of the dual resonance models corresponded to the normal modes of vibrations of a string. The interacting string theories were established (including open-closed interactions), and the role of the various consistency conditions (involving intercepts and spacetime dimensions) known and understood. Despite the fact that the dual model qua string theory idea was well in place in the early 1970s, it was then still considered tentative: a convenient model in which to think about the mathematical structure. It did not have a robust existence as a picture of string fields living in spacetime, for example. The string model provided a nice way of visualising processes that are rather difficult to handle in the operator approach. Hence, we should not be misled into thinking that string theory in anything like the modern sense (that is, a sense corresponding to the 'real world') was in operation in this initial phase.

References

1. Ademollo, M., D'Adda, A., D'Auria, R., Napolitano, E., di Vecchia, P., Gliozzi, F., et al. (1974). Unified dual model for interacting open and closed strings. *Nuclear Physics, B77*(2), 189–225.
2. Alessandrini, V. (1971). A general approach to dual multiloop diagrams. *Il Nuovo Cimento, 2*(2), 321–352.
3. Bardakçi, K., & Mandelstam, S. (1969). Analytic solution of the linear-trajectory bootstrap. *Physical Review, 184*, 1640–1644.
4. Bardeen, W. A., Bars, I., Hanson, A. J., & Peccei, R. D. (1976). Study of the longitudinal kink modes of the string. *Physical Review D, 13*(8), 2364–2382.
5. Brink, L., & Olive, D. I. (1973). The physical state projection operator in dual resonance models for the critical dimension of space-time. *Nuclear Physics, 56*(1), 253–265.
6. Brink, L., & Nielsen, H. (1973). A simple physical interpretation of the critical dimension of space-time in dual models. *Physics Letters B, 45*(11), 332–336.
7. Brower, R. C. (1972). Spectrum-generating algebra and no-ghost theorem in the dual model. *Physical Review D, 6*, 1655–1662.
8. Brower, R. C., & Thorn, C. B. (1971). Eliminating spurious states from the dual resonance model. *Nuclear Physics, B31*, 163–182.

9. Brower, R. C. (2012). The hadronic origins of string theory. In A. Capelli et al. (Eds.), *The birth of string theory* (pp. 312–325). Cambridge: Cambridge University Press.
10. Cao, T. Y. (2010). From current algebra to quantum chromodynamics. Cambridge: Cambridge University Press.
11. Castellani, E. (2012). Early string theory as a challenging case study for philosophers. In A. Cappelli, et al. (Eds.), *The birth of string theory* (pp. 63–78). Cambridge: Cambridge University Press.
12. Chang, L. N., & Mansouri, F. (1972). Dynamics underlying duality and gauge invariance in the dual-resonance models. *Physical Review D, 5*(10), 2535–2542.
13. Chew, G. F. (1970). Hadron bootstrap: triumph or frustration? *Physics Today, 23*(10), 23–28.
14. Chew, G. F. (1978). Dual topological unitarization: an ordered approach to Hadron theory. *Physics Reports, 41*(5), 263–327.
15. Fairlie, D. (2012). The analogue model for string amplitudes. In A. Capelli et al. (eds.), *The birth of string theory* (pp. 283–293). Cambridge: Cambridge University Press.
16. Fubini, S., & Furlan, G. (1965). Renormalization effects for partially conserved currents. *Physics, 1*, 229–247.
17. Fubini, S., Gordon, D., & Veneziano, G. (1969). A general treatment of factorization in dual resonance models. *Physics Letters, B29*, 679–682.
18. Fubini, S., & Veneziano, G. (1969). Level structure of dual resonance models. *Il Nuovo Cimento, A64*, 811–840.
19. Giles, R., & Thorn, C. B. (1977). Lattice approach to string theory. *Physical Review D, 16*(2), 366–386.
20. Gliozzi, F. (2012). Supersymmetry in string theory. In A. Capelli et al. (eds.), *The birth of string theory* (pp. 447–458). Cambridge: Cambridge University Press.
21. Goddard, P., & Thorn, C. B. (1972). Compatibility of the dual pomeron with unitarity and the absence of ghosts in the dual resonance model. *Physics Letters, 40B*(2), 235–238.
22. Goddard, P., Goldstone, J., Rebbi, C., & Thorn, C. B. (1973). Quantum dynamics of a massless relativistic string. *Nuclear Physics, B56*, 109–135.
23. Goddard, P. (2012). From dual models to relativistic strings. In A. Capelli et al. (eds.), *The birth of string theory* (pp. 236–261). Cambridge: Cambridge University Press.
24. Gomez, C., & Ruiz-Altaba, M. (1992). From dual amplitudes to non-critical strings: A brief review. *Rivista del Nuovo Cimento, 16*(1), 1–124.
25. Gotō, T. (1971). Relativistic quantum mechanics of one-dimensional mechanical continuum and subsidiary condition of dual resonance model. *Progress in Theoretical Physics, 46*(5), 1560–1569.
26. Green, M. B., & Schwarz, J. H. (1982). Superstring interactions. *Nuclear Physics, B218*, 43–88.
27. Gross, D. (2005). The discovery of asymptotic freedom and the emergence of QCD. *Proceedings of the National Academy of Science, 102*(26), 9099–9108.
28. Gross, D. J., Neveu, A., Scherk, J., & Schwarz J. (1970). Renormalization and unitarity in the dual-resonance model. *Physical Review D, 2*(4), 697–710.
29. Kaku, M., & Kikkawa, K. (1974). Field theory of relativistic strings. I. Trees. *Physical Review D, 10*(4), 1110–1133.
30. Kaku, M., & Kikkawa, K. (1974). Field theory of relativistic strings. II. Loops and pomeron. *Physical Review D, 10*(6), 1823–1843.
31. Kaku, M. (1988). *Introduction to superstrings*. New York: Springer.
32. Kalb, M., & Ramond, P. (1974). Classical direct interstring action. *Physical Review D, 9*(8), 2273–2284.
33. Kibble, T. (1999). Recollections of Abdus Salam at Imperial College. In J. Ellis, F. Hussain, T. Kibble, G. Thompson & M. Virasoro (Eds.), *The Abdus Salam memorial meeting* (pp. 1–11). Singapore: World Scientific Publishing.
34. Kraemmer, A. B., Nielsen, H. B., & Susskind, L. (1971). A parton theory based on the dual resonance model. *Nuclear Physics, B28*, 34–50.
35. Lovelace, C. (1971). Pomeron form factors and dual Regge cuts. *Physics Letters, B34*, 500–506.

36. Lovelace, C. (2012). Dual amplitudes in higher dimensions: A personal view. In A. Capelli et al. (Eds.), *The birth of string theory* (pp. 198–201). Cambridge: Cambridge University Press.
37. Mandelstam, S. (1974). Interacting-string picture of dual-resonance models. *Nuclear Physics, B64*, 205–235.
38. Mandelstam, S. (1974). Interacting-string picture of Neveu-Schwarz-Ramond model. *Nuclear Physics, B69*, 77–106.
39. Mandelstam, S. (1974). II. Vortices and quark confinement in non-Abelian gauge theories. *Physics Reports, 23*(3), 245–249.
40. Nambu, Y. (1970). Quark model and the factorization of the Veneziano amplitude. In R. Chand (Ed.), *Symmetries and quark models* (pp. 269–277). Singapore: World Scientific.
41. Nambu, Y. (1970). Duality and hadrodynamics. (Notes prepared for the Copenhagen High Energy Symposium). In T. Eguchi & K. Nishijima (Eds.), *Broken symmetry, selected papers of Y. Nambu* (pp. 280–301). Singapore: World Scientific.
42. Nambu, Y. (1974). Strings, monopoles, and gauge fields. *Physical Review D, 10*, 4262–4268.
43. Nambu, Y. (2012). From the S-matrix to string theory. In A. Capelli et al. (Eds.), *The birth of string theory* (pp. 275–282). Cambridge: Cambridge University Press.
44. Nash, C. (1999). Topology and physics—A historical essay. In I. M. James (Ed.), *History of topology* (pp. 359–415). Amsterdam: Elsevier.
45. Nielsen, H. (2012). The string picture of the Veneziano model. In A. Capelli et al. (Eds.), *The birth of string theory* (pp. 266–274). Cambridge: Cambridge University Press.
46. Nielsen, H. (1970). An almost physical interpretation of the integrand of the n-point Veneziano model. Unpublished manuscript: http://theory.fi.infn.it/colomo/string-book/nielsen_preprint.pdf
47. Nielsen, H., & Oleson, P. (1973). Vortex-line models for dual strings. *Nuclear Physics, B61*, 45–61.
48. Rebbi, C. (1974). Dual models and relativistic quantum strings. *Physics Reports, 12*(1), 1–73.
49. Redhead, M. L. G. (2005). Broken bootstraps—The rise and fall of a research programme. *Foundations of Physics, 35*(4), 561–575.
50. Sakita, B., & Virasoro, M. A. (1970). Dynamical models of dual amplitudes. *Physical Review Letters, 24*(20), 1146–1149.
51. Schwarz, J. H. (1972). Physical states and pomeron poles in the dual pion model. *Nuclear Physics, B46*(1), 61–74.
52. Susskind, L. (1969). Harmonic oscillator analogy for the Veneziano amplitude. *Physical Review Letters, 23*, 545–547.
53. Susskind, L. (1970). Structure of Hadrons implied by duality. *Physical Review D, 1*, 1182–1186.
54. Susskind, L. (1970). Dual-symmetric theory of Hadrons. 1. *Il Nuovo Cimento, A69*(3), 457–496.
55. Susskind, L. (1997). Quark confinement. In L. Hoddeson et al. (Eds.), *The rise of the standard model: Particle physics in the 1960s and 1970s* (pp. 233–242). Cambridge: Cambridge University Press.
56. Susskind, L. (2006). *The cosmic landscape*. New York: Back Bay Books.
57. Susskind, L. (2012). The first string theory: Personal recollections. In A. Capelli et al. (Eds.), *The birth of string theory* (pp. 262–265). Cambridge: Cambridge University Press.
58. Takabayasi, T. (1970). Relativistic quantum mechanics of a mechanical continuum underlying the dual amplitude. *Progress in Theoretical Physics, 44*(5), 1429–1430.
59. t' Hooft, G. (1974). Magnetic monopoles in unified gauge theories. *Nuclear Physics, B79*, 276–284.
60. Thorn, C. B. (1980). Dual models and strings: The critical dimension. *Physics Reports, 67*(1), 163–169.
61. Tryon, E. P. (1972). Dynamical parton model for Hadrons. *Physical Review Letters, 28*, 1605–1608.
62. Virasoro, M. (1970). Subsidiary conditions and ghosts in dual resonance models. *Physical Review D, 1*, 2933–2936.

63. Weinberg, S. (1977). The search for unity: Notes for a history of quantum field theory. *Daedalus*, *106*(4), 17–35.
64. Wheeler, J. (1994). Interview of John Wheeler by Kenneth Ford on March 28, 1994, Niels Bohr Library and Archives, American Institute of Physics, College Park, MD USA, www.aip.org/history/ohilist/5908_12.html

Chapter 5
Supersymmetric Strings and Field Theoretic Limits

> *You claimed to have solved a problem that many people*
> *including my colleagues at Berkeley have been trying to solve.*
> *I do not know who you are, and from what you have told me,*
> *I cannot tell if you have succeeded, but I will study it and let*
> *you know.*
>
> Stanley Mandelstam (to Pierre Ramond)

The first dual models were a strictly bosonic affair. A major challenge was therefore to make them more physically realistic by including fermions in the spectrum. In 1971 both Pierre Ramond[1] and, independently, John Schwarz and Andrè Neveu (though ultimately unworkable) Korkut Bardakçi and Martin Halpern, attempted to implement fermions into the dual model. This led to the concept of 'spinning strings', and pointed the way to a method for removing the problematic tachyon. In this chapter we describe these advances and the events leading up to them.

Ramond was at the National Accelerator Laboratory (NAL, now known as Fermilab), fresh from Syracuse, during this transitionary period from dual models to strings to an early form of superstrings (or 'spinning strings'): (1969–1971). There were two other 'dualists' at NAL at the same time as Ramond: Louis Clavelli (from Chicago) and David Gordon (from Brandeis). The three theorists[2] were hired in order to analyse the various experimentsthat would be conducted (once the equipment was

---

[1] The quotation above is how Ramond recalls Mandelstam's skeptical response to his claim to have generalised the Dirac equation to the Veneziano model [27, p. 6].

[2] In fact, there were five theorists in all, "The NAL Fives" [26, p. 362], hired by Robert Wilson, the then director of NAL—the others were Jim Swank and Don Weingarten, all housed at 27 Sauk Boulevard. Hoddeson, Kolb, and Westfall refer to this period as Fermilab's "experiment in theory" [16, p. 139], motivated by Wilson and Edwin Goldwasser's desire to generate dialogue between theory and experiment.

D. Rickles, *A Brief History of String Theory*, The Frontiers Collection,
DOI: 10.1007/978-3-642-45128-7_5, © Springer-Verlag Berlin Heidelberg 2014

up and running). Ramond and Clavelli were working on the problem of the previous chapter: generalizing the Veneziano model. Their approach involved a group-theoretical analysis with a view to incorporating anti-commuting oscillators.

Curiously, though the dual models (and the rapidly emerging string theory; though still hadronic) were not quite considered 'unscientific' at this stage (as is often argued today), there was still some animosity towards it for similar reasons to those raised today; primarily because it was already becoming very formal (and anti-phenomenological). As Clavelli describes it, the management (under the directorship of Robert Wilson) terminated their contracts prematurely in the fall of 1970, with the statement:

> ..we had hoped that considerably stronger interactions would develop between you and the experimental physicists than has been the case... It is pretty clear that our experiment has not been a total success, and it would be foolish to pretend otherwise [3, p. 3].

As Clavelli goes on to point out, not even Ramond's discovery of supersymmetry on the string worldsheet was considered sufficient progress (phenomenologically or theoretically) despite the fact that they had been informed by the deputy director, Edwin Goldwasser, that significant *formal* advances would be acceptable in lieu of the primary task of analysing data.[3] All three left NAL in August 1971. Ramond took up a position at Yale. Clavelli took up a position at Rutgers, alongside Claud Lovelace and Joel Shapiro. But Clavelli notes that even in the environment of Rutgers, there was a division between the dual theorists and other particle physicists:

> In the Fall of 71, a war of wall posters broke out between Lovelace and Bogdan Maglich. Maglich posted a challenge to the theorists to stop working on strings and tell him what they would see at Fermilab. Claud Lovelace responded with a picture suggesting that Maglich would see a jagged cross section while the rest of the world would see a smooth Regge behavior. Maglich had become well known for reporting that the A2 resonance had a pronounced dip in the center. This result, which was initially confirmed by another experiment and which had triggered a barrage of theory papers, later evaporated [3, p. 4].

However, a network was formed when Bunji Sakita relayed the Fermilab dualists' ideas to CERN in 1970.[4] But this Fermilab experience must have fostered some tribal 'them and us' instincts in the three theorists. Also at Rutgers, fresh from very productive post-docs at Berkeley and Maryland, was Joel Shapiro. Clavelli wrote a paper with Shapiro, at Rutgers in 1973, that in many ways anticipated the heterotic strings we meet in Chap. 9. The paper, on "Pomeron Factorization in General Dual Models", covered many issues, including the extra dimensions of the dual models which, as they say, "provoked overly much adverse reaction" [4, p. 491]. They argued that a 4-dimensional model (expressed in terms of the Pomeron's being a factorizing pole in 4D) could be constructed by employing 44 anti-commuting (Fermi) degrees

---

[3] Note that the full interpretation of supersymmetry on the worldsheet was to come later than Clavelli indicates.

[4] See Lou Clavelli's article "On the Early Tension between String Theory and Phenomenology": http://bama.ua.edu/~lclavell/papers/tension1.pdf. (Note that Clavelli was a student of Nambu's in Chicago.)

Fig. 5.1 Table of the factorization conditions for various models considered by Clavelli and Shapiro. Here we see laid out the consistency conditions for the several models then known. This analysis clearly shows the model dependence of the critical dimension (or "magic dimension"). *Image source* [4, p. 504]

Model	Reggeon vertex	Transformed symmetrized vertex	Loop equation	$S(w)$ in magic dimension	$S'(r^2)$	D (magic dimension)	α_0 (of ground state reggeon)	α_0 of pomeron
CV	V_0	$\tilde{V}_0$	(3.5)	$f^2(w)$	$f^2(r^2)$	26	1	2
NS	$V_0 k \cdot H$	$G=+1$ $\tilde{V}_0 k \cdot H$	(3.8)	$w^{\frac{1}{2}}\left[\dfrac{f(w)}{\phi_+(w)}\right]^2$	$r\left[\dfrac{f(r^2)}{\phi_+(r^2)}\right]^2$			2
		$G=-1$ $\tilde{V}_0 k \cdot \bar{H}$	(3.13)	$w^{\frac{1}{2}}\left[\dfrac{f(w)}{\phi_-(w)}\right]^2$	$\dfrac{ir^2}{16}\left[\dfrac{f(r^2)}{\phi_{0+}(r^2)}\right]^2$	10	$-\frac{1}{2}$	1
Scalar Mobius spinor ($G=+1$ pomeron)	$V_0 \displaystyle\prod_1^l H$	$\tilde{V}_0 \displaystyle\prod_1^l H$	(3.15)	$f^2(w)$	$f^2(r^2)$	$26-\frac{1}{2}D_f$	$1-\frac{1}{4}l$	2
Scalar Bardakci–Halpern	$V_0 \displaystyle\sum_1^{22} \bar{\psi}_i\psi_i$	$\tilde{V}_0 \displaystyle\sum_1 \bar{\psi}_i\psi_i$	(3.15)	$f^2(w)$	$f^2(r^2)$	4	0	2

of freedom in place of the residual dimensions—which they achieve by using 22 scalar quarks and 22 scalar anti-quarks.[5] The conditions leading to this choice can be seen in the table, reproduced from their article, in Fig. 5.1—note that the critical dimension is called the "magic dimension" in this chapter.

---

[5] This bears some resemblance to the so-called Frenkel-Kac mechanism for compactifying degrees of freedom on a lattice—a construction that was crucial in the development of the heterotic string theory (see Sect. 9.2).

5.1 Ramond's Dual Dirac Equation

In 1970 Pierre Ramond was working, largely on his own, at Yale on a way to extend dual theory to include fermionic degrees of freedom (free fermions). In fact, he had already published a paper with David Gordon[6] in 1969, while at NAL, on a "Spinor Formalism for Dual-Resonance Models" [13].[7] And, as mentioned above, he had also been pursuing a group-theoretical study of dual models with Clavelli. Ramond had initially been introduced to dual resonance models while working on the 3-Reggeon vertex with J. Nuyts and H. Sugawara in Trieste, during a 3-month appointment arranged by his supervisor, in the summer of 1969 [27, p. 2].[8]

Ramond describes having the initial insight for the dual model version of the Dirac equation during research visit at the Aspen Centre for Physics early in 1970:

> It was a wonderful stay. The town was in the afterglow of the hippie era, and my days were spent playing volleyball in Wagner Park, listening to music outside the music tent in the late afternoons, and in other nonscientific activities. In my spare time, I started thinking about the particle spectrum that had been extracted from the dual amplitudes. People had already found some sort of position operator $Q_\mu(\tau)$ which appeared in the vertex, and its derivative was like a generalized momentum $P_\mu(\tau)$. Indeed, if one pursued the analogy further, the inverse propagator looked like the square of that generalized momentum. This led me to think of a 'correspondence principle' by which simple notions of point particles were related to dual models. At last, a glimpse of simplicity! [26, p. 364]

Though he initially constructed a bosonic version of the correspondence principle linking particle theory and dual models, he later noticed that the same principle could be applied to the Dirac equation. This route to the dual theory of (free) fermions, then, proceeded by direct analogy with the point particle case. Ramond invoked a correspondence principle whereby operators in the point particle case are to be thought of as averages over internal motions of the hadronic system. First setting up the bosonic case, then, he generalizes the Klein-Gordon operator so that the solutions of the Klein-Gordon equation correspond to states of the free bosonic system. He recovers Virasoro's gauge conditions and infinite-dimensional algebra from this correspondence principle. With this to hand, he is then able to proceed to the fermionic case via the Dirac equation, using the analogous γ-matrices. His matrices will be averages $\Gamma_\mu(\tau)$ (where τ gives the cycles of the internal motions) over internal cycles that give the standard Dirac matrix: $\langle \Gamma_\mu(\tau) \rangle = \gamma_\mu$. For $\Gamma_\mu(\tau)$ he finds:

---

[6] This is the same Gordon responsible for the influential oscillator formalism paper with Fubini and Veneziano.

[7] Neither this nor Ramond's subsequent paper, introducing the dual Dirac equation, were couched in terms of the string picture. Note also that the Ramond model was a free theory.

[8] It's worth pointing out that Ramond had originally intended to study general relativity in Peter Bergmann's group, after receiving a four-year graduate fellowship to study at the University of Syracuse. He had been advised to join Sudarshan's particle physics group instead, later studying under him, though later switching to Balachandran (cf, [26, p. 361]).

$$\Gamma_\mu(\tau) = \gamma_\mu + i\omega_0\tau\delta_\mu + i\sqrt{2}\gamma_5 \sum_{n=1}^{\infty} [b_\mu^{(n)\dagger} e^{i\omega_n\tau} + b_\mu^{(n)} e^{-i\omega_n\tau}] \tag{5.1}$$

Ramond describes the next steps:

> To my surprise, this led to an algebra of a kind I had never seen: it contained both commutators and anticommutators, and was in essence the square-root of the Virasoro algebra. Tremendously excited, I barely ate and drank for weeks, as every derivation brought more conceptual clarity and more questions. There were some odd things. The generalization of the Dirac matrices led to fermionic harmonic oscillators $b_\mu^{(n)}$ and $b_\mu^{(n)\dagger}$ with spacetime four-vector indices, but they came with their own operators, F_n, of the right structure to decouple the negative norm states. I also realized that it was a truly novel algebraic structure, since I could now take the square root of any Lie algebra [26, p. 364].

The key feature to emerge from Ramond's analysis of this generalised Dirac equation was, then, an algebra which contained both the standard harmonic oscillator operators (viz. $[a_\mu^{(n)}, a_\nu^{(m)\dagger}] = -g_{\mu\nu}\delta^{n,m}$), but also anti-commuting operators satisfying:

$$\{b_\mu^{(n)}, b_\nu^{(m)}\} = \{b_\mu^{(n)\dagger}, b_\nu^{(m)\dagger}\} = 0 \tag{5.2}$$

$$\{b_\mu^{(n)}, b_\nu^{(m)\dagger}\} = -g_{\mu\nu}\delta^{n,m}, \quad (n, m = 1, \ldots) \tag{5.3}$$

The resulting structure is an example (the first) of a 'superalgebra'. This model would soon merge with Neveu and Schwarz's bosonic 'spinning' model to generate what is usually seen to constitute the earliest version of superstring theory—though at the time they called it the 'dual pion model,' and, as above, there is no mention of strings.[9] Neveu and Schwarz were able to construct the amplitude for creating $N - 1$ pions from a fermion line (a quark). A model was also found with boson emission from a fermion-antifermion vertex.

Korkut Bardakçi and Martin Halpern were also concerned with the lack of space-time fermions in the bosonic oscillator dual models. Their approach involved the introduction into the dual picture of spinorial (quark-like, spin one-half) operators satisfying mutual anticommutation relations:

$$(b_r(n), \overline{b}_s(m))_+ = \delta_{rs}\delta_{mn} \tag{5.4}$$

$$(d_r(n), \overline{d}_s(m))_+ = \delta_{rs}\delta_{mn}, \quad n \geq 0, \quad r = 1, 2, 3, 4 \tag{5.5}$$

---

[9] In his recollections [25, pp. 8–9] André Neveu notes how he first encountered Pierre Ramond quite by chance when they were crossing the Atlantic on the same ship, bound for Princeton and Fermilab respectively. Neveu was studying the Fubini-Veneziano paper on the factorization of dual models in the ship's library, and left momentarily, at which point Ramond entered, finding the paper: the same paper he was working through! As a result of this fortunate interaction, Ramond thought to send Neveu (and Schwarz, with whom Neveu was collaborating) a copy of his dual (free) fermion paper which inspired them to work out the details for the interacting theory. As Schwarz recalls, he and Neveu had been working on a new bosonic theory, but then noticed that their's and Ramond's "were different facets of a single theory containing our bosons and Ramond's fermions" [31, p. 11]. Moreover, the Ramond model suffered from a lack of manifest duality.

In addition were introduced local fields generalising Chan-Paton factors:

$$\overline{\psi}^I(z) = \sum_{p\in\mathbb{Z}+1/2} \overline{\psi}^I(p)z^{-p-1/2} \qquad (5.6)$$

$$\psi^I(z) = \sum_{p\in\mathbb{Z}+1/2} \psi^I(p)z^{-p-1/2} \qquad (5.7)$$

The resulting model was known as the 'dual quark model,' with $\{\psi, \overline{\psi}\}$ the dual quark fields.[10] As they have noted, however, their idea of using generalised Chan-Paton factors "was unsuccessful because it included negative-norm states" [1, p. 395]. Note again, as with Ramond's fermion model, that there was initially no string interpretation associated with this dual quark model: neither the terminology nor the conceptual idea of strings played a role at this stage. Hence, claims one often finds about these models marking the 'introduction of worldsheet fermions' or 'superstrings' and the like, must (historically speaking) be taken with a pinch of salt.

5.2 The Ramond-Neveu-Schwarz Model

While they were together at Princeton, in 1970, Neveu and Schwarz[11] enlarged the Fock space of the standard dual model by adding anticommuting creation and annihilation operators. They call the resulting structure a "new dual resonance model" and hoped to be able to establish a closer fit with the "real world of mesons" by adding spin degrees of freedom [22, p. 108].[12] Initially they produced only a bosonic model, though with interactions: [21]. Within a month of completing this paper they found a way to eliminate the $M^2 = -1$ (leading trajectory) tachyon of the Veneziano model: [22]. This model introduced a new tachyon, however, at the next trajectory up, $M^2 = -1/2$, which they took to be a pion (thus patching a perceived problem with Veneziano's model—see p. 42). They had hoped that the new tachyon problem could be resolved by finding a mechanism to shift any such particles onto their correct masses. Halpern and Thorn [15] tried (albeit unsuccessfully) to resolve the problem, presenting a method (called the "shifted Neveu-Schwarz (SNS) model" in [4, p. 491]) in which pion mass could be varied arbitrarily without affecting the otherwise

---

[10] This work also had mathematical implications. In fact, they had discovered in this work affine Lie algebras (independently of knowledge of the mathematical work then available) and constructed a concrete, fermionic representation of $\widehat{sl}(3)$. Clavelli produced a broadly similar construction during his time at NAL in his paper "New Dual N-Point Functions" (NAL Preprint: http://lss.fnal.gov/archive/1971/pub/Pub-71-009-T.pdf).

[11] Here I am indebted to Schwarz's own account presented in [32]. A more recent account can be found in [33].

[12] The Veneziano model is sometimes called the "orbital model" on account of the absence of spin degrees of freedom.

consistent nature of the framework. In which case one could set the mass to avoid the existence of tachyons. The approach involved the introduction of an additional spacelike dimension for the oscillators and momenta. Edward Corrigan recalls not being immediately convinced by this approach

> The fact that Neveu and Schwarz introduced a set of anticommuting operators to achieve this was very interesting, but seemed less compelling [than Ramond's approach[13]]; after all, at that time there were many ideas for modifying the basic amplitudes and there did not appear to be any obvious reasons why anticommuting operators should be part of a story involving mesons, unless the newly introduced operators were somehow to be associated with their constituent quarks [6, p. 380].

A little after Neveu and Schwarz's second paper, they realised that there were similarities ("a deep connection") between their model and Pierre Ramond's fermion model; they suspected that Ramond's could be embedded in their own, giving a theory of bosons and fermions. They managed to construct a vertex operator representing the emission of their pion from a fermionic state and then built amplitudes from this involving pairs of fermions and N-pions [23]. Charles Thorn [35] was then able to find a vertex function for the emission of fermions: [(excited meson) $\rightarrow$ (fermion-antifermion pair)].[14] Goddard and Waltz [10] argued that the leading trajectory of the NS model would have a Regge intercept of two—this was later proven by Clavelli and Shapiro [4].

The Neveu-Schwarz model included Ramond's fermion spectrum, integrating it with a bosonic spectrum. But it is important to note, as in the previous section, that there was no mention of 'strings'.[15] It was when Mandelstam extended his analysis of scattering by splitting and joining strings to the Ramond and Neveu-Schwarz models [19] and then the Neveu-Schwarz-Ramond model [20] that these models received an interpretation in terms of 'spinning strings'.[16]

In this sense, then, the Neveu-Schwarz-Ramond construction (with Thorn and Mandelstam's inputs) marks the birth of the very first superstring theory (superstrings 1.0, if you like—or perhaps, a 'beta version'), with strings possessing both bosonic

---

[13] Though, in a paper Corrigan co-authored with David Olive, they also note that while it is "tantalizing to think of the Ramond fermion as a quark or a baryon ... in fact it is probably neither, but just an important clue on the way to more physically realistic theories" [5, p. 750].

[14] Thorne also considered N pions and two fermions, recovering the spectrum of Neveu and Schwarz's model (from fermion-meson channels), in addition to the spectrum of Ramond's propagator (from the fermion-meson channels). Corrigan and Olive generalised this work by constructing a general dual vertex giving the transition of a Ramond fermion into a Neveu-Schwarz meson by the emission of a general excited fermion state [5].

[15] It was still, like the Veneziano model, a dual-resonance model. This initial disconnection from the (known) string interpretation (instead, employing operator methods) seems to have been essential for building solid results and moving the field forward in the earliest phase of dual model research—such an approach constitutes one of many such examples of pushing point-particle analogies as far as they will go (e.g. before specific string-specific issues arise).

[16] Iwasaki and Kikkawa had, however, given an earlier model of a *free* spinning string: [17].

and fermionic coordinates. Note that at the time they were published, there was no constraint imposed on the dimensionality of the spacetime, other than that it had to be some even number.[17]

5.3 Many Roads to Supersymmetry

It is now well-known that spacetime supersymmetry was discovered (at least) twice: once in the context of string theory (in the West) in 1973, by Julius Wess and Bruno Zumino (where they called it "supergauge symmetry in four dimensions"). The other, earlier, in the context of particle scattering, by Yuri Golfand and his student Evgeny Lichtman in 1971 [11] (where they referred to an "extension of the algebra of Poincaré group generators" and called supersymmetry itself "the spinor translation").[18]

Supersymmetry is described by a (non-Lie) algebra that extends the Poincaré algebra by anticommuting spinorial terms representing fermions. The algebra has a $\mathbb{Z}_2$ grading, splitting it into an odd fermionic part (anti-commutators) and an even bosonic part (commutators).[19] If a theory possesses supersymmetry then bosons and fermions can be rotated into one another so that both lie in a multiplet.[20] This also brings about a mutual cancellation between the contributions from fermions and bosons. As Ramond discovered, this algebra is a square root of the pre-generalised algebra, and a similar relationship can be established between all sorts of algebras.

This ignores the work of Jean-Loup Gervais and Bunji Sakita in 1971 [9], on the two-dimensional case in which they use the concept of 'supergauge,' taken from Neveu, Schwarz, and Ramond's work on dual models—a term later employed by

---

[17] The full understanding that $d = 10$ in these embryonic superstring theories came in several stages, as with the $d = 26$ result for bosonic strings, with evidence that the theory not only became simpler but also that it was required for the theory to be unitary and ghost-free (see [2, 30]).

[18] There was a now well-known incident involving the dismissal of Golfand during Russia's staff reduction campaign (a thinly disguised anti-semitic campaign). Golfand sent an appeal to Harald Fritzsch, which resulted in a letter being signed by many physicists, including several string theorists [see Gell-Mann Papers: Box 8, Folder 21].

[19] More precisely, following Golfand and Likhtman [12, p. 3], a linear space L is said to be $\mathbb{Z}_2$-graded if it possesses a subspace of vectors ^0L which are even, with another subspace of odd vectors, ^1L, and for which the whole of L is a direct product of these subspaces: $L = {}^0L \oplus {}^1L$.

[20] This rotation is often described by saying that the rotation happens within a spinorial extension of space, or superspace. One can view this in terms of operators Q (spinorial charges) acting on bosonic and fermion states as: $Q|fermion\rangle = |boson\rangle$ and $Q|boson\rangle = |fermion\rangle$. The Qs satisfy the commutation relations $\{Q_\alpha^i, Q_j^{\bar{\alpha}}\} = -2(\sigma^\mu)_\alpha^{\bar{\alpha}} \delta_j^i P_\mu$ (where P_μ is the energy-momentum operator—as Haag et al. [14, p. 258] pointed out, the appearance of such operators amongst the elements of the superalgebra implies that a "fusion" occurs between geometric (spacetime) and internal symmetries). The number N of supersymmetries leads to a classification of theories as follows: $N = 1$ is known as simple symmetry; $N > 1$ cases are known as extended supersymmetry; $N \leq 4$ is demanded by renormalizable gauge theories; and $N \leq 8$ is required for helicity-2 theories like supergravity. (Note that simple supersymmetry is required if one wishes to construct parity-violating theories.)

Wess and Zumino (who in fact were explicitly seeking four-dimensional versions of Gervais and Sakita's model— cf. [33], p. 6). It is often referred to as 'worldsheet supersymmetry,' though that terminology wasn't used in the original, which focused on new invariances of the generalized dual models understood as two-dimensional field theories on 'strips'.[21] Gervais and Sakita found a two-dimensional Lagrangian that was invariant under certain kinds of mappings that mixed scalar (bosonic) and spinor (fermion) fields.[22] Their paper also involves an explanation of the difference between fermions and mesons in terms of the boundary conditions at the ends of the strings (see [9, p. 634]).

5.4 Early Explorations of the Zero Slope Limit

One of the most remarkable early developments of dual models came from the study of their zero slope limits by Jöel Scherk in 1971. It was found that the standard classical field theories (of both Yang-Mills and, later, Einstein type) can be found to emerge from dual models in this limit. Hence, having begun life as a top-down approach to physics (a solution of analyticity and bootstrap conditions), the dual models appear to have the potential to merge with standard (local) quantum field theory. While one might be forgiven for thinking that this would have radically increased the general degree of belief in dual models,[23] the timing was unfortunate (as we discuss in the next chapter).

The basic idea of taking the zero slope limit is easy enough to see, as is the reasoning behind doing it and understanding the structure that emerges. Dual models were known to have three sectors, corresponding to three particle-types and depending on where the associated Regge trajectory's intercept was located. The three sectors are specified by the following leading trajectories:

1. $\frac{1}{2} + \alpha'$: fermionic sector
2. $1 + \alpha'$: mesonic sector
3. $2 + \frac{\alpha'}{2}$: pomeron sector

---

[21] The term 'supersymmetry' was introduced by Salam and Strathdee in 1974, in Trieste—it first appears in print, in hyphenated form as 'super-symmetry,' in [28].

[22] For an excellent survey of the history of supersymmetry, including its intersections with supergravity and superstrings, see: [7].

[23] Some did see the potential, of course. For example, Frampton and Wali write that the results are of "considerable interest because they provide a linkage between the hadronic models on the one hand, and Lagrangian field theories on the other" and that since "the latter is considerably more fully explored and understood than the former, we may hope to learn a great deal from the connection" [8, pp. 1879–1880]. Frampton and Wali discuss an interesting non-locality, resulting from the high-spins in hadron scattering experiments (as they say, expected from the string interpretation). They also suggest the possibility of utilising renormalization methods from dual models on the Lagrangians they study.

The idea is to then leave the intercepts (denoting spin) fixed while varying the α' parameter (of course, understood to determine a unit of length[24]) so that, as $\alpha' \to 0$, the masses of the states go to infinity giving one a decoupling in which only the massless states survive. Of course, for a theory of hadrons, having such massless states is a defect, one thought to be resolvable by some kind of symmetry breaking mechanism.[25]

In his first paper on the subject, Jöel Scherk argued that the (N-point generalised) dual resonance model reduced to the well-known ϕ^3 Lagrangian field theory in the zero-slope limit. That is, there is a deformation relationship between a (unitary) dual resonance model and a quantum field theory with Lagrangian:

$$\mathscr{L} = \frac{1}{2}(\partial^\mu \phi \partial_\mu \phi - m^2 \phi^2) + \frac{1}{6}\lambda \phi^3 \tag{5.8}$$

The specific reduction process involves keeping a parameter $\lambda = g/\alpha'$ fixed while varying the Regge-Mandelstam slope α' and the coupling constant g, sending both latter parameters to zero.[26] The anti-field theoretic principles, already eroded by the operator approach, by this point have been all but virtually eliminated—indeed, Yoneya spoke of this work as an "important step in understanding field theoretical foundations of dual models" [36, p. 1907].[27]

Dual resonance models could, it seemed, be understood as field theories of one-dimensional systems (with length $\sqrt{\alpha'}$). Here we see once again the importance of the Regge-Mandelstam slope, both in providing the intrinsic length of the strings and in providing an adjustable parameter responsible for examining the inter-theoretic relationship between quantum field theory and the dual resonance model. Such an equivalence (in the limit) is computationally useful since one can use the theory in the limit to probe aspects of the structure of dual models.[28]

Scherk's strategy involves a kind of inversion of Nielsen's relationship between standard Feynman graphs and his fishnet diagrams. In Nielsen's approach, one starts with fishnet diagrams and takes a large N limit (where N is the number of internal lines contributing to the mesh). Scherk considers a perturbative unitary expansion

---

[24] In a later paper on dual models, Scherk and John Schwarz, describe how the string picture makes the zero slope recovery of field theories intuitive: "the length of the strings is characterized by $\sqrt{\alpha'}$, where α' is the universal Regge slope parameter" so that "given this situation, it is not surprising that in the limit $\alpha' \to 0$ dual models reduce to field theory models of point particles" [29, p. 347].

[25] Of course, in the case in which dual models are taken to be models of gravitational (from closed strings) and gauge bosons (from open strings), then the massless states are necessary.

[26] Scherk credits Ray Sawyer with the discovery of this zero-slope limit method, and notes that it was also implicit in [18]. Yoneya [36] later argued that Scherk's result in fact clarified the relationship between the duality diagrams and Feynman diagrams, as discussed in [18].

[27] Neveu and Scherk, in 1971, write that the "dual-resonance models seem to be an approach to strong interactions which stands between field theories and the S-matrix" [24, p. 155]. Their reasoning is that on the one hand "one writes down phenomenological amplitudes with desirable physical properties without using a Lagrangian, but on the other hand, the factorizability of those amplitudes allows [one] to compute unitary corrections as in a field theory" (*ibid.*).

[28] Such a method amounts to a low-energy expansion.

of a dual resonance model as a function of the constants mentioned above, and with fixed λ. Taking the $\alpha' \to 0$ and $g \to 0$ limits amounts to taking a low energy limit (in which the external dynamical variables are very small relative to the mass of the ground state). Scherk identifies the Veneziano amplitude in such a scenario with the leading Born term of the limiting ϕ^3 theory.

In his slightly later paper with Neveu, Scherk briefly mentions gravitation, though only to point out that not only Yang-Mills fields, but also gravitational amplitudes (analysed according to Feynman rules) possess the property that the gauge group shows up by a transversality condition [24, p. 161]. We return to these in Chap. 7, in the context of the transition of dual models from a theory of hadrons to a more general fundamental theory of all known interactions.

Not long after these studies of limits appeared, a variety of (still highly influential) papers came out that attempted to start with the limiting field theories and worked out features of dual models, in the reverse direction as it were, or 'bottom-up,' thus further solidifying the links between ordinary field theories and dual models. [29]

5.5 Summary

By the early 1970s, dual resonance models had shed their bootstrapping past, and forged an intimate link with field theory. Tools from field theory had first enabled a transparent formalism to be constructed (the operator formalism), and then had led to various features that made dual models more physically realistic. By the time the zero-slope limits had been studied, the field theoretic links were very solid.

References

1. Bardakçi, K., & Halpern, M. B. (2012). The dual quark models. In A. Capelli et al. (Eds.), *The birth of string theory* (pp. 393–406). Cambridge: Cambridge University Press.
2. Brink, L., Olive, D. I., Rebbi, C., & Scherk, J. (1973). The missing gauge conditions for the dual fermion emission vertex and their consequences. *Physics Letters B, 45*(4), 379–383.
3. Clavelli, L. (2007). *On the early tension between string theory and phenomenology*. Manuscript: http://bama.ua.edu/~lclavell/papers/tension1.pdf
4. Clavelli, L., & Shapiro, J. A. (1973). Pomeron factorization in general dual models. *Nuclear Physics, B57*, 490–535.
5. Corrigan, E., & Olive, D. I. (1972). Fermion-meson vertices in dual theories. *Il Nuovo Cimento, A11*, 749–773.
6. Corrigan, E. (2012). Aspects of fermionic dual models. In A. Capelli et al. (Eds.), *The birth of string theory* (pp. 378–392). Cambridge: Cambridge University Press.
7. Di Stefano, R. (2000). Notes on the conceptual development of supersymmetry. In G. Kane and M. Shifman (Eds.), *The supersymmetric world* (pp. 169–271). Singapore: World Scientific.

---

[29] Here I have in mind examples like 't Hooft's large N expansion in which he proves that the topological structure of the perturbation series in $1/N$ (for a gauge theory with 'colour group' $U(N)$) matches that of dual models [34].

8. Frampton, P. H., & Wali, K. C. (1973). Regge-slope expansion in the dual resonance model. *Physical Review D, 8*(6), 1879–1886.
9. Gervais, J.-L., & Sakita, B. (1971). Field theory interpretation of supergauges in dual models. *Nuclear Physics, B34*, 632–639.
10. Goddard, P., & Waltz, R. E. (1971). One-loop amplitudes in the model of Neveu and Schwarz. *Nuclear Physics, B34*(1), 99–108.
11. Golfand, Y. A., & Likhtman, E. P. (1971). Extension of the algebra of Poincaré group generators and violation of P-invariance. *Journal of Experimental and Theoretical Physics Letters, 13*(8), 452–455.
12. Golfand, Y. A., & Likhtman, E. P. (1986). On $N = 1$ supersymmetry algebra and simple models. In P. West (Ed.), *Supersymmetry: A decade of development* (pp. 1–31). Bristol, UK: IOP Publishing.
13. Gordon, D., & Ramond, P. (1970). Spinor formalism for dual-resonance models. *Physical Review D, 1*(6), 1849–1850.
14. Haag, R., Łopuszański, J. T., & Sohnius, M. (1975). All possible generators of supersymmetries of the S-matrix. *Nuclear Physics, B88*, 257–274.
15. Halpern, M. B., & Thorn, C. B. (1971). Dual model of pions with no tachyon. *Physics Letters, 35B*(5), 441–442.
16. Hoddeson, L., A. W. Kolb, & Westfall, C. (2008). *Fermilab: Physics, the frontier, and megascience.* Chicago: University of Chicago Press.
17. Iwasaki, Y., & Kikkawa, K. (1973). Quantization of a string of spinning material Hamiltonian and Lagrangian formulations. *Physical Review D, 8*(2), 440–449.
18. Kikkawa, K., Sakita, B., & Virasoro, M. A. (1969). Feynman-like diagrams compatible with duality. I. Planar diagrams. *Physical Review, 184*, 1701–1713.
19. Mandelstam, S. (1973). Manifestly dual formulation of the Ramond-model. *Physics Letters, 46B*(3), 447–451.
20. Mandelstam, S. (1974). Interacting-string picture of Neveu-Schwarz-Ramond model. *Nuclear Physics, B69*, 77–106.
21. Neveu, A., & Schwarz, J. (1971). Tachyon-free dual model with a positive-intercept trajectory. *Physics Letters, 34B*, 517–518.
22. Neveu, A., & Schwarz, J. H. J. (1971). Factorizable dual model of Pions. *Physics Letters Nuclear Physics B, 31*, 86–112.
23. Neveu, A., & Schwarz, J. H. J. (1971). Quark model of dual Pions. *Physics Letters Physical Review, D4*, 1109–1111.
24. Neveu, A., & Scherk, J. (1972). Connection between Yang-Mills fields and dual models. *Nuclear Physics, B36*, 155–161.
25. Neveu, A. (2000). Thirty Years Ago. In G. Kane and M. Shifman (Eds.), *The supersymmetric world* (pp. 8–10). Singapore: World Scientific.
26. Ramond, P. (2012). Dual model with Fermions: Memoirs of an early string theorist. In A. Capelli et al. (Eds.), *The birth of string theory* (pp. 361–372). Cambridge: Cambridge University Press.
27. Ramond, P. (2000). Early supersymmetry on the Prairie. In G. Kane and M. Shifman (Eds.), *The supersymmetric world: The beginnings of the theory* (pp. 2–7). Singapore: World Scientific.
28. Salam, A., & Strathdee, J. (1974). Super-symmetry and non-abelian gauges. *Physics Letters, B51*(4), 353–355.
29. Scherk, J., & Schwarz, J. (1974). Dual models and the geometry of space-time. *Physics Letters, 52B*(3), 347–350.
30. Schwarz, J. H. (1972). Physical states and pomeron poles in the dual pion model. *Nuclear Physics, B46(1)*, 61–74.
31. Schwarz, J. H. (2000). Strings and the advent of supersymmetry: The view from Pasadena. In G. Kane and M. Shifman (eds.), *The supersymmetric world* (pp. 11–18). Singapore: World Scientific.
32. Schwarz, J. H. (2001). String theory origins of supersymmetry. *Nuclear Physics, B101*, 54–61.
33. Schwarz, J. H. (2012). Early history of string theory and supersymmetry. arXiv, 1201.0981v1.

34. 't Hooft, G. (1974). A planar diagram theory for strong interactions. *Nuclear Physics*, *B72*, 461–473.
35. Thorn, C. (1971). Embryonic dual model for pions and fermions. *Physical Review*, *D4*, 1112–1116.
36. Yoneya, T. (1973). Connection of dual models to electrodynamics and gravidynamics. *Progress of Theoretical Physics*, *51*(6), 1907–1920.

Part II
A Decade of Darkness: 1974–1984

Chapter 6
An Early Demise?

> *There was in fact a continuing low level of activity throughout*
> *these years but, for the most part, it had become reckless to work*
> *in the field without a very secure position or a very strong patron.*
>
> Louis Clavelli

An unfortunate (for string theory) series of events terminated the growing popularity that string theory was enjoying in the early 1970s. The sense of optimism in 1973, and early 1974, can be seen in many chapters. Consider the following:

- The opening line of Chodos and Thorn's 1973 article on 'Making the Massless String Massive' states that "[t]he massless relativistic string represents the first consistent relativistic extended model of the hadron" ([16], p. 509).
- Rebbi, in his survey of Dual Models and Relativistic Quantum Strings, from January 1974, writes: "[t]his new vision of the structure of fundamental objects and of the mechanisms of interaction is surely one of the major contributions that the theory of dual models and their string interpretations has brought to the theory of elementary particles" ([46], p. 62).
- Fubini, in his general introduction to a 1974 reprint volume on dual models, writes that: "theoreticians who are not in some way acquainted with the new features present in the dual models, strongly risk being unable to follow and to contribute with profit to future developments in strong interaction physics" ([18], p. 1).

Even Gell-Mann, in his summary talk of a very large conference on high-energy physics in September, 1972, saved for last "the most exciting theoretical work on the strong interaction, namely the attempt to construct a dual resonance theory of hadrons" ([19], p. 354). However, he is also quick to draw attention to various defects: difficulties in incorporating fermions, massless vector mesons, and the unphysical number of space-time dimensions demanded by consistency. The string picture is also identified by Gell-Mann which he views as a way of 'imitating' the dual scheme by a multilocal field theory "corresponding to the relativistic quantum mechanics of a string" (p. 355). This he describes as "an amusing point of view"—that David Olive's talk at the same conference was entitled "Clarification of the Rubber String

D. Rickles, *A Brief History of String Theory*, The Frontiers Collection,
DOI: 10.1007/978-3-642-45128-7_6, © Springer-Verlag Berlin Heidelberg 2014

Picture" [37] clearly added to the amusing-sounding character of hadronic string interpretations of dual models.[1]

By the end of 1973, Peter Goddard notes that David Olive, wrote to him that "[v]ery few people are now interested in dual theories here in CERN. Amati and Fubini independently made statements to the effect that dual theory is now the most exciting theory that they have seen but that it is too difficult for them to work with. The main excitement [is] the renormalization group and asymptotic freedom, which are indeed interesting" [22, p. 256]. Thus, in his history of research in the theory division at CERN, John Iliopoulos writes that the dual theory "from exaggerated heights of faith has fallen into totally unjustified depths of oblivion" ([28], p. 301). This much is purely hadronic, of course, but a somewhat cruel twist of fate meant that, what we now view as string theory's great breakthrough (to convert itself into a quantum theory of gravity, and other interactions: a potential unified theory of all interactions), went largely ignored by the particle physics community.

However, to say, as one often hears, that the creation of QCD led to the *immediate* demise of dual string theory as a theory of hadrons (offering an alternative to gluons) is, of course, an exaggeration.[2] For example, there was a conference on the string model of hadrons in 1975 (held in Durham, UK). But certainly, post-1973 work on string theory *qua* self-sufficient theory of hadrons dramatically slowed down, and had virtually ceased by the 80s (see Fig. 6.1): to pursue it further as a central subject would have been (almost) tantamount to professional suicide.[3] There were some good solid reasons behind this. Veneziano himself ([60]: 31) claimed that he persisted playing with string theory for a time on account of its interesting topological structure, only fully deserting it in 1974 when 't Hooft indicated how the same planar diagram structure could be generated from within QCD (given the generalisation to the $1/N_c$ expansion, where c, the number of colours, is taken to be large). For Veneziano this simply left no work for string theory to do.

The $1/N_c$ expansion, in the limit of large numbers of colours, has planar diagrams that correspond to free strings. Moreover, several physicists returned to the notion of a 'QCD string,' of the kind introduced by Nielsen and Oleson. In other words, within the broader church of string theory there was a split in which the hadronic string was pursued as a part of QCD itself (in the role of the vortex lines, like the Abrikosov flux tubes in a superconductor), and also pursued along the lines of

---

[1] The 16th International Conference On High-Energy Physics was remarkable in many ways. Not least, because it presents the state of the art in quantising non-Abelian gauge theories, with the latest work of 't Hooft and Veltman, Wilson, and others. It features important results on Bjorken scaling. But it is also dual model heavy. The conference marked an unstable equilibrium that, from the perspective of most of the attendees, could have shifted in any number of directions.

[2] Michael Redhead writes that "the bootstrap programme was not so much *refuted* as *overtaken* by the new fundamentalist approach involving truly basic constitutes like quarks and gluons" ([47], p. 573).

[3] I might also note that, in terms of the poor job prospects for dual theorists, this was very likely also a symptom of the 1973-5 financial crisis, which in the UK at least involved a very severe 'double-dip' recession leading to general hiring freezes. (John Schwarz has pointed out that the academic job market was already very bad for all physicists in 1972, adding that he was very fortunate that Gell-Mann 'discovered' him at that time (private communication).)

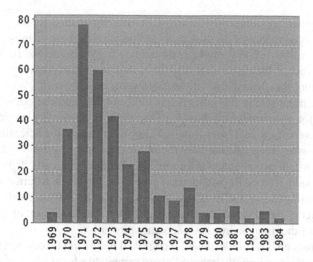

Fig. 6.1 Graph showing the number of publications on dual models from 1968, the year of Veneziano's original publication, to 1984, the year in which string theory is often said to have had its renaissance. The search (done using *Web of Science*) involved "dual models" or "dual reso-nance", with instances not relating to the modelling of hadrons (such as supergravity, gravity, and gauge bosons) removed from the data set. General papers looking at broad structural features of dual resonance models have been included. There was an evident boom in 1971 and a bust (not quite as dramatic) in 1976. Image source: Thompson-Reuters, *Web of Science*

the more fundamental superstring responsible for both gravitational and Yang-Mills forces. David Olive points out that the latter strand had prepared the dual-string theorists for the idea of QCD since they already "knew from the dual models that all fundamental interactions besides gravity had to be gauge mediated" so that their "reaction was a lack of surprise" [40, p. 352].

But it certainly seems accurate to say that the little work that was done on string theory and dual models during the decade 1974–1984 was carried out by those physicists that had already worked on the theory *prior* to 1974: it's very rare to see newcomers working in the area. Given the string theory renaissance that we observe in the next part, this obviously had a certain curious impact on the state of the professional aspects of string theory. The gap meant that any new generation of string theorists would have to learn tools from scratch (and tools that were very different from those found in orthodox field theories at the time).[4] The way string theory was 'sold' after the fallow period was very different to beforehand, and makes direct reference to certain apparently problematic features of the point particle field theories underpinning the standard model and so on.

However, it should not be forgotten that the dual resonance models were amongst the key driving forces behind the early understanding of quark confinement. The notion that the generalised dual amplitudes could be interpreted in terms of strings

---

[4] It was perhaps natural, then, that the first textbook [24] to appear on a specific approach to quantum gravity was devoted to string theory.

with quarks at their endpoints provides a qualitative explanation for the confinement. An early analogue model of Nielsen and Oleson derives these strings from a field theory as vortex lines (cf. [58]) and this model was highly influential for many years afterwards. Hence, it would be absurd to think that the S-matrix theory and the dual resonance theory that flowed from it were defective in some way.

The sharp initial rise in the influence of dual models and the embryonic string theories had not gone unnoticed outside of the field of elementary particle physics. Dual models were used, by Henry Small [52] of the Institute for Scientific Information, as a case-study for a new method of citation analysis: *co-citation* (a measure of the degree to which two chapters are coupled). The downfall came from two sides: inside and outside. It faced its own internal problems. However, externally, there was a strong competing theory: QCD. But we have to be careful that we are discussing the right string theory here, namely the hadronic theory. The extension of string theory to interactions other than hadrons is a separate matter.

However, though it was indeed well recognised that string theory was not as good a model for hadrons as initially believed, there remained a loyal following that found it hard to give up such a remarkable mathematical structure. In a plenary talk on dual models from the 17th International Conference on High Energy Physics in 1974 (one of the 'Rochester' conferences), David Olive refers to those enamoured of the mathematical beauty of dual theory (especially cast in the operator formulation) as "addicts" ([38], p. I-270). The focus on mathematical beauty is an approach associated with Dirac. For example, in response to a criticism of one of his chapters, Dirac writes:

> It is true that the ultimate goal of theoretical physics is merely to get a set of rules in agreement with experiment. But it has always been found that highly successful rules are highly beautiful and ugly rules are of only restricted use. In consequence, physicists generally have come to believe in the need for physical theory to be beautiful, as an overriding law of nature. It is a matter of faith rather than logic. ... If the theory fails to agree with experiment, its basic principles may still be correct and the discrepancy may be due merely to some detail that will get cleared up in the future ([17], p. 268).

One can see this same mindset in the present day controversy over the scientific status of string theory—something we return to in the final chapter.

6.1 Tachyons, Critical Dimensions, and the Wrong Particles

Even in the early 1970's, soon after their creation, the dual models of hadrons faced some serious *internal* problems.[5] As Ramond notes, with wry amusement:

> In 4 years, this S-matrix theory evolves into a Lagrangian (in two dimensions) formulation, but it ends up in 26 dimensions, with a tachyon and long range forces. It is no wonder that at its moment of greatest theoretical clarity, the Dual Resonance Model enters its first hibernation ([45], p. 50).

---

[5] By "internal" here I simply mean problems that either have to do with consistency issues, or issues that related to very general physical problems (rather than, e.g. the failure to deal with specific experimental results (a matter that will be discussed in the next section).

Interestingly, all of the basic mathematical consistency problems then understood would be resolved within the decade (of course, this is not to say that *new* problems didn't emerge), though the *physical* consistency troubles remained, as the next chapter shows, they were 'upgraded' from vices to virtues in the context of a new non-hadronic theory of strings. Olive and Scherk open their paper on the no-ghost theorem, from March 1973, with:

> The existing dual models seem to be more a self-consistent alternative to polynomial local field theories than a phenomenological approach to describe the real world of hadrons. The main advantage over field theories is that at each order of perturbation expansion an infinite number of particle states of any spin is included, while maintaining the properties of duality, Regge behaviour, positive definiteness of the spectrum of resonances (absence of ghosts), and perturbative unitarity. The drawback is that this set of conditions is so constraining that it can be realised only for the unphysical values of the number of dimensions of space-time, and at the expense of having, in general, tachyons ([41], p. 296).

The first problem is simple and stark: in order to be consistent string theories demanded more dimensions than the four we observe: the string theories have a "magic" critical dimension of 26 (or 10). This is, of course, immediately falsified by direct experience! Though compactification methods could be devised, without any dynamical mechanism coming from the theory itself, such a scheme would be clearly far too *ad hoc* to take seriously (as had been the case with earlier compactification schemes). In the case of hadronic physics there was no conceivable reason as to why the theory should only work in such a number of dimensions.

A further problem is equally simple to see: the strong interaction binds hadrons together at close quarters, it is not a long range force. Yet there are massless particles of many kinds in the spectrum of the dual models. These particles coupled in the same way as the graviton and the Yang-Mills gauge bosons, a fact that would prove crucial in the evolution of hadronic string theory into a unified theory of all interactions.

There are also, strictly speaking, issues that lie on the boundary of internal and external. For example, in 1973 Lars Brink and Holger Nielsen show that QCD has no tachyon in its spectrum, while dual string models do, as a result of the zero-point energy of the quantum string. The tachyon problem was eventually resolved by the application of supersymmetry (as we see in Sect. 7.3).[6]

The next section describes the 'external' problems that led to a decrease in confidence with respect to dual models of hadrons. It isn't completely clear which set of problems posed the biggest threat to the dual string model of hadrons. Certainly, the problem of spacetime dimensions and tachyons were known while the theory was being pursued in its most intense phase of development, and did not appear to slow it down. Mathematical consistency problems,[7] perhaps, are seen as problems that can

---

[6] Note that Korkut Bardakçi [10] had argued that the tachyon problem simply pointed to the fact that the dual resonance model was using the wrong ground state and that a Goldstone-Higgs-type spontaneous-symmetry breaking mechanism of the gauge symmetry would lead to the correct ground state, eliminating the tachyon in the process. Indeed, he writes that "the existence of a scalar tachyon in the model turns out to be an advantage, rather than a defect!" (p. 332).

[7] Of course, this depends to a large extent on how one partitions one's class of problems into 'mathematical' and 'physical'. Such partitions might well allow for some exchanges between them.

be overcome, whereas physical problems (disagreement with a solid experiment, say) are rather more stubborn, with any potential accommodation being viewed as 'fudging' or introducing ad hoc elements.

6.2 Hard Scattering and Charmonium

In his Nobel lecture, David Gross recounts how he was led, relatively early, to a theory of point-like entities on the basis of experiments:

> These SLAC deep-inelastic scattering experiments had a profound impact on me. They clearly showed that the proton behaved, when observed over short times, as if it was made of point-like objects of spin 1/2 ([25], p. 9102).

He had been working on dual resonance models at CERN, and at Princeton (with Neveu, Scherk and Schwarz) for two years, but lost faith:

> At first I felt that this model, which captured many of the features of hadronic scattering, might provide the long sought alternative to a field theory of the strong interactions. However, by 1971 I realized that there was no way that this model could explain scaling, and I felt strongly that scaling was the paramount feature of the strong interactions. In fact, the dual resonance model led to incredibly soft behavior at large momentum transfer, quite the opposite of the hard scaling observed (ibid.).

String theory and QCD had very little overlap with respect to the features responsible for the quick uptake of the latter—especially so for the hard scattering events. One of string theory's great virtues is its non-local interaction structure, meaning that stringy interactions were 'smeared' out rather than occurring at a single space-time point.[8] This leads to excellent UV properties (soft scattering amplitudes), but obviously does not give the required hard scattering in the UV region, as mentioned, giving exponential decay in the case of fixed angle scattering.[9]

String theory might well have been pursued at the same intense rate despite its internal shortcomings and its obvious empirical problems had there been nothing else on the table. But the existence of a competitor eliminated this possibility. Quantum chromodynamics suffered from none of the severe internal problems plaguing string

---

(Footnote 7 continued)
For example, the constraint demanding $d = 10$ was initially viewed as a mathematical obstruction to be solved mathematically, but later was given a physical spacetime interpretation so that the solution to the problem of why we observe only four had to also involve physical structures (e.g. spontaneous compactification).

[8] Dual string model amplitudes decay exponentially in the Bjorken limit (i.e. when $s, t \to \infty$ and $s/t = $ const). Hence, it is seemingly unable to deal with the phenomenon of Bjorken scaling, which instead requires point-like entities—though recent work using the gauge/string duality suggests that string theory is able to accommodate such hard scattering, giving hard amplitudes (see [42]).

[9] Note, however, that in the immediate aftermath of the SLAC deep-inelastic collision results, the quark model also faced a serious empirical adequacy problem in that it could not explain why, given the point-like collisions (that appeared to fit experiment) quarks were not knocked free of the proton.

theory, and it could account for the hard, point-like events that string theory struggled with, thanks to its asymptotic freedom.[10] The quark theory was given a realistic, material interpretation as the various experiments at SLAC and Brookhaven began to indicate that just such entities were needed to explain the appearance of scaling at high energy—later data required the specific quantum numbers assigned to quarks. However, one must be careful not to over-exaggerate the speed with which QCD and the quark picture became firmly established. This process took some time to achieve definitive confirmation.[11] As David Gross explains in his Nobel lecture:

> The experimental situation developed slowly, and initially looked rather bad. I remember in the spring of 1974 attending a meeting in Trieste. There I met Burt Richter who was gloating over the fact that $R = \sigma_{e^+e^- \to hadrons}/\sigma_{e^+e^- \to \mu^+\mu^-}$ was increasing with energy, instead of approaching the expected constant value. This was the most firm of all of the scaling predictions. R must approach a constant in any scaling theory. In most theories one cannot predict the value of the constant. However, in an asymptotically free theory the constant is predicted to be equal to the sum of the squares of the charges of the constituents. Therefore, if there were only the three observed quarks, one would expect that $R \to 3[(1/3)^2 + (1/3)^2 + (2/3)^2] = 2$. However, Richter reported that R was increasing, passing through 2, with no sign of flattening out. Now many of us knew that charmed particles had to exist. Not only were they required, indeed invented, for the GIM mechanism to work, but as C. Bouchiat, J. Illiopoulos, and L. Maini, and independently R. Jackiw and I, showed, if the charmed quark were absent, the electroweak theory would be anomalous and non-renormalizable. Gaillard, Lee, and Rosner had written an important and insightful paper on the phenomenology of charm. Thus, many of us thought that since R was increasing, probably charm was being produced. In 1974, the charmed mesons, much narrower than anyone imagined (except for Appelquist and Politzer), were discovered, looking very much like positronium, and easily interpreted as Coulomb bound states of quarks. This clinched the matter for many of the remaining skeptics. The rest were probably convinced once experiments at higher energy began to see quark and gluon jets. The precision tests of the theory—the logarithmic

---

[10] The history of QCD has been studied well enough already to warrant a further treatment here. I refer the reader to [12] for more details.

[11] The hard-scattering (deep-inelastic) experiments had been carried out at SLAC between 1967 and 1973 (under the leadership of Jerome Friedman, Henry Kendall and Richard Taylor: the SLAC-MIT Collaboration)—the "deep inelastic" terminology refers to the fact that the energies are able to probe beyond the resonance region (scattering involving very large momentum transfer). In these experiments electrons were scattered from protons (a liquid hydrogen target) to reveal hard, point-like constituents within the proton: the precise details can be found in the original DOE R&D report: http://www.osti.gov/accomplishments/documents/fullText/ACC0173.pdf. Originally, not understood to correspond to quarks (whose fractional charge, $\pm\frac{1}{3}e$ or $\pm\frac{2}{3}e$, still struck many as physically dubious), they had been labeled 'partons' by Feynman—this model assumed that the particles were free in the deep-inelastic region, thereby delivering the scaling result. The experimental equivalence of partons and quarks required several theoretical and experimental developments. As Michela Massimi argues, some of the reluctance to side with the quark model, over the parton model of Feynman, can be explained by the fact that the quark model was initially constructed on primarily theoretical grounds, "independently of the deep inelastic scattering experiments that ... were giving evidence for partons" ([31], p. 45). One could make the point rather more simply by noting that Feynman's parton model was *tailored* to the SLAC results, whereas the quark model had preceded them—John Polkinghorne finds an "earthiness" in parton modelling, not found in the quark model ([44], p. 127). Note that in the same paper, Massimi argues that we should be careful in speaking of quarks and partons as being really two names for the same object: they had very different theoretical presuppositions associated with them.

Fig. 6.2 Cross section for hadron production against centre-of-mass energy, from Richter's data generated at SPEAR [Stanford Positron Electron Asymmetric Rings]. The sharp spike is a result of the long lifetime of the particle. Image source: [48], p. 288

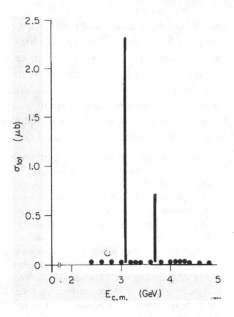

deviations from scaling—took quite a while to observe. I remember very well a remark made to me by a senior colleague, in April of 1973 when I was very excited, right after the discovery of asymptotic freedom. He remarked that it was unfortunate that our new predictions regarding deep-inelastic scattering were logarithmic effects, since it was unlikely that we would see them verified, even if true, in our lifetime. This was an exaggeration, but the tests did take a long time to appear. Confirmation only started to trickle in 1975–1978 at a slow pace ([25], pp. 9107-7).

One of the reactions studied in the early hadron accelerator experiments is $e^+e^- \rightarrow q\bar{q}$ (involving electron-positron collision beams). Recall the tube-like behaviour of the colour field as quark and anti-quark pair separate. The tension in the string embodies potential energy $V(r)$, which is kept at a constant per unit tube length r: more tube means more energy. When this energy reaches a certain threshold $2m_q$ there is a probability for $q\bar{q}$ pair creation (that is, for the creation of a new hadron). This process, called "hadronization", generates jets of hadrons parallel to the initial quark-anti-quark pair. This reaction was used as a tool to investigate hadron production. Such a meson was discovered independently in 1974 at both Brookhaven National Laboratory (by Samuel Ting's group) and at the Stanford Linear Accelerator Collider [SLAC] (by Burton Richter's group), showing up as a very significant peak in the cross-section for $e^+e^- \rightarrow$ hadron production (see Fig. 6.2). The announcements were published back to back in *Physical Review Letters*: [8, 9]. The former labeled the particle a 'J particle' while the latter labeled it a 'ψ particle'.[12] The particle is a bound state consisting of a quark and an anti-quark, as above. Ting's group's paper

---

[12] Note that Richter had studied under James Bjorken while at Stanford. Richter gives a very nice presentation of his discovery of the ψ (for 'psion') in his Nobel lecture: [48].

speculated that their J might be one of the charmed particles (i.e. a bound state of a quark/anti-quark pair: the charm quantum number. The charmed particle pair, $c\bar{c}$, was named 'charmonium' by Thomas Applequist and David Politzer, though it was coined by Alvaro De Rújula—see ([43], p. 92) in which Politzer also notes that the name was rejected by the editors of *Physical Review Letters*.

Writing of the quick change in the fortunes of gauge theories and the rise of QCD, John Iliopoulos writes that "[b]elieving in gauge theories before 1972 required an act of faith" yet by 1976 gauge theories had accumulated strong experimental support so that the "study of their physical consequences became the dominant research theme everywhere" ([28], p. 315). Geoffrey Chew, writing two years earlier, in 1970, was still defending his 'anti-fundamentalist' view of hadrons against the approach that searches for "basic building blocks" ([15], p. 23). He describes his approach as follows:

> The bootstrapper seeks to understand nature not in terms of fundamentals but through self-consistency, believing that all of physics flows uniquely from the requirement that components be consistent with one another and with themselves. No component should be arbitrary. Now by definition a "fundamental" component is one that is arbitrarily assignable; thus, to a bootstrapper, the identification of a seemingly fundamental quark would constitute frustration (ibid.).

Chew believed that the mathematical principles determining the consistency of the structure would *uniquely* pin down a single S-matrix that would approximate observed hadron physics, leaving no arbitrariness whatsoever. Of course, this would strike a chord with string theorists. Chew is describing what Steven Weinberg calls a "logically isolated theory":

> In a logically isolated theory every constant of nature could be calculated from first principles; a small change in the value of any constant would destroy the consistency of the theory. The final theory would be like a piece of fine porcelain that cannot be warped without shattering. In this case, although we may still not know why the final theory is true, we would know on the basis of pure mathematics and logic why the truth is not slightly different ([63], p. 189).

At the root of the division between the bootstrap approach and standard (reductive) quantum field theory approaches, then, is the question of arbitrariness. Exactly this issue would be raised after string theory takes off once again, in the mid-1980s, when the standard model's parameters are viewed as too numerous for such a model to provide a good understanding of how nature works, and why it is the way it is. Moreover, as string theory is rescaled, to become a theory of gravitons and gauge bosons (with a significantly smaller string scale), Chew's vision of mathematical consistency guiding the construction of the theory becomes non-optional. As Weinberg put it, in his interview for a BBC radio programme on string theory (from the 1980s):

> This is physics in a realm which is not directly accessible to experiment, and the guiding principles can't be physical intuition because we don't have any intuition for dealing with that scale. The theory has to be conditioned by mathematical consistency ([61], p. 221).

However, that comes later. At this phase in the life of string theory (and dual models), experiment rules the roost, and the hadronic strings quite clearly possess exponential decay for large transverse momenta that cannot be made to fit with experiments. QCD perturbation theory is, quite simply, not string-like. It can deal with the experiments.[13]

After all the novel work, pushing for a revolution in physics as a way of understanding hadrons, no such conceptual revolution was needed. Chew's "ultimate frustration" was realised:

> I would find it a crushing disappointment if in 1980 all of hadron physics could be explained in terms of a few arbitrary entities. We should then be in essentially the same posture as in 1930 when it seemed that neutrons and protons were the basic building blocks of nuclear matter. To have learned so little in half a century would to me be the ultimate frustration ([15, p. 25]).

Although there are obviously aspects of QCD that were not known in the 1930s, the resulting framework was essentially the same as the quantum field theory introduced by Heisenberg and Pauli. As Steven Weinberg put it, "revolution is unnecessary" ([62], p. 17).

6.3 Dual Strings and QCD Strings

That QCD possesses asymptotic freedom implies that at large distances it gets complicated in much the same way as standard field theories get complicated at small distances. In other words, QCD has infrared divergences just as potentially catastrophic as the ultraviolet divergences found in field theories without asymptotic freedom. In this case one finds that the perturbation theory is not able to tell us what we will observe at such large distances (corresponding to experiments conducted at the kinds of scale we can probe).[14] One of these low energy features, that we are well aware of, is that we don't observe individual quarks (nor do we observe anything with colour quantum numbers): they appear to be *confined* in pairs or triples within microscopic volumes.[15] Hadrons are colour neutral. As Susskind and Kogut point out, this has to be due to an infrared divergence in the self-energy of colour-bearing objects ([53], p. 348). This leaves open the opportunity for string theory to play a role, for it achieved good qualitative success with the lower energy phenomenology, for example the Regge recurrences from the apparently rotational sequences found in experiments. From the perspective of the quark model these trajectories (if continued onto infinite energies and spins) imply that quarks can never be liberated.

---

[13] Interestingly, recent work involving the 'gauge/gravity' duality has attempted to recover such 'hard scattering' behaviour from warped geometry in a dual gravity theory—see, e.g., [42].

[14] This idea of new objects at the non-perturbative level that cannot be seen at the perturbative level arises again in the final chapter, when we look at the physics of D-branes.

[15] The property of confinement (and several other properties) have yet to be given a convincing mathematical derivation from QCD, though there is plenty of evidence from computer simulations. The problem is considered to be important enough to be amongst the Clay Mathematics Institute's million dollar 'Millennium Prize Problems': [29].

In 1974, Nambu [33], not so long after suggesting the string model of hadrons, suggested incorporating it directly into QCD to try and account for the Regge behaviour in this context:

> On the phenomenological side, the string model seems to give a good overall description of hadron dynamics, so most of the theories that have been proposed for the confinement mechanism aspire to realize the string as a flux of gauge fields ([35], p. 372).

Of course, this harks back to the quantized vortices of Nielsen-Oleson in which quarks were viewed as magnetic monopoles [36]. The idea is to treat the hadrons as made of confined quarks, bound by flux tubes.[16] These hadrons have high angular momentum values, as revealed in the experimental Regge plots, and so a centrifugal force sends them apart. As Mandelstam points out ([30], p. 272), at low values of angular momentum, the length of such flux tubes is comparable with its thickness, and the theory reduces to the MIT 'Bag Model' (in which fields are constrained to lie in a finite region of space). In this sense, the string model of hadrons provides a neat qualitative account of the 'soft' processes (the Regge phenomenology, along with duality), while the quark model provides an account of 'hard' processes (deep-inelastic scattering): they are complimentary rather than competing.

There was, however, a problem in making this link work: Nielsen and Oleson's original flux tubes were without ends, being either closed on themselves or else infinitely long. Dual strings, on the other hand, had a finite length fixed by α'—though the closed-string/Pomeron connection was known by the early 1970s.[17] What Nambu added to this picture was the idea that, to be finite (like dual strings) the Nielsen-Oleson tubes would have to terminate on monopoles, to 'capture' the flux that terminates at the end points—this is roughly similar to the way open strings must end on D-branes in the context of the modern understanding of superstring theory.[18]

This turns on a kind of 3-way analogy between confined quarks, dual strings, and a monopole/anti-monopole pair in a superconductor. Confinement comes about from the fact that the energy of the monopole system increases proportionally to the distance between them (as mentioned earlier, the more flux tube there is, the more energy there is). This system is analogous to a meson viewed as a pair of quarks

---

[16] Recall that the example depends on an analogy between dual strings and magnetic flux tubes in a superconductor, with a Meissner-type effect creating the tube, caused by the pressure of a superfluid it sits within and displaces.

[17] A more concrete understanding of the claim that the Pomeron must have vacuum quantum numbers can be given once one has a closed string interpretation since one need simply note that quark quantum numbers flow along the edges of the surface (in the fashion of Harari-Rosner diagrams), yet this is not possible in the case of a surface with no edges: hence, the Pomeron must have only vacuum quantum numbers (cf. [39], p. 139).

[18] Nambu later considered the case of *electric* confinement, borrowing from the work of Kalb and Ramond [34]. (Though note that 'electric' and 'magnetic' here are being used by analogy with the situation in electromagnetism: the systems of interest occur in the context of strong interactions). Unfortunately, the model was only an Abelian model, and was unable to account for the additional confinement of gluons that one finds in QCD.

sitting at the ends of a string.[19] The quarks are, then, viewed as carriers of 'magnetic charge' which are bound (permanently) by string bonds. As Kenneth Wilson puts it, the "confinement of quarks is caused by the strings they are attached to; quarks may separate from each other, but at a cost of creating more string" ([65]. p. 332).[20]

This basic idea was developed considerably in subsequent years, as can be found in a meeting on "Extended Systems in Field Theory" held at the Ecole Normale Supérieure in Paris in 1975 (organised by Gervais and Neveu). Dual strings were still very much on the table at this conference. Still later work was carried out by Charles Thorn: [59]. Thorn expressed the hope, in 1979, that the dual string approach to quark confinement might be able to enable researchers to build a bridge between the disparate scales of hadronic physics given by the Regge slope on the one hand and in deep-inelastic scattering on the other.

In the 1970s, 't Hooft's investigations into $U(N)$ Yang-Mills theories (in the large N limit) pointed to a precise mathematical relationship with dual resonance models, on account of the shared topological structure of the planar diagrams that result: the topological structure of the perturbation series in $1/N$ is the same as that computed in the dual model picture [56].[21] He managed to prove some interesting results in the

---

[19] Kerson Huang gives an exceptionally clear presentation of this 3-way analogy in Chap. 19 of [27]. (Note that 't Hooft's work on monopoles was also inspired by the paper of Nielsen and Oleson.)

[20] Note that in this chapter Wilson introduces strings as elements of his construction of lattice gauge theories. These strings are rigid, unstretchable "string bits" which are created and destroyed by string-bit operators. Wilson had already pointed out, in his landmark paper on the confinement of quarks [64], that his structure was "reminiscent of relativistic string models of hadrons" in that the strong-coupling expansion of his lattice gauge theory has the same general structure as the string models—this paper introduced the concept of 'Wilson loop' (an integral part in the development of the loop quantum gravity approach and very many other areas of physics). Wilson's relationship with the S-matrix is curious, and rather unorthodox. He had trained in the methods of renormalisation group theory (and also computational techniques) during his doctoral years at Caltech (under Gell-Mann), but his thesis had been on the Chew-Mandelstam theory. He was trying to program numerical solutions to Mandelstam's bootstrap approximation. However, early on, as he says, he had "come to the presupposition that S-matrix theory was going no place [and] that field theory was the only way to go" ([7]). He eventually came to dual strings while trying to make physical sense of the mathematical formulation he had of lattice gauge theory: " I did that, initially, just so I could have something that I was confident I could understand. Then I found myself faced with this problem that the lattice theory is something that has a simple strong coupling limit. It was the first experience in my life when I found that I could do the mathematics (the mathematics of solving the theory for strong coupling) but I couldn't figure out the physics. I just couldn't get any kind of concept in my mind as to what all the results meant when I did the strong coupling expansion. And I spent a full year just building a sense of the physics, working partly from the ideas of [Leonard] Susskind, partly [J.] Kogut-Susskind, partly just Susskind on strings, to build an ability to relate, to build a model physical world in which the strong coupling expansion made sense. That was a very different kind of experience from the experience that I had before, where the physics was not in question, it was a question of getting mathematical approximations to a known physics" ([7]). I mention this to point out the importance of string models in guiding key aspects of the construction and development of QCD. These aspects are often sidelined (or completely ignored) in discussions of the history of QCD and the standard model. They also quite clearly serve to dampen the notion that string theory was dropped in 1973.

[21] 1974 was clearly an *annus mirabilis* for 't Hooft: together with Martinus Veltman, he computed the one-loop divergences in vacuum general relativity (bringing together earlier work on dimensional

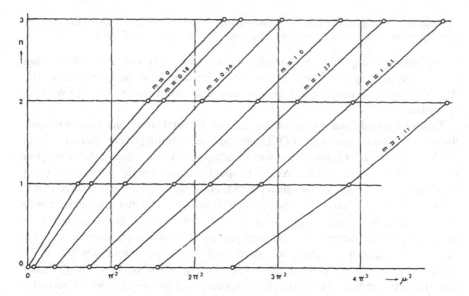

Fig. 6.3 The Regge trajectories derived from 't Hooft's two-dimensional meson model built from a quark/anti-quark pair. Image source: [55], p. 469.

case of two-dimensional (one space and one time) gauge theories [55]. In particular, he showed in detail, for a two-dimensional model of mesons, how the gauge field's interactions correspond to those of a quantized dual string.[22] He was also able to derive a physical mass spectrum which can approximately reproduce the straight Regge trajectories (see Fig. 6.3).

't Hooft describes the reasoning behind this work as follows:

> I thought what I have to do is first of all is try to see if I can rearrange perturbation expansion for the strong interactions such that I can show why the quarks do not come out, which is why the $N \to \infty$ limit was interesting once you got diagrams. They looked a little like string theory diagrams. And, string theory put quarks at the ends of a string, which would prohibit them from coming out. So, I thought that the large-N theory was a good indication as how to deal with the strong interactions such that you can show quark confinement. But then, I couldn't unfortunately solve those high, those planar diagrams. They're still too complex. So even the $N \to \infty$ limit of QCD is not a solvable limit. Except, and that was a nice thing, in two dimensions. And that was after asymptotic freedom had been argued about by Gross and Wilczek and Politzer that I could show in two dimensions QCD just does everything that you expect it to do. Because, the large N limit then works out beautifully. It gives you

---

(Footnote 21 continued)
regularisation and the Feynman rules for gauge theories) [54], he introduced magnetic monopoles (without Dirac strings) [57], and also introduced the tool of using the $N \to \infty$ limits (where N is the number of colours in a gauge theory) used to link quark theory to the dual string theory [55].

[22] Four-dimensional gauge theories proved to be far more complicated. These links were eventually probed in the context of the 'gauge/string' dualities associated with the Maldacena conjecture: on which, see Sect. 10.4. There we find that the same limiting procedure (the 't Hooft limit') is utilised, in which $N \to \infty$ and $g^2 N$ is kept fixed.

the spectrum of mesons right away . . . and it shows that the mesons are confined and there, this time, you can compute everything (Interview with the author, pp. 67–68).

The interesting thing about this two-dimensional case, of course, is that the 1-space is essentially a string, with the time dimension allowing the string to propagate. This idea of focusing on the two-dimensional theory would inspire much later work, as we shall see.

To summarize the main point of this section: the dual string picture was extraordinarily fruitful in the context of QCD and the understanding of confinement.[23] The role it played shows, I think, that it was in fact essential for many of the conceptual insights that emerged in QCD. Again, to speak of the 'demise' of the dual string model does a disservice to the impact it had. However, the conflicts between QCD and string theory are still there, and subsequent progress in superstring perturbation theory (in flat space) increasingly drove a wedge between QCD and string theory (as a more general framework). The superstring at the critical dimension $d = 10$ has a zero-mass graviton, six extra dimensions and extra supersymmetries that are, to say the least, not easily reconcilable with QCD. Those still pursuing string theory, as a structure mapping onto the fundamental nature of the physical world, turned to a version theory of gravity interacting with matter. Meanwhile, as Brower notes, the string was rendered less fundamental in its role in the description of strong interactions: "The QCD flux tube was viewed merely as an effective low energy or long distance description no more fundamental than flux tubes in a superconductor". ([11], p. 322).[24]

6.4 Soldiering on

Whilst we have seen that the basic idea of a string theory persisted through the emergence of QCD, and beyond, it is fair to say that the idea that string theory might be used to construct a quantum theory of gravity (and other interactions) was not quickly taken up. However, it is quite wrongheaded to say that this was caused by QCD's rise to fame. QCD was a theory of hadrons, while the transformed string theory (discussed in the next chapter) was a theory that promised far more. The two simply were not competitors, and those interested in strong interactions were unlikely to be much impressed with quantum gravity. Hence, aside from those continuing to work on the hadronic string (which featured peculiar 'strong gravitons' and 'strong

---

[23] One might look at these attempts to recover QCD effects from string models as a factor in the shift from the once prevalent operator approach to the string picture.

[24] I should point out, also, that the fact that many of the same people were shifting seamlessly between the dual string theory and elements of what would become QCD shows that they would not have recognised a 'split' at the time: it was a case of applying whatever tools were at hand to the problem of understanding strong interactions—this can be seen quite clearly in the papers of, e.g., 't Hooft and Wilson mentioned above.

photons,' rather than genuine versions thereof[25]), one should view the transformed version as an entirely different theory. Given the 20 order of magnitude shift in the distances that separate the hadronic and transformed versions, this seems to be an entirely appropriate stance!

As is often the case in episodes in the history of science, the dual model approach to hadrons did not immediately surrender, and its practitioners were strongly enamoured of its mathematical structure. In fact, as I intimated above, much of the 'hard work' (in terms of mathematics) was carried out using the operator approach, with the string picture reaping the benefits, as it were, but also providing a nice way to conceptualise and visualise the results, giving a physical, dynamical explanation for otherwise mysterious aspects of the dual models, as mined using the operator approach. In this sense, both were necessary for a proper understanding of the dual models.

The most orthodox approaches attempted to recover QCD from dual theory either by some kind of limiting procedure, or by showing the equivalence between the theories. Schwarz and Scherk themselves proposed a dual field-theory of quarks and gluons. However, the notion that there *were* 'dark years' for string theory, is something of an exaggeration. The output was indeed leaner than the unusual explosion of interest that followed Veneziano's work, but that can be explained by the restoration of relative normalcy of focus in the field of strong interactions to a specific approach. There was a steady flow of important and interesting work on string theory throughout the 70s and early 80s, with many key results being generated during this period.

There was a curious parallel integration of ideas concerning supersymmetry flowing into both the hadronic string models and their non-hadronic extensions. The latter we consider in more detail in the next chapter, but the former contain some interesting developments—developments that stretch out until the 80s. An important cluster of papers in this respect come from the workshop of Ademollo et al. [2–4].[26] Here the authors attempted to milk the new aspects of QCD from the dual string models (and achieve more *realism*) by adding more structure to their worldsheets (cf. [21], p. 453). But more to the point, this work constitutes part of a detailed effort to construct and understand supersymmetric string theories and includes variables that would later be used to construct to the Green-Schwarz superstrings (see Sect. 8.3).[27]

---

[25] One can find the strong graviton terminology (referring to 'strongly interacting spin-2 particles') appearing at the cusp of the period in which QCD begins to replace dual models (both operator and string formulations) and in which 'unified' and non-hadronic versions of dual models begin to take form—see, e.g., [1], p. 191. The strong photon concept can be found earlier: see, e.g., [20], p. 377.

[26] A group of such size that Lars Brink has referred to it as "the Italian football team" (see his talk: http://hep.caltech.edu/ym35/presentations/Brink.pdf; see also [49], p. 201).

[27] For more along these lines, from the same period, see: [13, 14]—note that [13], by Chang, Macrae, and Mansouri, contains what I believe is the first instance of the term "super-string" (in this case in the hyphenated form: see p. 59). There was some earlier work, by Hopkinson and Tucker [26], in which new degrees of freedom were added to dual string models to get out para-statistics (in an independent work, [6] called such structures "parastrings"). John Schwarz also developed an earlier 'realistic' model in which quark statistics were included: [50].

Finally, it is worth mentioning Murray Gell-Mann's influential role in keeping at least one string theorist,[28] John Schwarz, working on strings. Gell-Mann had taken his sabbatical (from 1971–1972) in CERN, where there was a small group of string theorists working at the time. Schwarz and Neveu's paper on the dual pion model had made it over there and was generating some interest, which Gell-Mann duly picked up on. This resulted in an invitation to Schwarz, initially to give a series of talks at Caltech. Shortly thereafter, in 1972, Schwarz discovered that he would not be given tenure at Princeton. A little after this he was offered a research associate position at Caltech. Gell-Mann's relationship with string theory is perhaps a little puzzling. On the one hand he is an ardent conservative in his physics, and very rigid in his attachment to experiment. He seems to have not been impressed by fancy mathematics.[29] And yet he was an early and faithful advocate of string theory (including the period in which dual models and string theory were not pursued so actively). However, with the potential benefit so high (a theory of all forces and particles with no arbitrary parameters) then one can rationally allow a large risk to achieve it (where the risk is using up lots of resources in the knowledge that it might well be false).

6.5 Summary

Summarising the developments that have led to the current research landscape, Nambu writes:

> String theory traces its origin to the Veneziano model of 1968. It also happens that the Weinberg-Salam model was born about the same time. The latter has led to the successful Standard Model. The descendants of the former, on the other hand, are still struggling to be relevant to the real world in spite of their enormous theoretical appeal. Indeed there exist two pathways in the development of theoretical particle physics since its beginnings in the Thirties. I will call them the quantum field theory and the S-matrix theory respectively. In its historical lineage, the Standard Model belongs to the former, whereas the superstring theory belongs to the latter. Even though the former has turned out to be the Royal Road of particle physics, this was not entirely clear before its final triumph, and the latter has also played very important contributing roles which continue to this day ([32], p. 275).

Looking back at the records, with the exception of the radical position of Chew, one finds not forked paths, but intersecting paths. The idea of absolute separation is far too simplistic. One way of seeing this, it to look at the trajectories of those responsible for the creation of QCD, who were perfectly at ease switching between

---

[28] Cf. Interview with John H. Schwarz, by Sara Lippincott. Pasadena, California, July 21 and 26, 2000. Oral History Project, California Institute of Technology Archives. Retrieved [2nd jan, 2012] from the World Wide Web: http://resolver.caltech.edu/CaltechOH:OH_Schwarz_J.

[29] For example, in a letter to Wick he writes that "mathematical rigour bores me" (March 21, 1962 [Box 22.23, Gell-Mann papers: Caltech]).

the S-matrix approach and the more orthodox field theoretic picture (and, indeed, several other approaches).[30]

In contrast to standard historical accounts of this period, I have tried to show in this chapter that the notion of an 'early demise' of string theory at the hands of QCD is perhaps not quite the right interpretation of events. String theory was very much bundled up with the creation of QCD, and in particular with the conceptualisation of colour confinement. Rather than a demise, I argue that it's far more accurate to speak of an *integration* of the hadronic string within QCD. However, once this integration was achieved, QCD could largely take care of itself, and string theory (so conceived) would become a proper part of QCD, rather than a field of its own. However, it is certainly true that as a complete, self-contained description of strong interactions, the dual string model succumbed to QCD. If one is basing this on 'having dominion' over strong interactions, then the standard tale is true. However, I don't think we are forced to follow that tale. Hence, dual string theory did not die then, but the notion of what string theory might be used to *represent* did diverge. This will be the topic of the next chapter. We will see again, however, that the story is not quite so simple as is often supposed. We can say, by way of prospectus, that the splintering of string theories (hadronic and non-hadronic) occurred partly on account of a certain enamourment that certain physicists had with the mathematical structure of string theory, and the thought that it *could* do so much more. Moreover, the expansion of the potential domain of the framework of string theory occurred *before* QCD was crowned.

References

1. Ademollo, M., D'Adda, A., D'Auria, R., Napolitano, E., di Vecchia, P., Gliozzi, F., et al. (1974). Unified dual model for interacting open and closed strings. *Nuclear Physics*, *B77*(2), 189–225.
2. Ademollo, M., Brink, L., D'Adda, A., D'Auria, R., del Guidice, E., di Vecchia, P., et al. (1976). Dual string with $U(1)$ colour symmetry. *Nuclear Physics*, *B111*, 77–110.
3. Ademollo, M., Brink, L., D'Adda, A., D'Auria, R., del Guidice, E., di Vecchia, P., et al. (1976). Dual string models with non-abelian colour and flavour symmetries. *Nuclear Physics*, *B114*, 297–316.
4. Ademollo, M., Brink, L., D'Adda, A., D'Auria, R., del Guidice, E., di Vecchia, P., et al. (1976). Supersymmetric strings and colour confinement. *Physics Letters*, *62B*(1), 105–110.
5. Andersson, B. (1998). *The Lund model*. Cambridge: Cambridge University Press.
6. Ardalan, F., & Mansouri, F. (1975). Interacting color string models. *Bulletin of the American Physical Society*, *20*(4), 552.
7. Ashrafi, B., Hall, K., & Schweber, S. (2002). Interview with Kenneth G. Wilson, 6 July 2002: Part III. http://authors.library.caltech.edu/5456/1/hrst.mit.edu/hrs/renormalization/Wilson/Wilson3.htm
8. Aubert, J. J., et al. (1974). Experimental bbservation of a heavy particle *J*. *Physical Review Letters*, *33*(23), 1404–1406.

---

[30] I might add to this case for intersecting paths that there are some Monte Carlo simulations (especially the so-called 'Lund Model' [5]) that invoke the string picture, viewing a gluon as "a kink on the string stretched between the quark and antiquark" ([51], p. 233)—cf. [23], p. 5.

9. Augustin, J.-E., et al. (1974). Discovery of a narrow resonance in e^+e^- annihilation. *Physical Review Letters*, *33*(23), 1406–1408.
10. Bardakçi, K. (1974). Dual models and spontanous symmetry breaking. *Nuclear Physics*, *B68*, 331–348.
11. Brower, R. C. (2012). The hadronic origins of string theory. In A. Capelli et al. (Eds.), *The birth of string theory* (pp. 312–325). Cambridge: Cambridge University Press.
12. Cao, T. Y. (2010). From current algebra to quantum chromodynamics: A case for structural realism. Cambridge: Cambridge University Press.
13. Chang, L. N., Macrae, K., & Mansouri, F. (1975). A new supersymmetric string model and the supergauge constraints in the dual resonance models. *Physics Letters B*, *57*(1), 59–62.
14. Chang, L. N., Macrae, K., & Mansouri, F. (1975). Geometrical approach to local gauge and supergauge invariance: Local gauge theories and supersymmetric strings. *Physical Review D*, *13*(2), 235–249.
15. Chew, G. F. (1970). Hadron bootstrap: Triumph or frustration? *Physics Today*, *23*(10), 23–28.
16. Chodos, A., & Thorn, C. (1974). Making the massless string massive. *Nuclear Physics*, *B72*, 509–522.
17. Dirac, P. A. M. (1954). Logic or beauty? *The Scientific Monthly*, *79*(4), 268–269.
18. Fubini, S. (1974). The development of dual theory. In M. Jacob (Ed.), *Dual theory*. Physics Reports Reprint Book Series, Vol. 1 (pp. 1–6). Amsterdam: Elsevier Science Publishers.
19. Gell-Mann, M. G. (1972). General status: Summary and outlook. In J. D. Jackson & A. Roberts (Eds.), *Proceedings of the 16th international conference on high-energy physics* (pp. 6–13) Sep 1972, Batavia, Illinois (pp. 333–355). Publications office, National Accelerator Laboratory, P.O. Box 500, Batavia, Ill. 60510.1973. 4 vols.: http://www.slac.stanford.edu/econf/C720906/
20. Del Giudice, E., Di Vecchia, P., & Fubini, S. (1972). General properties of the dual resonance model. *Annals of Physics*, *70*(2), 378–398.
21. Gliozzi, F. (2012). Supersymmetry in string theory. In A. Capelli et al. (Eds.), *The Birth of string theory* (pp. 447–458). Cambridge: Cambridge University Press.
22. Goddard, P. (2012). From dual models to relativistic strings. In A. Capelli et al. (eds.), *The birth of string theory* (pp. 236–261). Cambridge: Cambridge University Press.
23. Gomez, C., & Ruiz-Altaba, M. (1992). From dual amplitudes to non-critical strings: A brief review. *Rivista del Nuovo Cimento*, *16*(1), 1–124.
24. Green, M. B., Schwarz, J. H., & Witten, E. (1987). Superstring theory, volume 1: Introduction. Cambridge: Cambridge University Press.
25. Gross, D. (2005). The discovery of asymptotic freedom and the emergence of QCD. *Proceedings of the National Academy of Science*, *102*(26), 9099–9108.
26. Hopkinson, J. F. L., & Tucker, R. W. (1973). Dual models with para-field excitations. *Physics Letters*, *47B*(6), 519–522.
27. Huang, K. (2007). *Fundamental forces of nature: The story of gauge fields*. New Jersey: World Scientific.
28. Iliopoulos, J. (1996). Physics in the CERN theory division. In J. Krige (Ed.), *History of CERN*, Vol. III (pp. 277–326). Amsterdam: Elsevier.
29. Jaffe, A., & Witten, E. (2006). Quantum Yang-Mills theory. In J. Carlson, A. Jaffe, A. Wiles (Eds.), *The millennium prize problems* (pp. 129–152). Cambridge: American Mathematical Society.
30. Mandelstam, S. (1983). The current theory of strong interactions and the problem of quark confinement. In A. van der Merwe (Ed.), *Old and new questions in physics, cosmology, philosophy, and theoretical biology* (pp. 265–274). Springer.
31. Massimi, M. (2004). Non-defensible middle ground for experimental realism: Why we are justified to believe in colored quarks. *Philosophy of Science*, *71*(1), 36–60.
32. Nambu, Y. (2012). From the S-matrix to string theory. In A. Capelli et al. (Eds.), *The birth of string theory* (pp. 275–282). Cambridge: Cambridge University Press.
33. Nambu, Y. (1974). Strings, monopoles, and gauge fields. *Physical Review D*, *10*(10), 4262–4267.

34. Nambu, Y. (1976). Magnetic and electric confinement of quarks. *Physics Reports*, *23*(3), 250–253.
35. Nambu, Y. (1979). QCD and the string model. *Physics Letters*, *80B*(4–5), 372–376.
36. Nielsen, H., & Oleson, P. (1973). Vortex-line models for dual strings. *Nuclear Physics*, *B61*, 45–61.
37. Olive, D. I. (1972). Clarification of the rubber string picture. In J. D. Jackson, A. Roberts, & R. Donaldson (Eds.) *16th international conference on high-energy physics, v. 1: Parallel sessions: Strong interactions* (pp. 472–474). National Accelerator Laboratory. Batavia, IL. [http://www.slac.stanford.edu/econf/C720906/]
38. Olive, D. I. (1974). Dual models. In J. R. Smith & G. Manning (Eds.), *Proceedings of the XVII International Conference on High Energy Physics* (pp. 269–280). Chilton: Rutherford Appleton Laboratory.
39. Olive, D. I. (1974). Further developments in the operator approach to dual theory. In M. Jacob (Ed.), *Dual theory* (pp. 129–152). Physics Repots Reprint Book Series—Volume 1. Oxford: Elsevier Science Publishers.
40. Olive, D. I. (2012). From dual fermion to superstring. In A. Cappelli et al. (Eds.), *The birth of string theory* (pp. 346–360). Cambridge: Cambridge University Press.
41. Olive, D., & Scherk, J. (1973). No-ghost theorem for the Pomeron sector of the dual model. *Physics Letters*, *44B*(3), 296–300.
42. Polchinski, J., & Strassler, M. J. (2002). Hard scattering and gauge/string duality. *Physical Review Letters*, *88*, p. 031601.
43. Politzer, D. (2004). The dilemma of attribution. In T. Frängsmyr (Ed.), *The Nobel Prizes 2004* (pp. 85–95). Stockholm: Nobel Foundation.
44. Polkinghorne, J. (1989). *Rochester roundabout*. New York: Longman.
45. Ramond, P. (1987). The early years of string theory: The dual resonance model. In R. Slansky & G. B. West (Eds.), *Theoretical Advanced Study Institute Lectures in Elementary Particle Physics 1987: Proceedings* (pp. 501–571). World Scientific.
46. Rebbi, C. (1974). Dual models and relativistic quantum strings. *Physics Reports*, *12*(1), 1–73.
47. Redhead, M. L. G. (1998). Broken bootstraps—the rise and fall of a research programme. *Foundations of Physics*, *35*(4), 561–575.
48. Richter, B. (1976). From the psi to charm—the experiments of 1975 and 1976. In S. Lundqvist (Ed.), *Nobel lectures, physics 1971–1980* (pp. 281–310). Singapore: World Scientific Publishing Co., 1992.
49. Schwarz, J. H. (1987) The future of string theory. In L. Brink et al. (Eds.), *Unification of fundamental interactions* (pp. 197–201). Physica Scripta, The Royal Swedish Academy of Sciences. World Scientific, 1987.
50. Schwarz, J. H. (1974). Dual quark-gluon theory with dynamical color. *Nuclear Physics*, *B68*, 221–235.
51. Sjöstrand, T. (1983). The Lund Monte Carlo for e^+e^- jet physics. *Computer Physics Communications*, *28*, 229–254.
52. Small, H. (1973). Co-citation in the scientific literature: A new measure of the relationship between two documents. *Journal of the American Society for Information Science*, *24*, 265–269.
53. Susskind, L., & Kogut, J. (1976). New ideas about confinement. *Physics Reports*, *23*(3), 348–367.
54. 't Hooft, G. & M. Veltman. (1974). One loop divergences in the theory of gravitation. *Annales de l'Institut Henri Poincaré*, *20*, 69–94.
55. 't Hooft, G. (1974). A planar diagram theory for strong interactions. *Nuclear Physics*, *B72*, 461–473.
56. 't Hooft, G. (1974). A two-dimensional model for mesons. *Nuclear Physics*, *B75*, 461–470.
57. 't Hooft, G. (1974). Magnetic monopoles in unified gauge theories. *Nuclear Physics*, *B79*, 276–284.

58. 't Hooft, G. (2000). The glorious days of physics: Renormalization of gauge theories. In A. Zichichi, (Ed.), *From the Planck length to the Hubble radius* (pp. 434–454). Proceedings of International School of Subnuclear Physics, Erice 1998, Subnuclear Series, Vol. 36. World Scientific.

59. Thorn, C. B. (1979). Quark confinement in the infinite-momentum frame. *Physical Review D, 19*(2), 639–651.

60. Veneziano, G. (2012). Rise and fall of the hadronic string. In A. Capelli et al. (Eds.), *The Birth of string theory* (pp. 17–36). Cambridge: Cambridge University Press.

61. Weinberg, S. (1988). Steven Weinberg. In P. C. W. Davies & J. Brown (Eds.), *Superstrings: A theory of everything?* (pp. 211–224). Cambridge: Cambridge University Press.

62. Weinberg, S. (1977). Search for unity: Notes for history of quantum field theory. *Daedalus, 106*, 17–35.

63. Weinberg, S. (1992). *Dreams of final theory*. New York: Pantheon Books.

64. Wilson, K. (1974). Confinement of quarks. *Physical Review D, 10*, 2445–2459.

65. Wilson, K. (1976). Quarks on a lattice, or, the colored string model. *Physics Reports, 23*(3), 331–347.

Chapter 7
Theoretical Exaptation in String Theory

> *[W]e must have both felt that it must be good for something,*
> *since it was just such a beautiful, tight structure.*
>
> John Schwarz

In a reprint volume on *Dual Theory*, published in 1974[1], David Olive had this to say about the status of the dual theory and its newly discovered potential for describing more than just hadrons[2]:

> The whole motivation of the dual resonance theory was in connection with strong interaction physics. Now we have seen the remarkable fact that in the (we hope) unlikely event of this being wrong, the theories of the other interactions, weak, electromagnetic, gravitational, appear as different special cases of the same dual theory. The most optimistic point of view is that we are on the way to a unified theory of all the interactions, but if not, we still have the most general and powerful theory yet found in the sense of generalizing all previously known theories of interest [28, p. 150].

One might have thought that this clear statement of the potential unifying power of the dual models might have led to much frenzied work in unpacking the details. Olive himself thought of the new point of view as a "conceptual revolution" [29, p. 35]. However, his efforts to motivate the unified dual models fell largely on deaf ears, with the majority of physicists finding it too risky a project.

---

[1] Note that this was the first volume of the *Physics Reports Reprint Series*, aimed at providing overviews of rapidly changing fields, in their early stages. This first volume spans exactly the transitionary period, charting the development of dual models from a pure hadron theory, potentially to a unified theory, and from a dual theory without a really clear physical grounding, to a theory of relativistic quantum strings. It also leads up to the cusp of string theory's existence as a description of strongly interacting systems—though as the quotation from Olive (written in 1974) below shows, it wasn't completely clear cut that the string models would *not* provide a good model of hadrons.

[2] Olive was referring to the then new results of Yoneya and Scherk and Schwarz, as well as earlier related results on the content of the zero-slope limits of dual models. We will discuss these below. What Olive didn't refer to in his account was the rescaling that was also required to obtain a theory able to reproduce the predictions of general relativity at observed energies, rather than some theory whose fundamental excitations lived at the hadronic scale.

D. Rickles, *A Brief History of String Theory*, The Frontiers Collection,
DOI: 10.1007/978-3-642-45128-7_7, © Springer-Verlag Berlin Heidelberg 2014

Nevertheless, the most curious feature in the history of string theory has to be this transition that occurs in its function, from a description of the forces binding protons and neutrons to a description of gravitational and other interactions. This is perhaps the most extraordinary case of 'theoretical exaptation' in the history of physics.[3] One can see from Olive's remarks that the seed of grand unification had already been planted by 1974. In a similar vein, Scherk and Schwarz write of the still relatively new results on zero-slope limits:

> [A] scheme of this sort might provide a unified theory of weak, electromagnetic, and grav-itational interactions. The gauge bosons and leptons would be identified with open strings and the graviton with the closed string [30, p. 347].

This, in essence, corresponds to the barest modern understanding of string theory, though later work would show that the links between open and closed string descriptions (and therefore between gauge and gravity) were far more complicated.[4]

There are, in fact, similarities between this shift within string theory and the way in which gauge theory developed and transformed in its early days, including the belief that the theory was too beautiful not to be useful for something. Recall that gauge theory was devised by Hermann Weyl in 1918 (in the context of classical field theory) as a way of unifying gravitation and electromagnetism, with his principle of *eich-invarianz* connecting the electromagnetic potentials ψ_i and g_{ik}. The theory involved the idea that parallel transported vectors experience a path-dependent change of length of: $\exp(\gamma \int_C A \cdot dx)$. It turned out to be dysfunctional in this environment for solid experimental reasons, as had been pointed out by Einstein. But it was later revived in the new 'quantum environment', involving not the non-integrability of length measurements, but of *phase*. Just as Vladimir Fock and Fritz London simply changed real into complex numbers—so that electromagnetic potentials were reinterpreted as linked with the components of the quantum wave-function Ψ—so string theory just had to adjust the value of the string constant to get a theory of quantum gravity. Of course, this simple modification has dramatic consequences with respect to the physical interpretation of the theory. Various elements of theoretical structure are impacted on; not least the critical spacetime dimension of the theory, which can be viewed through the lens of the dynamical nature of geometry of general relativity. This itself then suggests an entirely new range of tools, concepts, and techniques that can be employed in the further development of the theory.

---

[3] "Exaptation" is, of course, a term from evolutionary biology introduced by Steven Jay Gould and Elizabeth Vrba [20], referring to the shift in function of some trait or aspect of physiology over time, so that it is 'co-opted' for another use. To the best of my knowledge, it has not previously been employed for use in the context of historical studies of scientific theories, but I think it entirely appropriate here.

[4] Recall that given the restrictions on the intercept $\alpha(0)$, to either 1 or 2, then we will get a pair of Regge trajectories, each with infinitely many particles lying on it, but the former trajectory will contain a massless spin-1 (vector) particle while the latter will contain a massless spin-2 particle (identified as a pomeron in the earliest phases).

This chapter will explore this transitionary phase of string theory's history[5], taking the story up to the early 1980s, at which point the notion that string theory might offer a mathematically consistent 'unified quantum theory' was fully known, if still not yet fully understood or commonly pursued. The next leap forward (the subject of the first chapter in Part III) was the isolation of a phenomenologically suitable model.

7.1 The Role of the Scherk Limit

A vital piece of structural knowledge that was required by the idea that the function of string theory could be shifted was the notion that the dual models reduced to field theories in specific limits, namely those for which $\alpha' \to 0$ (the zero-slope or 'Scherk' limit.[6] The method of transformation involved the Mandelstam-Regge trajectory slope, modifying it from approximately $1/GeV^2$ to $10^{-38}/GeV^2$.[7] In terms of length, the shift is one of 20 orders of magnitude, from $l_s \sim 10^{-13}$ cm (the scale of hadrons) to $l_s \sim 10^{-33}$ cm (the Planck length, at which quantum gravitational effects become non-negligible). In terms of string tension, given that it goes as $1/l_s^2$, we find a shift of 40 orders of magnitude. As the slope is reduced, the masses of any initially massive particles increases, going to infinity in the zero slope limit. Only the massless states survive this limit and these correspond to the known classical field theories. As we have already seen, Jöel Scherk was responsible for figuring this out. The initial suggestion for thinking about what happens when the slope goes to zero seems to have come from Roland Omnès, during Scherk's Doctorat d'Etat lecture.[8]

Scherk's final papers were on supergravity, and in particular on dimensional reduction and spontaneous compactification, and the idea of using the compact dimensions as physical resources. Indeed, following the discovery that dual models reduce to Einstein gravity, Scherk appears to have increasingly diverted his attention to gravity.[9] Schwarz had been visiting the Ecole Normale Supérieure in Paris one year before Scherk died. During that year they worked on their paper entitled "How to Get Masses from Extra Dimensions" (see [34] for the published version).

---

[5] What Yuri Manin has called a "romantic leap" [26, p. 60]. Gomez and Ruiz-Altaba call it "a healthy extrapolation" [19, p. 5].

[6] As we saw earlier, Scherk had initially derived a ϕ^3 quantum field theory, but later work with Neveu showed that a Yang-Mills theory resulted.

[7] In this Scherk (small-string) limit (where zero slope $\equiv$ infinite tension) the spin-2 massless mode (the graviton) persists and couples in a generally covariant manner, as in Einstein's theory of general relativity cf. [13, p. 373].

[8] As Omnès points out, this was a kind of examination, for which he was one of the jury members (private communication). Gervais recalls Omnès' remark occurring over lunch in the cafeteria of Orsay [15, p. 410].

[9] However, it seems Scherk was not completely comfortable with supergravity. As David Olive recalls, "I remember Jöel Scherk complaining later that he felt obliged to work on supergravity whereas his real conviction lay with string theory" [29, p. 355]. We can guess that these were career-based obligations and peer-pressure.

The zero slope limit was the central device that enabled the dual string model to morph into the superstring theory we know today, with the problematic massless particles given a realistic interpretation. The compactification techniques he devised (which we return to later), to reduce critical to observed dimensions, are central to the generation of phenomenologically acceptable physics from superstrings. As Schwarz wrote at the Second Aspen Winter Conference on Physics, in 1987, "two of the most troubling features of string theory for application to hadronic physics could be turned into virtues if the goal was changed" [33, p. 269]—the zero slope technique and compactification models were central to this new-found virtuous status.

One might also mention that the softness of the scattering amplitudes, that had posed empirical problems with the hard-scattering experiments on hadrons, would also serve as a further virtue in this case since it tames the otherwise fatal ultraviolet divergences of gravitational interactions. However, the relationship between the divergences and non-locality of strings took longer to fully understand.

7.2 Dual Models of Everything

By 1975, Tamiaki Yoneya was able to write:

> By its string formulation, the dual-resonance theory has been acquiring a unified and clear physical picture. In particular, we are now able to treat interacting reggeons and pomerons, from the outset, by considering the interaction among open and closed strings [41, p. 440].

As we have discussed already, one of the (initially) embarrassing features of the dual model was that in, what was interpreted as the closed string sector of the general framework there was a spin-2 particle which was forced, by the gauge invariance required by the absence of ghosts, to be massless. It became clear to at least a handful of people that this particle had the properties required by the graviton (the carrier of gravitational force), and that given this it would be forced to behave in a generally covariant fashion. David Olive recalls that the idea that the dual models might therefore provide a unified framework for gauge and gravitational interactions was discussed as far back as 1971:[10]

> The price that the Dual Resonance Model has to pay for consistency with fundamental principles is that it looks increasingly less like a theory of strong interactions and more like a unified theory. Not only does it possess massless gauge particles but also massless gravitons.

---

[10] In fact, Keiji Kikkawa (just as he was preparing to leave for his new position at CUNY) and Hikaru Sato considered the compatibility of gauge boson interactions with the dual resonance model in 1970 [24], though not using the varying slope method, which had yet to be introduced. They were concerned with the incorporation of the electromagnetic and the weak interactions in the dual resonance scheme (there is no mention of gravitation). Bars, Halpern, and Yoshimura [3] also considered a unified theory of all non-gravitational interactions in a way that was, as they acknowledge, heavily influenced by Neveu and Scherk's earlier work on the connection between dual models and Yang-Mills fields—see also [2] in which Bardakçi and Halpern consider what they call "M-models" (essentially an early gauge theory of hadrons) that are also tightly bound to Neveu and Scherk's work.

Of course the same was true of the dual fermion theory (if indeed it does really exist) and it had the innate advantage of possessing fermions. As I remember, this idea of unification of gauge and gravitational interactions was much discussed by the community in CERN Theory Division in the year 1971–1972 even though this was before the discovery of asymptotic freedom and the formulation of the Standard Model [29, p. 352].

There were several independent generalisations of the dual models to gravity and other non-hadronic interactions—note that the title of this subsection, "Dual Models of Everything," is borrowed from Green, Schwarz, and Witten's textbook [22, §1.2]. The root of these alternative applications of dual models was, later on, the troublesome spectrum of massless particles, including massless spin-1 and spin-2 particles. The massless spin-2 case is especially interesting since, as has been known since the late 1930s (thanks to Wolfgang Pauli and Markus Fierz[11]), it corresponds to the expected features of a gravitational force carrying particle. However, initially the particle was not treated as having anything to do with gravity, and so was named the 'pomeron' instead.[12]

Schwarz and Scherk are usually credited with instigating the gravitational application of dual models. Yet, as Schwarz points, gravity was at that stage simply not in the toolkit of most particle physicists. Schwarz (and, one can guess, Scherk) learned general relativity later, as a result of the potential application of dual models to gravity:

We knew that that was an issue, but it wasn't our problem; we were trying to understand the strong interactions. And in those days physics was much more compartmentalized than it is now. The first thing that people who were brought up in particle physics were taught was that you can forget about gravity, because if you just look at the force between two protons, or even between an electron and a proton, the gravitational force compared to, say, the electric force, is smaller by ten followed by 38 zeros or something. It was just fantastically negligible. So we were taught to forget about gravity. It had nothing to do with our problem. Particle physicists wouldn't talk about gravity. I mean, if anyone tried to, they'd be viewed as a crackpot. It wasn't part of the problem (http://resolver.caltech.edu/CaltechOH:OH_Schwarz_J, p. 27).

Hence, though we now think of the 'killer app' of string theory as its consistent implementation of quantum gravity, no one in the dual model community was concerned with that problem at the time: certainly not the particle physicists, who would have been the natural audience, given the concepts and methods employed. The killer app simply didn't take at its inception and it is important to reiterate that this was *not*

---

[11] It is interesting to note that Alton Coulter brought out a paper [7] explaining the relationship between massless spin-2 fields and gravitational theory the same year as Scherk's first paper on the zero slope limit, 1971. Coulter mentions the Fierz and Pauli result, though he proposes a modified version of the theory based on the physical components of the spin-2 field, rather than potentials.

[12] There was something of a battle of names over this Regge intercept second hadron: Mandelstam, Chew, and other central figures can be found calling the particle the 'Pomeranchon'. One can find also a more extended version, 'Pomeranchukon,' to fit more of Pomeranchuk's name in! Throughout the late sixties and early seventies all of these names were utilised. Later 'Pomeron' became the accepted term—Gribov introduced the Pomeron concept, though credit is usually given to Gell-Mann (via Geoff Chew) with coining the term 'Pomeron' to refer to the vacuum pole (i.e. a pole with vacuum quantum numbers).

the result of QCD being the stronger theory.[13] To compare Scherk and Schwarz's modified dual string model with QCD is to compare apples and oranges: very different fruit. To 'sell' strings, they needed the research landscape to alter in such a way that string gravity was well adapted to it. John Schwarz is quite rightly credited as being one of the main researchers keeping string-gravity alive while this change happened—though, Schwarz was also part of the 'refashioning' of the wider research landscape too, as I shall explain in the next chapter.

Hence, at the time of their initial attempt to forge a new path for the dual models, they were not especially interested in the conflict between quantum theory and general relativity:

> [I]t wasn't a problem that we were particularly concerned about. However, when Scherk was here in '74, at some point in our deliberations we said, "Just for the fun of it, let's see whether this massless spin-2 particle behaves in the right way to give the standard gravitational force of the Einstein theory of general relativity." And having posed the question, it wasn't actually very hard to answer by invoking some appropriate theorems and making the case that indeed that was right. [.] And the reason we found this exciting was that we knew that string theory was going to give a consistent quantum theory. [.] And it became clear to both of us, immediately, that this was the way to make a consistent quantum theory for gravity. So we figured that we'd just tell the world and they'd all get excited and start working on it (http://resolver.caltech.edu/CaltechOH:OH_Schwarz_J, p. 28).

Given that the framework promised a consistent framework for quantum gravity one might have expected the quantum gravity community to jump on it. But, as Schwarz goes on to note, "Nobody took it seriously—not even the relativists or the people who had been working on string theory before. Nobody!" (ibid., p. 29).[14] To reiterate, the string theory of gravity and non-hadrons, was a new theory that started more or less in 1973/4. Like most new theories, it takes time to generate interest. One can look to other quantum gravity proposals (Roger Penrose's twistor framework, for

---

[13] For this reason, I think Schwarz is mistaken in assigning the 'blame' over the delayed uptake of string theory as a fundamental theory of all interactions to "the stigma associated with its origins [in S-matrix theory]" [32, p. 5]. The theory was, at the time, still plagued by a tachyon, and still had the curiosity of the additional spacetime dimensions (despite the potential dynamical explanation in a gravitational context). On the latter, in 1974, Fubini associated the varying of the number of spacetime dimensions with "science fiction" [14, p. 5], —though he did not dismiss the approach; but rather thought that the $d = 26$ and $d = 10$ results "should suggest further study on the rôle of dimensionality in the general structure of physical theories" (ibid.). One might add to this that gravitational physics (*prima facie* a natural habitat for the newly transformed string theory) was still not long out of a transformation of its own, from a field that had become synonymous with 'crackpot science' to one based on solid experimental and observation evidence. At this stage of its development, string theory might well have looked like just another unified field theory, which is precisely what the relativity community was keen to avoid.

[14] This is a slight exaggeration. I suspect that had they pitched string theory more along the lines of what the general relativity and quantum gravity community were used to, engaging with their concerns, they might have fared better. I might add that Schwarz and Scherk did receive an 'honourable mention' (though along with 29 others!) in the 1975 Gravity Research Foundation essay competition for their paper "Dual Model Approach to a Renormalizable Theory of Gravitation" (a strong year, in which Roger Penrose took first prize and Julian Schwinger second).

example) to see a similar phenomenon.[15] What is curious about the dual models of non-hadrons, of course, is that they were discovered via dual models of hadrons, and probably would not have been otherwise—here perhaps is a difference from the case of Weyl's gauge theory: there was an *immediate need* (rather than basic survival) driving that case of exaptation.

The introduction of gravity suggested to Scherk and Schwarz that the problem of the mismatching dimensions might be given a dynamical explanation: general relativity allows for features of space–time to be determined by equations of motion, so perhaps the determination of the space–time dimension is not so bad after all!

> You see, before it had been a problem. When we were just doing strong interactions, it didn't make sense. But in gravity, the geometry of space and time is determined by the equations of the theory. So it became a possibility that the equations of the theory would require that six of the dimensions, for some reason, would curl up into some invisible little ball or something, and then it could be perfectly consistent with observation. It wouldn't make sense to give that kind of a story if you were just doing strong interactions, but in a theory of gravity, that kind of story made sense. We certainly understood that (ibid., p. 30).

This dynamical feature of space–time geometry became known as "spontaneous compactification", and can be found in Scherk's work with Cremmer, from 1976 [8, 9]. However, this is related to 'dimensional reduction' which goes back farther.

Directly influenced by Neveu and Scherk's earlier work [27][16] showing that Yang-Mills field theories can be given a tree approximation using the zero slope technique, Tamiaki Yoneya had independently realised that the low-energy behaviour of dual models was equivalent to Einstein gravity too, from the scalar amplitudes of the Virasoro-Shapiro model (again, with $\alpha' \to 0$ and fixed $g\sqrt{\alpha'}$). Since he explicitly refers to quantum gravity along the lines of the "Gupta-Feynman" [38, p. 951], perturbative approach[17], we can infer that the links Feynman and others had drawn between Yang-Mills theories and general relativity were behind the extension to gravitation in this case. Yoneya explicitly interprets the massless spin-1 and spin-2 states, required by the no-ghost condition, as a photon and a graviton [39, p. 1907].[18]

---

[15] Rather interestingly, Chang and Mansouri mention in 1971: [4, p. 2541], some potential overlap between Penrose's twistors and dual string models of hadrons, in their discussion of the introduction of spin degrees of freedom onto the time-evolved string's two-dimensional surface (they didn't use the worldsheet terminology).

[16] In his reminiscences, Yoneya recalls that he received a preprint of the Neveu-Scherk paper in 1972, from M. Minami (a fellow dual theorist from the Research Institute of Mathematical Sciences in Kyoto)—Minami was offering comments on an earlier preprint of Yoneya's (on the nature of the gauge principle in open string models) that had in fact been rejected. Yoneya switched to the closed strings of the Virasoro-Shapiro model, finding gravity. However, he claims to have been "prejudiced against general relativity" on account of the dominance of the S-matrix programme, so his idea languished. It was the resurgence of interest in gauge theories that spurred him on into revisiting his idea.

[17] That is, the approach in which the metric tensor is split into two parts, $g_{\mu\nu} = \delta_{\mu\nu} + \kappa h_{\mu\nu}$ (where $\kappa^2 = 16\pi G_{Newton}$), with the Lagrangian then expanded in powers of κ.

[18] Of course, we now associate these with open and closed string descriptions respectively, but Yoneya only mentions strings in a brief appendix of his paper. This highlights the fact that even into the mid-1970s there was a parallel operator algebraic approach that was capable of discovering

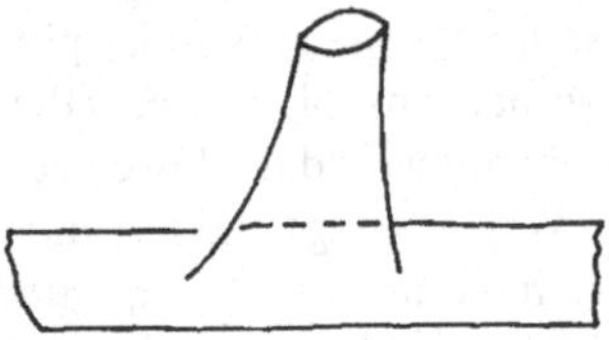

Fig. 7.1 Emission of a *closed*
from an *open* string according
to Ademollo et al. *Image
source* [1, p. 193]

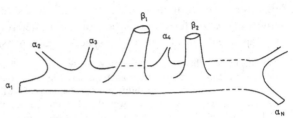

Fig. 7.2 A combination of
'*open* → *open*' transitions
(described by the vertex
$V_\alpha(z, k)$) and '*open* → *closed*'
transitions (described by the
vertex $\omega_\beta(z, \bar{z}, k)$). *Image
source* [1, p. 196]

One can also find a clear statement of the existence of a graviton "coupling universally with the energy momentum tensor of the string" in a 1974 paper of the Ademollo et al. collaboration [1, p. 191],—that is to say, the 'strong graviton' just *is* a graviton. They use this universality property to develop a scheme for coupling open and closed strings. However, they make no attempt to rescale the physics to describe a gravitational physics coupling according to Newton's constant, and are primarily concerned with constructing a *unified* model capable of incorporating both open and closed strings, in interaction, thus bringing together the generalised Veneziano model and the Shapiro-Virasoro models. The basic vertex, $\omega_\beta(z, \bar{z}, k)$, for the emission of a closed from an open string is achieved by treating the closed string interaction as an external field (see Fig. 7.1).

This vertex can be combined with the vertex, $V_\alpha(z, k)$, of the original Veneziano theory (for open → open transitions) to write down complex amplitudes, such as that depicted in Fig. 7.2.

Before leaving this topic, mention should be made of a further, quite distinct, attempt to forge a connection between dual string theory and general relativity, by Takabayasi [36], this time based on an analysis of general covariance in string theory and a formal analogy between this and general relativity. Takabayasi bases his approach on the geometric string model of Nambu-Gotō that he had played a role in. However, the connection in this case is a purely formal one, involving an overlap of mathematical formalism, and there is no suggestion that gravitation is involved in, what for Takabayasi are still hadronic strings.

(Footnote 18 continued)
many of the features that we often associate with the more geometrical, string picture. He made the connection to closed strings explicit in his 1975 paper on dual string models and quantum gravity [41].

7.3 The GSO Projection and 'Real' Superstrings

One of the most serious flaws with the dual string models had been the persistent presence of a tachyon located at the lowest mass state. For example, in the 26 dimensional Veneziano model, we find the following spectrum containing $M^2 = -1$:

Mass^2	J (spin)
-1	0
0	2 (also: spinless dilaton and 2-form)
1	≤ 4
2	≤ 6
$\vdots$	$\vdots$

This was finally and fully resolved in 1976, within the supersymmetric model of Gliozzi, Scherk, and Olive, in their paper "Supersymmetry, Supergravity Theories and the Dual Spinor Model" [17]. The method involved the imposition of a certain chiral projection that suppressed ('truncated') a large sector of the states, including that containing the tachyon, so that the ground state (for the bosonic sector: NS) instead comprises a massless graviton, a massless scalar, and a massless antisymmetric tensor. The NSR sector (also containing left-handed Majorana fermions), is also tachyon-free and has a massless spin $3/2$ state (then called a "hemitrion" rather than a gravitino[19]) and a massless scalar (see Fig. 7.3).[20]

---

[19] As is so often the case, Murray Gell-Mann was behind this earlier naming scheme. Peter van Nieuwenhuizen describes, in his Dirac lecture from 1994, how he and Gell-Mann browsed through dictionaries searching for a "venerable name" for the massless spin 3/2 particles. They settled on 'hemitrion' since it means 'half-3'. Alas, as is also so often the case, the editors of *Physical Review* were not keen on the name, and suggested their own: "massless Rarita-Schwinger particle". He notes that Sidney Coleman and Heinz Pagels coined the current name of 'gravitino' [37, pp. 14–15].

[20] There were some relevant earlier steps along the way to this result. For example, Clavelli and Shapiro [5] made a detailed study of G-parity states in the NS-model, in their paper on Pomeron factorization, showing that both odd and even G-parity states contribute. They also consider projection operators onto the G-parity states and the possible "cancellation of the tachyon pole" (noting that there is no such possibility from the positive-G-parity part: p. 505). Fairlie and Martin performed a "systematic replacement of the factors in the one-loop integrals for the original model by factors incorporating the anticommuting elements ψ_i" (where $i = 1, ..., N$ are linked to the N Koba-Nielsen variables and form an anticommuting Grassmann algebra, $\psi_i\psi_j + \psi_j\psi_i = 0$: [10, p. 375]; see also [11]). They were able to show that odd G-parity Pomerons disappeared (cf. Mandelstam's review of dual-resonance models, in which he also explicitly notes how one could use this approach to "exclude off g-parity particles from the Pomeron sector of the N.S.R. model" [25, p. 348]. Note that these issues were dealt with at length in David Martin's PhD *Investigations into Dual Resonance Models*, completed under Fairlie at Durham between 1971–1974: http://etheses. dur.ac.uk/8277/1/8277_5278.PDF. Michael Green [21] also made inroads on similar problems in 1973. In 1978, in an interesting review of the 'spinning string theory from a modern perspective,' John Schwarz writes that "[i]t was clear from the beginning that one could restrict the [NSR] model

Fig. 7.3 A Chew-Fraustchi
plot of the spectrum of states
of the supersymmetric NSR
model (for *closed strings*)
showing the elimination
of the $M^2 = -1$ (tachyonic)
state and suggesting supersymmetry between the bosons
and fermions (i.e. an equal
number of bosons and fermions located at each mass level)
[17, p. 281]

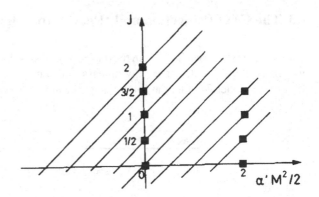

The truncation also generates a spacetime supersymmetric spectrum of states,
associating one-to-one at each mass level, bosons and fermions in ten dimensions[21],
so that (in the NSR theory) there is at each mass-level an equivalence between the
number of physical states in the bosonic and fermionic sectors (which points to the
existence of supersymmetry in the full 10-dimensional theory).[22] This work produced
what would later be called 'Type I' superstrings (where the 'I' refers to the number

to the subspace of even "G parity" particles, which is free from tachyons ... [but] this restriction
was not advocated partly because of our commitment to hadronic interpretation, partly because of
other hopes for eliminating the tachyon, and partly because of the concern that fermionic coupling
would restore the odd-G states through duality" [31, pp. 433–434].

[21] Of course, as we saw earlier, it was already known that there existed a two-dimensional worldsheet
supersymmetry (as they mention in their paper), but not yet a spacetime (or 'target space') supersymmetry. An earlier version of the paper submitted to *Physics Letters B* contains the abstract: "We
find that the spinor dual model is locally supersymmetric not only in the two-dimensional surface
spanned by the string, but also with respect to the embedding space–time" [16, p. 282].

[22] In fact, they hedge somewhat by writing that "this model has a good chance of being supersymmetric" [16, p. 266], noting later that a full proof would demand a definition of the supersymmetric
transformations that exchange the NS (bosonic) and R (fermionic) states (ibid., pp. 267–268). Their
own proof involved an identity that had already been proven by Carl Jacobi in 1829 (his *Aequatio
identica satis abstrusa*)—yet more evidence for Fairlie's remark that "String theory is more like
something dragged out of the nineteenth [century]!". This involves another instance of the serendipity resulting from 'turning to the maths books' for an expected answer. Gliozzi writes: "I knew I
had to look for some identity involving Jacobi theta functions. I took from the shelf of mathematical book ... a copy of Whittaker and Watson ... [and] it magically opened on page 470, where the
following exercise is proposed:

'Shew that : $\dfrac{1}{2}\left[\displaystyle\prod_{n=1}^{\infty}(1+q^{2n-1})^8 - \prod_{n=1}^{\infty}(1-q^{2n-1})^8\right] = 8q\prod_{n=1}^{\infty}(1+q^{2n})^8.$'   (7.1)

It was exactly the sought after formula!' [18, p. 455]. That is, the formula describes (with its signs
capable of describing bosons [NS] and fermions [R]) their level-by-level equality. As Gliozzi puts
it: "[t]he left-hand side is the relevant part of the generating function of the NS physical states
after removing the odd G-parity sector, while the right-hand side is the analogous function for
the Ramond states, after projection on the Majorana-Weyl spinors; the factor 8 comes from the
degeneracy of the ground state fermion" (ibid.).

of supersymmetries) and marks the birth of the modern understanding of (consistent) superstrings *qua* supersymmetric strings. Of course, the 'dual spinor model' in this paper simply refers to an embryonic version of superstrings and highlights the fact that string theory (even in its exapted form) was still connected by an umbilical cord to the old dual resonance models.

As Gliozzi remembers it, he began discussing these ideas (initially with Scherk) leading to the GSO result while "under the influence" of the recent work on supergravity, that was taking place in the offices next door to his [18, p. 545].[23] His idea was to extend this work to RNS [Ramond-Neveu-Schwarz] strings. Using the Scherk limit they found that the RNS theory defined a $d = 10$ supergravity theory. Applying the kinds of compactification techniques Scherk had developed with Cremmer, they were able to show that pure supergravity in $d = 10$ generates supergravity coupled to matter in four $d = 4$. The massless spin 3/2 particle mentioned above was the signal that supersymmetry was involved (since such a particle only consistently couples to supersymmetric matter), so that each physical state in the NS-sector should be partnered with a physical state in the R-sector.[24]. It was then the fact that such a partnership breaks down for both the tachyon[25] and the NS-subsector satisfying $\alpha' M^2 = n - 1/2$ (the odd-G-parity sector, where the G-parity operator combines charge conjugation and a 180° rotation about the second axis of isospin space), that formed the basis for the projecting out of such sectors:

> we [Gliozzi and Scherk] discovered that this sector transformed a right-handed fermion into a left-handed fermion, therefore it decoupled altogether if the right-handed fermions were projected out using Weyl spinors. Moreover the fermion-fermion and the fermion-antifermion states had the same spectra as bosonic bound states. In order to avoid infinite degeneracy of the bosonic spectrum we were led to require that the fermions satisfy also the Majorana condition. The resulting projected model, as tachyons had been removed, was the first example of a totally consistent string theory. Only later, thanks to the contribution of David Olive, we realized that the requirement of the Majorana-Weyl condition is very constraining and is possible only if d is 2 modulo 8 [18, pp. 454–455].

In sum: half of the fermion states and the odd G-parity (boson) states are removed, leaving the bosonic and fermionic spectra evenly-balanced (and recovering, in a natural way, $d = 10$ for the RNS model, as a result of the joint imposition of Majorana and Weyl restrictions on the Dirac spinors).

This work heralded (though after a brief 'intermission') the beginning of a new wave of dimensional reduction in string theories. In this case it included a link between the compact manifold and the low energy (four dimensional) properties that

---

[23] Indeed, in [17] the work explicitly aims to tie dual model research to that taking place in supergravity theories (see p. 254).

[24] One finds that the transverse (physical) Fock spaces associated with the Neveu-Schwarz and Ramond theories both decompose into a pair of invariant subspaces (chiral projections) under the transverse subgroup, $SO(8)$, of the Lorentz group in $d = 10$, $SO(9,1)$.

[25] Initially the tachyon elimination was a major motivation, along with the derivation of supergravity, but later work on higher-loop (non-tree level) amplitudes, placed the GSO projection even more centrally in the superstring programme, revealing that the truncation it enforces is in fact required in order to preserve unitarity and modular invariance (see [35, p. 285], [23]).

remain after the compactification. In particular, the preservation of supersymmetry in higher dimensions depended on features of the manifold, with a torus leaving invariant *all* of the supersymmetry of the higher dimensional theory. However, it wasn't until Michael Green and John Schwarz's work on new superstring theories, from 1980 onwards, that a version of string theory with explicit spacetime supersymmetry was constructed.

7.4 Summary

We have seen how the Scherk limit, discovered during the heyday of dual models, was utilised as a tool for converting the function of dual models from strong interactions (with 'strong photons' and 'strong gravitons') to a theory of non-hadrons (electrodynamics, Yang-Mills theory, and gravitation). This shift in function led to new (positive) ways of viewing what were previously viewed as insurmountable problems: the presence of massless particles in the dual model spectra, and the requirement of 26 or 10 dimensions of spacetime, both demanded by consistency. The remaining problem of the tachyon was also finally ironed out by following connections with supergravity, with spacetime supersymmetry offering a mechanism for controlling the theory. The next chapter looks at the steady rise of string theory work in the early '80s, followed by the dramatic shift in factors triggered by Green and Schwarz's anomaly cancellation proofs.

References

1. Ademollo, M., D'Adda, A., D'Auria, R., Napolitano, E., di Vecchia, P., Gliozzi, F., et al. (1974). Unified dual model for interacting open and closed strings. *Nuclear Physics*, *B77*(2), 189–225.
2. Bardakçi, K., & Halpern, M. B. (1974). Dual M-models. *Nuclear Physics*, *B73*(2), 295–313.
3. Bars, I., Halpern, M. B., & Yoshimura, M. (1973). Unified gauge theories of hadrons and leptons. *Physical Review D*, *7*(4), 1233–1251.
4. Chang, L. N. F., &Mansouri. (1972). Dynamics underlying duality and gauge invariance in the dual-resonance models. *Physical Review D*, *5*(10), 2535–2542.
5. Clavelli, L., & Shapiro, J. A. (1973). Pomeron factorization in general dual models. *Nuclear Physics*, *B57*, 490–535.
6. Clavelli, L. & Halprin, A. (eds.). (1986) *Lewes string theory workshop*. Singapore: World Scientific.
7. Coulter, C. A. (1971). The mass-zero spin-two field and gravitational theory. *Il Nuovo Cimento*, *7*(2), 284–304.
8. Cremmer, E., & Scherk, J. (1976). Spontaneous compactification of space in an Einstein-Yang-Mills-Higgs model. *Nuclear Physics*, *B108*, 409–416.
9. Cremmer, E., & Scherk, J. (1976). Spontaneous compactification of extra space dimensions. *Nuclear Physics*, *B118*, 61–75.
10. Fairlie, D. B., & Martin, D. (1973). New light on the Neveu-Schwarz model. *Il Nuovo Cimento A*, *18*(2), 373–383.
11. Fairlie, D. B., & Martin, D. (1974). Green's function techniques and dual fermion loops. *Il Nuovo Cimento A*, *21*(4), 647–660.

12. Frampton, P. H., & Wali, K. C. (1973). Regge-slope expansion in the dual resonance model. *Physical Review D, 8*(6), 1879–1886.
13. Freund, P. G. O., Oh, P., & Wheeler, J. T. (1984). String-induced space compactification. *Nuclear Physics, B246,* 371–380.
14. Fubini, S. (1974). The Development of Dual Theory. In M. Jacob (Ed.). *Dual theory, physics reports reprint book series* (Vol. 1, pp. 1–6). Amsterdam: Elsevier Science Publishers.
15. Gervais, J-L (2012). Remembering the dawn of relativistic strings. In A. Capelli et al. (Eds.). *The birth of string theory* (pp. 407–413). Cambridge: Cambridge University Press.
16. Gliozzi, F., Scherk, J., & Olive, D. I. (1976). Supergravity and the dual spinor model. *Physics Letters B, 65*(3), 282–286.
17. Gliozzi, F., Scherk, J., & Olive, D. I. (1977). Supersymmetry, supergravity theories and the dual spinor model. *Nuclear Physics, B122,* 253–290.
18. Gliozzi, F. (2012). Supersymmetry in string theory. In A. Capelli et al. (Eds.). *The birth of string theory* (pp. 447–458). Cambridge: Cambridge University Press.
19. Gomez, C., & Ruiz-Altaba, M. (1992). From dual amplitudes to non-critical strings: A brief review. *Rivista del Nuovo Cimento, 16*(1), 1–124.
20. Gould, S. J., & Elizabeth, S. V. (1982). Exaptation—a missing term in the science of form. *Paleobiology, 8*(1), 4–15.
21. Green, M. B. (1973). Cancellation of the leading divergence in dual loops. *Physics Letters B, 46,* 392–396.
22. Green, M. B., Schwarz, J. H., & Witten, E. (1987). Superstring theory, volume 1: Introduction. Cambridge: Cambridge University Press.
23. Kawai, H., Lewellen, D. C., & Henry Tye, S.-H. (1986). Classification of closed fermionic-string models. *Physical Review D, 34*(12), 3794–3805.
24. Kikkawa, K., & Sato, H. (1970). Non-hadronic interactions in the dual resonance model. *Physics Letters, 32B*(4), 280–284.
25. Mandelstam, S. (1974). Dual-resonance models. *Physics Reports, 13*(6), 259–353.
26. Manin, Y. (1989). Strings. *Mathematical Intelligencer, 11*(2), 59–65.
27. Neveu, A., & Scherk, J. (1972). Connection between Yang-Mills fields and dual models. *Nuclear Physics, B36,* 155–161.
28. Olive, D. I. (1974). Further developments in the operator approach to dual theory. In M. Jacob (Ed.). *Dual theory, physics reports reprint book series* (Vol. 1, pp. 129–152). Amsterdam: Elsevier Science Publishers.
29. Olive, D. I. (2012). From dual fermion to superstring. In A. Cappelli et al. (Eds.). *The birth of string theory* (pp. 346–360). Cambridge: Cambridge University Press.
30. Scherk, J., & Schwarz, J. (1974). Dual models for non-hadrons. *Nuclear Physics, B81*(1), 118–144.
31. Schwarz, J. (1978). Spinning string theory from a modern perspective. In A. Perlmutter & L. F. Scott (Eds.). *New frontiers in high-energy physics* (pp. 431–446). New York: Plenum Press.
32. Schwarz, J. (1985). Superstrings: The first fifteen years of superstring theory. Singapore: World Scientific.
33. Schwarz, J. H. (1987). Superstrings—an overview. In L. Durand (ed.). *Second aspen winter school on physics* (pp. 269–276). New York: New York Academy of Sciences.
34. Scherk, J. & Schwarz, J. H. (1979). How to get masses from extra dimensions. *Nuclear Physics B, 153,* 61–88.
35. Seiberg, N., & Witten, E. (1986). Spin structures in string theory. *Nuclear Physics, B276,* 272–290.
36. Takabayasi, T. (1974). General-covariant approach to relativistic string theory. *Progress of Theoretical Physics, 52*(6), 1910–1928.
37. van Nieuwenhuizen, P. (1994). Dirac lecture: Some personal recollections about the discovery of supergravity. *News from the ICTP, 78,* 10–15.
38. Yoneya, T. (1973). Quantum gravity and the zero-slope limit of the generalized virasoro model. *Lettere al Nuovo Cimento, 8*(16), 951–955.

39. Yoneya, T. (1974). Connection of dual models to electrodynamics and gravidynamics. *Progress of Theoretical Physics*, *51*(6), 1907–1920.
40. Yoneya, T. (1975). Interacting fermionic and pomeronic strings: Gravitational interaction of the Ramond Fermion. *Il Nuovo Cimento*, *27*(4), 440–458.
41. Yoneya, T. (1975). Dual string models and quantum gravity. In H. Araki (ed.). *International Symposium on Mathematical Problems in Theoretical Physics, Lecture Notes in Physics 39* (pp. 180–183). Springer.
42. Yoneya, T. (2012). Gravity from strings: Personal reminiscences. In A. Capelli et al. (Eds.). *The birth of string theory* (pp. 459–473). Cambridge: Cambridge University Press.

Chapter 8
Turning Point(s)

Interest in the theory of superstrings as a fundamental theory of
matter reached near hysterical proportions ... as a consequence
of the significant work of Green and Schwarz showing that such
theories are anomaly free and probably finite.

L. Clavelli and A. Halprin, 1986

A common myth of string theory has it that string theory was simply ignored until the famous anomaly cancellation result of Green and Schwarz in 1984. This result is said to be the origin of "the first superstring revolution". It is the goal of this chapter to tame this myth a little, showing that research on the subject was steadily increasing up to 1984, with several important developments between 1981 and 1983, while admitting that there is certainly much truth to the claim that Green and Schwarz's anomaly cancellation paper [38] *triggered* a very large increase in the production of papers on the subject, including a related pair of papers that between them had the potential to provide the foundation for a realistic unified theory of both particle physics and gravity.[1] The anomaly cancellation paper of Green and Schwarz amounted to a challenge to produce a string theory embodying a particular (more phenomenologically appealing) symmetry group. The so-called 'heterotic string theory' that fitted the bill, will be the subject of the next chapter.

---

[1] If we take revolutions in the Kuhnian sense [44] (i.e., involving the production of a successor theory incommensurable with the older theory), then Green and Schwarz were clearly operating more in the 'normal science' mode than 'crisis mode': the anomalies were one of many problems that had been faced in the development of the theory up to that point. Though the *response* to their work might have been very dramatic, the work itself was part of a smoother story, and so strictly speaking the concept of revolution doesn't seem to be applicable here.

D. Rickles, *A Brief History of String Theory*, The Frontiers Collection,
DOI: 10.1007/978-3-642-45128-7_8, © Springer-Verlag Berlin Heidelberg 2014

8.1 Supergravity, Cousin of Superstrings

While string theory was somewhat underwhelming the community of high-energy physicists during the late 1970s, supergravity appeared rather more promising. Indeed, while string theory is absent from virtually all of the conferences on 'grand unification' and similar themes, before the early 1980s, supergravity can be found in abundance. Many of those that had (and still were) working on string theory had in fact transitioned into supergravity research. Indeed, I don't think it is too unfair to say that the existence and professional support of supergravity through the mid- to late seventies were a lifeline for string theorists, whether they liked it or not. Moreover, many serious technical obstacles to string theory had been worked out in the context of supergravity research, and many new concepts were introduced into string theory as a result of work in supergravity. A notable example, already mentioned in the previous chapter, is the work of Gliozzi, Olive, and Scherk (the GSO projection). Importantly, the kind of 'mind-stretching' being accomplished in the supergravity community would make for a community more receptive to the *prima facie* radical ideas that were yet to come from string theory. I have in mind ideas such as 11 dimensional supergravity with compactification schemes, and so on—which, of course, provided future tools as well as future open minds.

The first example of an $N = 1$ supergravity model in four spacetime dimensions (rather than superspace) was that of Sergio Ferrara, Daniel Freedman, and Peter van Nieuwenhuizen, in 1976 [19]. This was really a toy model, describing only the graviton and its superpartner, the gravitino. Eugéne Cremmer, Bernard Julia, and Jöel Scherk [13] later discovered supergravity in eleven dimensions (the first supergravity theory to be formulated in $D > 4$—cf. [54, p. 135]). They left open two tasks at the close of their paper: (1) the reduction down to four dimensions, and (2) the reduction to ten dimensions to recover the zero slope limit of the closed string dual model. The former was achieved by Cremmer and Julia in [14] while the later was completed by Chamseddine [10].

One can see the rise in research of supergravity in Fig. 8.1, beginning following the first papers in 1975, until 1988 at which point superstring theory was more secure and supergravity research became far less popular, due to the growing realisation that it would remain forever non-renormalizable because of the local degrees of freedom.

An important part of the supergravity approach was the fact that it too introduced additional space-time dimensions, in this case 11. One of the tools that it brought in was the old Kaluza-Klein mechanism for dimensional reduction.[2] It was discovered in this context that the maximum number of spacetime dimensions for a supersymmetric theory with spin-2 particles is 11.[3] Recall that the Kaluza-Klein mechanism

---

[2] Freund and Rubin explain the procedure thus: "One essentially accounts for a seemingly complicated theory with many force and matter fields in a low-dimensional space-time, in terms of a simple geometrical theory in a space-time of higher dimensionality. The extra dimensions are assumed to compactify with a very small characteristic size" [24, p. 233].

[3] The reason for this restriction is that higher dimensions would lead to massless particles of spins greater than 2, and therefore a conflict with quantum field theory.

Fig. 8.1 Citation graph showing the number of papers published on supergravity between 1975 and 1988. There is a clearly a spike in the early 1980s, which coincides with a tandem increase in the study of string theories. *Image source* Thompson-Reuters, *Web of Science*

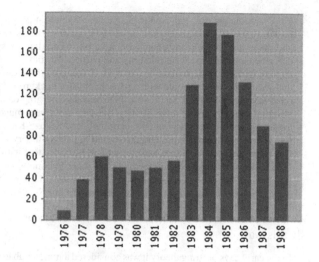

was part of a programme to unify electromagnetism and gravity. The idea was to extend spacetime to five dimensions with the electric charge being associated with the 5th momentum component in this higher-dimensional spacetime. The electromagnetic potential is associated with the $g_{5\mu}$ component of the 5×5 metric tensor. Geometrically, the picture involves compactifying the 5th coordinate onto a circle, from which one gets the one-dimensional $U(1)$ symmetry group associated with electromagnetism.[4] Part of the problem with this kind of 'geometrical unification' in its earliest phase of development was the problem of incorporating spinorial matter, providing it with a geometrical foundation. Of course, in uniting bosons and fermions supersymmetry overcomes this problem.

This is a clear example of what I had in mind in the previous chapter and above when I discussed the modification of the research landscape that was required in order to make non-hadronic string theory better adapted. In general, in fact, there was a greater disposition towards unification (even 'grand' unification) as a legitimate way of doing physics. John Schwarz describes some of this emerging alteration in the research environment:

> The point I wanted to make first was that in this period, 1980 through 1984, Michael and I published quite a few papers, and in each case I was quite excited about the results, and I

---

[4] In the context of grand unified theories, where we require $SU(3) \otimes SU(2) \otimes U(1)$ in addition to gravity, the method is more of the same: $SU(2)$ is associated with the two-dimensional space of S^2 (the sphere) and $SU(3)$ is associated with the four-dimensional space $\mathbb{C}P^2$ (complex projective space). The number 11 arises from this by simple dimension counting: $1 + 2 + 4 + 4 = 11$ (with the second instance of 4 coming from the usual spacetime dimensions). On the number 11 Abdus Salam once wrote that "as a number, [it] has the merit, that to my knowledge, nothing mystical has ever been associated with it" [53, p. 143]. Werner Nahm [47] was responsible for proving that $D = 10$ is the highest number of spacetime dimensions posessing supersymmetry representations with spins of 1 or lower and $D = 11$ is the highest number for supergravity theories, with spins of 2 or lower.

think he probably was, too. And in each case, we felt that people would now get interested, because they could see how exciting the subject was. But there was still just no reaction, and the culture had already changed quite a bit from what I described in the early days: I described a culture in which relativity and particle physics did not communicate. That had already changed, independent of our work. I mentioned the development of supergravity; that was a case where people with a particle physics background had been building general relativity into the theory. So by now there was an emerging community which was sort of bridging relativity and particle physics. But they were very much committed to working in the framework of what they called supergravity and Kaluza-Klein theory. And so the bottom line is that this was quantum field theory and not string theory. So on that basis I felt that it was misconceived, because it wasn't going to be consistent with quantum mechanics. And I interacted with these people, because I had more in common with them than anyone else; we spoke at the same meetings and so on. They were all aware of what Michael and I were doing, but none of them got particularly interested in it.[5]

Wolfgang Lerche, Dieter Lüst, and Bert Schellekens put the point a little more explicitly this way:

> In the early days of string theory it was considered a major embarrassment when it turned out that string theories could only be formulated consistently in 26 or 10 dimensions. The revival of interest in the subject was for a small, but not unimportant part due to a change in attitude towards extra dimensions, namely the acceptance of the idea that they can be compactified. This idea dates back to the first half of this century, but received serious attention only during the last ten years, after the end of the first string era. When strings were reconsidered one initially attempted to compactify their field theory limits, with the help of the technology developed during the past decade. More recently the attention has slowly shifted towards more "stringy" compactifications [45, p. 477].

Likewise, Yuri Manin makes a similar point about the restructuring that had to take place in physics in order for superstring theory (with its previously outlandish extra dimensions) to find receptive minds:

> This romantic leap of twenty orders of magnitude makes the situation in modern theoretical physics extremely bizarre and poses new problems of relating theory to phenomenology of low energy (former high energy physics). Psychologically this leap was prepared for by a decade of Grand Unification models based upon Yang-Mills fields with a large gauge group and a bold extrapolation of the high-energy behaviour of coupling constants of strong and electro-weak interactions [46, p. 60].

In fact, at the time, there were some very grand claims made on behalf of supergravity, not so distinct from the later claims made of string theory. For example, Abdus Salam spoke of the theory as realising Einstein's dream about a unified field theory:

> A supersymmetric theory of gravity would thus realize Einstein's dream of elevating the 'base wood' of (fermion) matter on the right-hand side of his equation $R_{\mu\nu} - \frac{1}{2}g_{\mu\nu}R = T_{\mu\nu}$, to the status of (spin-2 bosonic) 'marble' of gravity on the left-hand side [53, p. 144].

But though the supergravity programme was pursued with zeal, there remained the basic problem of its ultraviolet behaviour, despite the slightly improved behaviour

---

[5] Interview with John H. Schwarz, by Sara Lippincott. Pasadena, California, July 21 and 26, 2000. Oral History Project, California Institute of Technology Archives. Retrieved [2nd jan, 2012] from the World Wide Web: http://resolver.caltech.edu/CaltechOH:OH_Schwarz_J.

relative to standard quantizations of general relativity (thanks to the cancellations brought about by supersymmetry[6]). According to superstring theorists of the time, and eventually the supergravity theorists, the problems would be ineradicable so long as the theory understood its fundamental objects to be point-like quanta, with the problematic vertices that go along with that picture (local quantum theory): it is well-behaved in the infrared, but not so well-behaved in the ultraviolet.[7] This became widely accepted, and one can find Michael Duff writing in 1988 that "[m]any of the supergravity theories that we used to study a few years ago are now known to be merely the field theory limit of an underlying string theory" [17, p. 189].[8]

8.2 Polyakov's New Perspective on String Theories

An important development in the understanding of the mathematical properties of string theories came from Alexander Polyakov's application of functional integration to string models in 1981. This opened up a connection between string perturbation theory and the classical theory of Riemann surfaces so that the space of states of a string is given by the space of states of a Riemann surface. One can then use the symmetry properties of string theory (specifically diffeomorphism and conformal invariance) to 'count' the number of string states using Riemann surfaces: there will be equivalence of states when Riemann surfaces are conformally identical and diffeomorphic.[9] This factoring out of multiply represented states (giving the classical

---

[6] This basic feature of supersymmetry was shown by Bruno Zumino in 1974 (see [69]). The mechanism is, as he puts it, the "compensations among contributions involving different fields of the supermultiplet" [70, p. 535]. This induces cancellations of divergences among Feynman diagrams. (Lars Brink has claimed that Zumino's interest in dual models was stimulated by his attendance of a talk by Jöel Scherk on his zero-slope limit idea: [8, p. 475].

[7] For an excellent, near-exhaustive collection of the formative papers in supergravity, from its origins until 1985 see the 2 volume collection *Supergravities in Diverse Dimensions*, edited by Salam and Sezgin [54].

[8] The title of this article, "Supermembranes: The First Fifteen Days", is clearly an amusing reference to John Schwarz's two-volume edited collection *Superstrings: The First Fifteen Years*. In the early to mid-1990s the eleven-dimensional form of supergravity was found to be more intimately related to superstrings, and formed a key component of the transformation in the understanding of string theory as part of a deeper structure known as M-theory. Schwarz later expressed a similar view, that supergravity theories were in some sense *secondary* to strings: "I always felt that the supergravity theories didn't really make much sense by themselves, because they weren't consistent theories. Only inside string theory, where the quantum mechanics was under control, did this really make sense" (Interview with John H. Schwarz, by Sara Lippincott. Pasadena, California, July 21 and 26, 2000. Oral History Project, California Institute of Technology Archives. Retrieved [2nd jan, 2012] from the World Wide Web: http://resolver.caltech.edu/CaltechOH:OH_Schwarz_J).

[9] Quotienting out the group of diffeomorphisms (connected to the identity) and conformal transformations from the domain space of integration (in the evaluation of the partition function) implies that the integration is over *Teichmüller space*. Note, however, that this is the case only for surfaces of genus 0—the family covered by Polyakov. In general, for genus $g \neq 0$, one must also quotient out by the action of the mapping class group, which involves performing the integration over moduli

moduli space of Riemann) is necessary for doing the path integral to make sure one is not over-counting (much as was the case with DHS duality). The new perspective provided by Polyakov opened up several new possibilities, include the applicability of standard field theoretic tools (which vastly improved the prospects for 'selling' string theory to the wider particle physics community[10]), as well as opening up alternative possibilities for studying string theory in curved target spaces, thus making the link to general relativity more apparent (this time improving the prospects for selling string theory to general relativists).

Polyakov's primary interest was non-Abelian gauge theories and a host of curious topological features that arise therein (such as instantons[11]). Thus, string theory was employed, more or less, as a tool by Polyakov to study QCD (see [51, p. 248]). His view of strings was firmly rooted in the QCD-string concept: "[e]lementary excitations in gauge theories are formed by the flux lines (closed in the absence of charges) and the time development of these lines forms the world surfaces" [48, p. 207]. The proposal was then to extend the study of sums over random paths, to sums over random surfaces. Deep connections with mathematics (and statistical physics) were forged from this initial step. Not least the development of conformal field theory. This was studied in terms of the kinds of 2D Riemann surfaces corresponding to the worldsheets traced out by strings. It was suggested, by Daniel Friedan and Stephen Shenker [26], that the two-dimensional worldsheet picture was the fundamental one, encoding the important conformal structure of the theory, with spacetime concepts being derivative.[12] (Gomez and Ruiz-Altaba later called string theory "the mother of conformal field theories" [32, p. 6].)

One of the crucial aspects of Polyakov's work, that has increased in importance in recent years, is the notion that one can *modify* the critical dimension by adding what is called a Liouville mode to 'embody' the central charge. This work takes place in the

---

(Footnote 9 continued)
space of parameters describing deformations of the surface's conformal structure. (see [42] for an elementary discussion of these issues.)

[10] As did the fact that Polyakov's approach forged new links between string theoretic ideas and other areas of physics, such as statistical physics and the physics of low-dimensional systems.

[11] The terminology of "instanton" was due to 't Hooft, though he notes that the editor of the journal, *Physical Review*, to which he submitted the paper with the newly coined term did not like it, suggesting in its place: "non-abelian solitonic pseudo particle solution" (interview with the author; see also: [62, p. 299]).

[12] As Goddard notes [29, p. 331], there is something a little atavistic about this approach since one of the key early moves in superstring theory (as we have seen) was the geometrical description involving this two-dimensional worldsheet embedded in an ambient spacetime. However, as Friedan and Shenker make clear, their aim is to avoid hitching string theory to some particular spacetime background, which they see as "unnatural" given the claim of string theory to provide a solution to the problem of quantum gravity: "[t]he structure of spacetime should be a property of the ground state" [26, p. 287]—this was perhaps a dig at the existing attempts to construct a string field theory (which involved a fixed spacetime metric or the light-cone gauge). Interestingly, Friedan and Shenker close their paper by pointing to similarities between their approach and the original bootstrap approach of Chew, though in their case with the constraints given in terms of consistency conditions on the partition function. The approach involved the notion of a "Universal Moduli Space" which they hoped would reduce down to a unique theory by imposing the right constraints.

context of the string theoretical description of the gauge fields found in Yang-Mills theories. Polyakov argued that there is an equivalence between such string theories and a Liouville theory, possessing a curved fifth dimension. In more detail, the critical dimension is usually understood to be required to cancel the conformal anomaly. Polyakov noted that it can also be canceled 'subcritically' by introducing a 'dilaton' worldsheet field as a Liouville mode (i.e., satisfying the Liouville equations). The role of Weyl invariance is of 'critical' importance (if you'll pardon the pun...). The violation of Weyl invariance at the quantum level, when $c \neq$ critical, can thus be cancelled.

As he notes, at $D = 26$ "one could quantize the theory without bothering about the conformal anomaly, as has been done in dual models" but, for "$D < 26$ in order to get proper quantization we must examine the quantum Liouville theory" [48, p. 210]. For example, for the genus 0 case, the partition function Z_0 is given by the functional integral:

$$Z_0 = \int \mathscr{D}\phi(\sigma, \tau) \exp\left(-\left(\frac{26 - d}{48\pi}\right) \int d^2(\sigma, \tau) \frac{1}{2}(\partial_\mu \phi)^2 + m^2 e^\phi)\right). \quad (8.1)$$

The theory that results when $D < 26$ (that is, for the *non-critical* scenario) resembles a two-dimensional $\phi(\sigma, \tau)$ field theory, that is both renormalizable and integrable. For string theory in four-dimensions (non-critical string theory), it becomes necessary to solve the Liouville theory to get the scattering amplitudes out, where the Liouville action is the exponent from above: $-\left(\frac{26-d}{48\pi}\right) \int d^2(\sigma, \tau) \frac{1}{2}(\partial_\mu \phi)^2 + m^2 e^\phi)$.

Polyakov employs a slightly modified version of the action constructed by Brink et al. [6],[13] which itself extends the original Nambu-Gōto action:

$$F = \int (\sqrt{g} g^{ab} \partial_a x \partial_b x + \mu \sqrt{g}) + \cdots d^2 \xi. \quad (8.2)$$

What Polyakov found was that when quantized the theory 'grows' an additional dimension. As he puts it: "This result implies that the natural habitat for the random surface in D-dimensional x-space is $D + 1$ dimensional (x, ψ) space" [50, p. 321].

By focusing squarely on the conformal invariance of string theory Polyakov's work suggested a new way of viewing string theories, providing much-needed injection of interest. In 1982, Tataru-Mihai wrote that "[a]fter a period of relative lethargy, the string theory experienced a sudden resurrection due to Polyakov's papers" [60, p. 80]. Hence, the perception that string theory lay dormant until 1984 is not wholly accurate.

---

[13] Deser and Zumino independently came up with an action with the same properties [16]. Actually, "independently" here perhaps needs qualifying a little: as Brink mentions in his recollections [8, pp. 479–481], there were multiple interactions between the two groups.

8.3 Green and Schwarz's New Superstring Theories

Michael Green and John Schwarz had begun collaborating in 1979, initially focusing on space-time supersymmetry within the GSO-projected RNS superstring theory. This was developed into a complete theory of supersymmetric strings (superstrings) [34–37]. In one of their first papers on the subject [34] they work in various motivations for studying string theory into the text, keen to make it clear that string theory was not just a mathematical exercise, but a framework with potential physical applications. This led to the classification of superstring theories still in use today (i.e., Type I, Type IIA, Type IIB), and included the development of entirely new theories.

The GSO formalism was for them not transparent enough to provide a firm foundation for the study of issues of divergences and the explanation and firm proof of the supersymmetry (offering strong evidence, but nothing conclusive, for the latter). They wished to provide "another step towards a proof of the conjectured supersymmetry of the ten-dimensional string theory" [34, p. 504]. Their initial approach involved the construction an alternative spectrum-generating algebra, which allows them to construct the supersymmetry operator transforming physical open-string states, "and prove that it is a spinor satisfying the requisite anticommutation relations" (ibid.).

By using a new set of oscillators and a light-cone gauge formalism, they were able to construct a physical Fock space (with no need to project out unphysical states). This made the computation of loop amplitudes much more tractable. Moreover, manifest spacetime supersymmetry (absent in the GSO approach) can be achieved in a light cone formalism—though this itself comes at the expense of manifest Lorentz invariance, itself then prompting the search for a covariant approach.

The method involves the introduction of a new set coordinates $\theta^A(\sigma, \tau)$ on the worldsheet, in addition to the standard $X^\mu(\sigma, \tau)$. The $\theta^A(\sigma, \tau)$ are Grassmann coordinates transforming as spacetime spinors (connected to the fermionic degrees of freedom) and worldsheet scalars (where $A = 1, 2$ for $N = 2$ theories).[14]

The kinds of supersymmetry algebra possible in ten dimensional spacetime, and with $J \leq 2$, leads to a classification scheme for the various (consistent) superstring theories into (at this stage) Types I and II. In type I theories there are 16 supercharges, while in type II theories there are 32. Type II theories further branch into type kinds, IIA and IIB. IIB theories have supercharges of the same chirality—as do Type I theories. IIA supercharges are partitioned into a pair of classes of both kinds of chirality, with 16 of each. This system of classification had been yet another gift from supergravity, though, as mentioned, the naming convention had been applied by Green and Schwarz to the superstring case in the above papers. Their focus was on Type II theories, which they were able to prove 1-loop finite in 1981 [33]—the Type I theory was known to be 1-loop renormalizable.

---

[14] The idea of employing Grassmann variables as a useful tool in the study of supersymmetric theories was suggested earlier, by Salam and Strathdee [52].

In their reminiscences about the advances achieved in their early collaboration, Green and Schwarz write[15]:

> Each of these developments persuaded us that string theory had a compelling consistency that spurred us to investigate it further, but they did not seem to arouse much interest in the theory community. It was understood in 1982 that Type I superstring theory is a well-defined ten-dimensional theory at tree level for any $SO(n)$ or $Sp(n)$ gauge group. However, in every case it is chiral (i.e., parity violating) and the open-string (gauge theory) sector is anomalous. Evaluation of a one-loop hexagon diagram exhibits explicit nonconservation of gauge currents, which is a fatal inconsistency. The only hope for consistency was that inclusion of the closed-string (gravitational) sector would cancel this anomaly without introducing new ones. An explicit computation was required to decide for sure.

Much of the initial interest with the newer form of superstrings came from consideration of the loop amplitudes for toroidal compactifications,[16] in which case one has supergravitational limits (as both compact radii and the Regge slope go to zero), which gives $N = 8$ supergravity and $N = 4$ super-Yang-Mills theory. Hence, it was the links to supergravity that were deemed important. It is important to note, also, that not only were the Type I strings anomalous, but the Type II strings were not phenomenologically impressive enough to gather attention. In many ways, then, it is perhaps not surprising that these new forms of superstring theory were not taken up by the community.

8.4   Shelter Island II, Aspen, and Anomalies

All physicists (and historians of physics) know the importance of the first Shelter Island conference. It was here that renormalization theory was presented.[17] Less well-known is the reunion that took place, also at the Rams Head Inn on Shelter Island, this time in 1983. Murray Gell-Mann was chosen to open the discussion and gave an overview of the various topics to come. He describes string theories (and 'super' theories in general) as "a realm where everything is very hopeful, very beautiful, exceedingly promising, probably renormalizable or maybe even completely finite, free of parameters—but with no obvious connection with experiment!" ([28, p. 16]).

Witten's talk in 1983, at Shelter Island II [65], indicated quite clearly that he had been thinking hard about string theories. In considering modifications to field theory (a "drastic modification to Riemannian geometry": p. 267), he notes that the new superstring theories would make "very attractive candidates." In the same place he also refers to string field theory and discusses what their future might bring. Hence,

---

[15] The following is taken from their online article "The Early Years of String Theory at the Aspen Center for Physics," available at: http://www.aspenphys.org/aboutus/history/sciencehistory/stringtheory.html.

[16] Developed as part of a collaboration between Green, Schwarz and Lars Brink [7].

[17] Silvan Schweber's provides a detailed historical treatment in [58]. See also his 2011 Pais Prize Lecture, "Shelter Island Revisited": http://www.aps.org/units/fhp/newsletters/spring2011/schweber.cfm.

even before the anomaly cancellation result, people were beginning to take notice once again. Witten had, of course, already acquired a strong reputation by this stage.

Hence, perhaps the name "revolution" to describe what occurred with the anomaly cancellation is too big a stretch. If anything, the challenge of finding consistent theories in the face of Witten's observation of gravitational anomalies could also be viewed as a 'trigger event'. Motivated by the promise of a finite quantum theory of gravity alone, Peter Freund, Philhal Oh, and James Wheeler considered models for string-induced compactification of space [25]. Their analysis was done in the background of the, still not yet published, results of Witten on gravitational anomalies.

Paul Frampton and Thomas Kephart had also done extensive work on anomalies in Kaluza-Klein theories, and higher-dimensional theories—see, e.g., [22]. They also considered the cancellation of the hexagon anomaly in $D = 10$ superstring theories [21], prior to Green and Schwarz's work on the subject. They write "Only zero-mass fermions contribute to the anomaly when the local symmetry is exact; because of this one may evaluate the anomaly in the zero-slope limit. The open-string theory reduces in this limit to supersymmetric Yang-Mills theory and is hence anomalous for any choice of internal symmetry group" and then add, in a footnote, "The only possible loophole in this argument would be if the infinite sum over massive fermions gives a nonvanishing anomaly precisely cancelling the anomaly of the massless fermions. We regard this as extremely unlikely" ([21, pp. 81–82]).[18]

Still prior to the Green-Schwarz anomaly cancellation result in 1984, Paul Townsend [63] had considered anomaly cancellations, this time in chiral supergravity. Townsend noted that chiral anomalies, of the sort considered by Green and Schwarz can occur in currents coupled to gauge fields in any even dimension so that simple supersymmetric gauge theories in both $d = 6$ (mod 4) and $d = 10$ are afflicted and therefore inconsistent as quantum field theories.[19] The absence of *gravitational* anomalies was also considered, and viewed as a heavily restrictive principle of theory construction, radically reducing the number of quantum field theories able to satisfy the restriction. Townsend had also suggested that compactification onto $K3$[20] allows the theory to maintain its chirality under spontaneous compactification [63, p. 284].

Anomalies were clearly on the agenda in the first years of the 1980s. Superstrings had already made something of a slight comeback, but they were, it has to be said, hardly centre stage. Yet Clavelli and Halprin wrote in the preface to their volume from

---

[18] Hence, the general expectation was strongly against string theory being anomaly free. In the preface to his book *Dual Resonance Models and Superstrings* ([23, pp. 5–6]), Paul Frampton writes that it was Lars Brink, in August 1984, who informed him that Green and Schwarz had shown that $O(32)$ is anomaly free for open superstrings just months after he (Frampton) had announced at the ICTP in Trieste (with Green and Schwarz in the audience) that all open superstrings were potentially anomalous. In fact, Schwarz recalls Frampton projecting a slide depicting a tombstone with the inscription: "type I superstring theory" (private communication).

[19] Given these problems with $D = 10$ supergravity, John Ellis speaks of Green and Schwarz as coming "to the rescue with the superstring" [18, p. 596].

[20] $K3$ stands for 'Kummer's third surface,' a four-dimensional, compact Riemannian manifold with zero isometries but possessing a self-dual Riemann tensor (with SU(2) holonomy)—crucially, $K3$ supports a Ricci flat metric. Candelas and Raine [9] argued that quantum effects might be necessary to force the four-dimensional non-compact manifold to be flat at the Planck scale.

the Lewes String Theory Workshop (6–27 July 1985) that "[i]nterest in the theory of superstrings as a fundamental theory of matter reached near hysterical proportions ... as a consequence of the significant work of Green and Schwarz showing that such theories are anomaly free and probably finite" [12].[21] The result in question—a resolution of the hexagon gauge anomaly in Type I superstring theory—opened up the possibility that realistic and consistent superstring theories could be established that were capable of providing realistic, unified accounts of *all* forces (though not, in fact, via the Type I strings): a 'theory of everything,' according to a common way of understanding that phrase.

As Peter Galison notes, the anomaly cancellation requirement is clearly serving as a strong guiding constraint on the construction of possible theories: "superstring theory can be constrained by the condition that the dreaded anomalies must be cancelled" [27, p. 378]. If a theory passes this requirement then, evidently speaking, the theory has gained some support just as sure as if an empirical test had been passed. However, we must not forget that underlying this constraint there *is* a further empirical principle: that in order to provide a model of physical reality, the theories must be chiral. In this case, it is in parity-violating theories that the perturbative anomalies may occur (the IIA theory is non-anomalous, but is also non-chiral). In other words, the importance of the mathematical constraint of anomaly cancellation to some extent 'piggybacks' on the importance of the physical constraint of chirality.[22]

Anomalies, in the kinds of case that concern us, are a result of mixing symmetries with quantization procedures, including regularization/renormalization techniques.[23] Regularizing divergent integrals is done by introducing a momentum cutoff,

---

[21] In his note in *Nature*, from 1986, on superstrings and supersymmetry, Alvaro De Rújula likewise writes (echoing David Olive's earlier remarks about the mathematical addiction of some dual theorists): "Only two years ago almost all elementary particle physicists considered the subject of 'superstrings' abstruse and irrelevant, perhaps because the number of space-time dimensions in which string theories can be consistently defined is somewhat unrealistic—either 10 or 26 or an astonishing 506. But in September 1984 Michael Green and John Schwarz published a paper on the cure of certain diseases (called anomalies) in these elaborate theories. Overnight, most particle physicists shelved whatever they were thinking about (mainly 'supersymmetry' and 'supergravity') and turned their attention to superstrings" [15, p. 678]. However, De Rújula suggests a cautious approach, which heralds a more general skeptical attitude in various sectors of physics: "So far, none of those fashionable subjects has proved to have any convincing relationship with physical reality, yet they have the irresistible power of addiction. Such gregarious fascination for theories based almost exclusively on faith has never before charmed natural philosophers, by definition" (ibid.). From what I have indicated above, however, there is clearly a large slice of rhetoric in this passage.

[22] Michael Green stresses the importance of chirality (in the four-dimensional world) as a physical constraint on *any* theory which, by implication, forces the higher-dimensional theories generating the lower-energy physics to be chiral also [41, p. 135].

[23] See [20] for an excellent discussion of both the history of the study of anomalies and an account of their role in the interplay between mathematics and physics. (I might note that the front cover of the proceedings volume from the 1983 Shelter Island II meeting has a large picture of the hexagon diagram corresponding to the anomaly that Green and Schwarz later resolved for (certain) superstring theories.)

Fig. 8.2 The closed loop triangle graph from Stephen Adler's analysis of the axial-vector vertex in spinor QED, leading to anomalies. *Image source* [2, p. 2428]

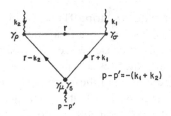

which can spoil a classical symmetry.[24] Sometimes the subsequent renormalization will remove the effects of the cutoff imposed, but when it doesn't then one faces an anomaly. Absence of such anomalies is essential for the consistency of gravitational and gauge theories. Hence, the key question is: 'do the symmetries of the classical theory survive the process of quantization?' Given the standard scheme for understanding symmetries in field theories, due to Noether, according to which conserved quantities (current or charge) are associated with the generators of symmetries, a further question is whether conserved quantities survive too. Probably the most famous example of an anomaly is the Adler-Bell-Jackiw anomaly.[25] Adler showed that the axial-vector vertex in QED faces an anomaly (non-conservation of the axial-vector current, though not the vector current) that arises from the presence of closed loop triangle graphs. For this reason the anomaly is often labeled the 'triangle anomaly'— the triangle graphs constituting the lowest non-trivial order (Fig. 8.2).

The precise nature of the problem has to do with the coupling between gauge bosons and chiral fermions in quantum field theories. Gauge bosons have different interaction strengths relative to left-handed and right-handed fermions. In calculations of scattering involving an incoming gauge boson, decaying into two other gauge bosons via an intermediate state with the structure of a triangle loop of fermions, one will pick up a (non-renormalizable) divergence because of the difference in coupling strength relative to the left and right-handed fermions in the loop. In such cases the theory is said to have a chiral anomaly (in this case a triangle anomaly).

Schwarz[26] puts the same kind of anomaly problem (though involving a hexagon graph) in string theory in the following way:

These theories have the same kinds of Yang-Mills gauge symmetries that you find in the Standard Model of elementary particles, which is one of the attractive features. However,

---

[24] This can also be seen at work in the context of path-integral quantization, where the anomalies arise due to features of the measure, which fails to preserve symmetries of the classical theory.

[25] Due, independently, to Stephen Adler [2] (during a visit to the Cavendish laboratory in Cambridge) and John Bell and Roman Jackiw [4] (working at CERN), in 1969. Adler had originally expected to be able to prove the finiteness of the axial-vector vertex in spinor QED, but this wasn't the case. In their paper, Bell and Jackiw had further argued that the anomaly could be removed by imposing regulators, something Adler argued against, pointing out that the anomaly could not be eliminated without thereby destroying either gauge-invariance, unitarity, or renormalizabilty.

[26] Interview of John Schwarz, by Sara Lippincott, Pasadena, California, July 21 and 26, 2000. (Oral History Project, California Institute of Technology Archives. Retrieved [2nd jan, 2012] from the World Wide Web: http://resolver.caltech.edu/CaltechOH:OH_Schwarz_J).

there was a danger that the quantum corrections would destroy those symmetries. And if they did, it would lead to a mathematical inconsistency of the entire theory. This kind of inconsistency would only arise in a situation where the theory was not left-right symmetrical—not mirror-symmetrical—and there were certain string theories that had that feature and others that didn't. So for example, the Type IIA theory was mirror-symmetrical, so certainly it wasn't going to have these anomalies. But the Type IIB theories and the Type I theories were not mirror-symmetrical, and it appeared as if they would have these anomalies and therefore be inconsistent. So you might say, "Well, why not just live with the IIA?" [...] But the thing is, we knew that nature is not mirror-symmetrical. So having an asymmetrical theory was a good thing, if we could make sense of the theory.

The problem is, more explicitly, that for quantum field theories (with chiral fields) in a special number of dimensions ($2 + 4n$, where n is an integer), the spacetime symmetries were destroyed at the quantum level. The supergravity approaches had faced such gravitational anomalies and, as Schwarz mentions, there were already examples of consistent superstring theories that didn't face the anomalies in question. The Type-IIA theory (of closed, oriented, with $N = 2$ supersymmetry) strings suffered no such fate since it is non-chiral. However, such a theory was not applicable to real-world, low energy scenarios. Moreover, if *some* string theories are inconsistent then one might think that the framework as a whole is thrown into jeopardy, or at least highly questionable. The Type-I theory (of non-orientable, open and closed, with $N = 1$ supersymmetry) strings *did* appear to suffer (Fig. 8.3).

Partly as a result of discussions with Edward Witten,[27] Green and Schwarz had begun to focus on anomalies. During the 1984 summer in Aspen,[28] they had initially focused their efforts on the more general problem of anomaly cancellations in supersymmetric gauge theories coupled to gravity.[29] They had in fact circulated a preprint entitled "Anomaly cancellations in supersymmetric $D = 10$ gauge theory

---

[27] Witten had recently written a paper with Luis Alvarez-Gaumé, "Gravitational Anomalies" [1], which demonstrated that anomalies affected a wider class of field theories than Yang-Mills gauge theories, and could affect theories of gravity too. It was in this paper that the IIB supergravity was shown to be anomaly free, and the expectation was that the same would hold for IIB superstring theory too.

[28] Schwarz had a close link with the Aspen Center for Physics, and had been present every summer but one since its inception. He was in fact the treasurer of the Center between 1982 and 1985. See http://www.aspenphys.org/aboutus/history/sciencehistory/stringtheory.html for Green and Schwarz's reminiscences of string theory at the Aspen Centre for Physics. A video of a lecture by Schwarz on "String Theory at Aspen" can be found here: http://vod.grassrootstv.org/vodcontent/11112.wmv. Dennis Overbye has written a brief article on the history of the Center: "In Aspen, Physics on a High Plane" (*New York Times*, August 28, 2001). A more recent article, written to celebrate the 50th anniversary of the Center, is [64].

[29] Schwarz had been aware of the potential problems posed by the hexagon graphs in 1982. He raises the problem in his *Physics Reports* review paper on superstring theory, stating that "inconsistencies are anticipated for class B gauge groups in SST I" (where SST I is Type I superstring theory), namely the "anomalous breakdown of Lorentz invariance for the hexagon loop diagram" [55, p. 313]—he notes that the hexagon graph in $D = 10$ is analogous to the triangle graph in $D = 4$. In his talk at the 2nd Shelter Island conference, one year later, he had referred to it as a "potentially fatal problem" [56, p. 222]. However, referring to the fact that the anomalies are controlled by short distance behaviour, that string theory is so well equipped for, he goes on to say: "I am optimistic that string theories can be free from bad anomalies" (ibid.).

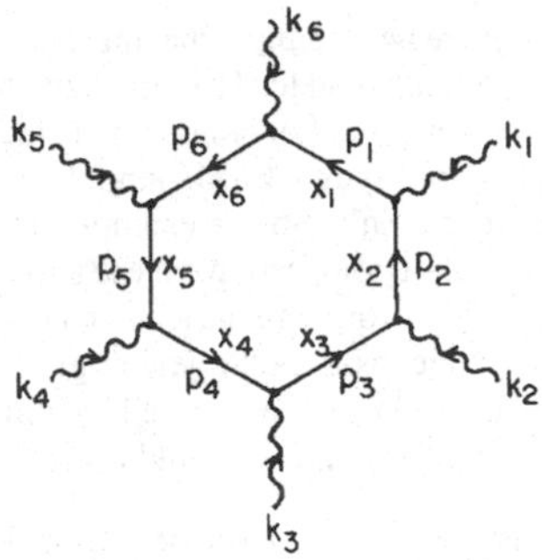

Fig. 8.3 Hexagon loop diagram, with six external particles, responsible for generating a chiral anomaly in Type I string theory. *Image source* [39, p. 101]

require $SO(32)$", quickly withdrawing it and replacing it with the preprint "Anomaly cancellations in supersymmetric gauge theory and superstring theory" ([38]; cf., [37, p. 357]). This latter version included an $E_8 \otimes E_8$ gauge group cancellation result.[30] The anomaly cancellation occurred for two very special gauge groups (both semi-simple Lie groups of dimension 496): $SO(32)$ or $E_8 \otimes E_8$.[31] Green and Schwarz focused on the $SO(32)$ case since the $E_8 \otimes E_8$ appeared intractable to them. While the open string theories could account for $SO(32)$, $E_8 \otimes E_8$ did not yet appear in a string theory and so could be seen as pointing to the existence of an entirely new

---

[30] Schwarz notes that it was during conversations with Dan Friedan and Steve Shenker that they switched from the more recently developed 'Green-Schwarz formalism' (employing the supersymmetric light-cone gauge formalism) for superstrings to the earlier 'Ramond-Neveu-Schwarz formalism' which they had believed to be inferior (Friedan and Shenker are acknowledged in [39]). Jean Thierry-Mieg wrote on how to extend the method employed by Green and Schwarz to $E_8 \otimes E_8$, [61] (the original preprint is [LBL-18464, Oct 1984]—however, Schwarz mentions that, during a talk at Berkeley, he had already informed Thierry-Mieg that he and Green were working out the $E_8 \otimes E_8$ case (private communication)). While Peter Goddard and David Olive [30] were independently studying how toroidal compactification generated non-Abelian symmetries in string theory via the affine Kac-Moody algebras generated by string-like vertex operators, they also uncovered the fact that in 16 dimensions there exist just two even, self-dual lattices: $\Gamma_8 \otimes \Gamma_8$ and Γ_{16} (which correspond, as root lattices, to the $E_8 \otimes E_8$ and $SO(32)$ cases)—see Goddard and Olive's edited collection [31] for an excellent survey of this and related themes.

[31] Having dimension 496 is crucial for the cancellation to occur: 'physically,' one needs 496 left-handed spin 1/2 fields in the matter sector to make the problematic anomaly term vanish—of course, this is 484 more than are required by the standard model (a photon, 3 weak vector bosons, and 8 gluons: cf. [15, p. 678]), and as a result, compactification will end up having to do much of the work in explaining how these 12 gauge bosons fall out. The group E_8 (the largest exceptional Lie group) has dimension 248, so two copies give the desired dimension. (The surplus structure 'beyond the standard model' is still there, of course, and much of the work in the years following the construction of the $E_8 \otimes E_8$ theory involved breaking the symmetries down to match observation.) As Becker, Becker, and Schwarz claimed in their recent string theory textbook, the anomaly analysis also permits the groups $U(1)^{496}$ and $E_8 \otimes U(1)^{248}$ [5, p. 9]. However, these do not seem to be applicable to string theories. Initially, it was believed that $E_8 \otimes E_8$ was not realisable in a string theory. (However, Schwarz informs me that Washington Taylor has since shown that these other groups are not allowed by anomaly cancellation.)

type. We see in the next chapter how the $E_8 \otimes E_8$ ('heterotic') string was constructed to play this role. This turns out to involve closed strings only.[32]

Let us quote at length from Schwarz's personal account[33] of the steps leading to the discovery of the cancellation:

> We were working in Aspen in the summer of 1984. We had to compute a particular amplitude, which we did. So we found the anomaly associated with a certain Feynman diagram of string theory—the so-called hexagon diagram, because it had six lines coming out of it. We got a certain answer. And then we realized that there was a second diagram that would also contribute to the anomaly, which is one that's kind of twisted, like a Möbius strip. And it gave another formula for the anomaly of the same structure as the first one. I remember quite clearly that we were going to a seminar at that time, and before the seminar I mentioned to Michael that maybe these two different contributions to the anomaly might cancel for a particular choice of the symmetry group. And at the end of the seminar—I don't remember what the seminar topic was or who gave it—Michael said to me, "SO(32)." [...]
>
> So, remember, we had this Type I theory that we had defined for an infinite class of symmetry groups. So now the conclusion was that out of this infinite class, only one of them would be consistent—all the others would be inconsistent, because it's only for this one that the anomaly cancels. And that is the correct result. But we just had one piece of evidence for it at that time, so we weren't yet totally convinced that it was not some accident, something which was not meaningful. So we had to do more tests and studies to really nail it down. What we did was to analyze the low-energy approximation to the theory, and in that low-energy approximation we could also study the gravitational anomalies, because these hexagons and Möbius strips hadn't analyzed the gravitational part, just the Yang-Mills part. So with this low-energy approximation, we could also analyze gravitational anomalies using the formulas from this paper of Witten and Alvarez-Gaumé. We just copied the formulas out of their paper and plugged in our numbers, and we saw that it worked beautifully. So we were very excited and started talking to people about it.[34]

Schwarz also describes his impressions of the immediate aftermath of the paper:

---

[32] Note that Green and Schwarz wrote that it "seems likely that $E_8 \times E_8$ superstrings exist" [40, p. 25]. This way of putting it—that they had simply yet to be formulated—amounted to a prediction, and one that would be proven correct within the year.

[33] John Schwarz interview with Sara Lippincott, Pasadena, California, July 21 and 26, 2000. (Oral History Project, California Institute of Technology Archives. Retrieved [2nd jan, 2012] from the World Wide Web: http://resolver.caltech.edu/CaltechOH:OH_Schwarz_J).

[34] In fact, Schwarz recalls that an 'informal' presentation of the results preceded their formal presentation: "[B]efore we had a chance to make any formal presentation of it, the Physics Center had what they called a physics cabaret....physicists acting and having fun for the benefit of other physicists. ... In [a] mid-seventies skit, at some point Murray jumped out of the audience, ran up on the middle of the stage, and said, 'I figured out the theory of everything,' and he starts going on and on and getting louder and louder. And then two guys dressed in white coats came up, grabbed him, and carried him off the stage. [Laughter] Well, there hadn't been such a cabaret for ten years, but now in 1984 they were going to have a second one, and the idea arose to have the same skit again. But Gell-Mann wasn't there. So I was asked whether I would play this role. ... [W]hen my time came at this cabaret, I ran up on the stage and said, "I figured out how to do everything. Based on string theory with a gauge group SO(32), the anomalies cancel! It's all consistent! It's a finite quantum theory of gravity! It explains all the forces!" And then the guys in the white coats came and carried me off. [...] Everyone just assumed it was a spoof [laughter], just like it was ten years earlier. But the funny thing is, that was actually our first announcement of our results."

[B]efore we even finished writing it up, we got a phone call from Ed Witten saying that he had heard from people who had been in Aspen—he hadn't been in Aspen himself—that we had a result on cancelling anomalies. And he asked if we could show him our work. So we had a draft of our manuscript at that point, and we sent it to him by *Fed Ex* ... and he had it the next day. And we were told that the following day everyone in Princeton University and at the Institute for Advanced Study, all the theoretical physicists (and there were a large number of them) were working on this.

So overnight it became a major industry ... at least in Princeton—and very soon in the rest of the world. It was kind of strange, because for so many years we were publishing our results and nobody cared. Then all of a sudden everyone was extremely interested. It went from one extreme to the other: the extreme of nobody taking it seriously, to the other extreme—which I thought was just as misconceived—of many people thinking that we were very close to describing all the experimental data. And I knew it wasn't that easy [laughter], so I wasn't that optimistic, even though I thought we were on the right track. I think most people have a more sober appraisal today than they did then, after fifteen-plus years of additional struggle and still not being there. People recognize that it's not such an easy problem.

However, as I suggested above, it wasn't so much the cancellation result itself that was of high importance—though it was in fact very important, and part of an ongoing investigation by many other physicists—rather, it was the *conditions* of the cancellation, which invoked interesting gauge groups that particle physicists could work with. As Witten pointed out in a paper published shortly after Green and Schwarz's first anomaly cancellation paper, $E_8 \otimes E_8$ can accommodate low energy physics ("physics as we know it": [66, p. 355]) since it has a maximal $E_6 \otimes SU(3)$ subgroup which enables one to make realistic models.[35]

There were various challenges to be met in the aftermath of Green and Schwarz's result. In particular, the challenge to find a consistent superstring theory which possesses as $E_8 \otimes E_8$ gauge group. Also important was the task of finding

---

[35] Witten also speculates on the possible shadow-like world associated with the other E_8: "it is amusing to speculate that there may be another low energy world based on the second E_8. The two sectors communicate only gravitationally. If the symmetry between the two E_8's is unbroken, it may be that half the stars in the vicinity of the sun are invisible to us, along with half the mass in the galactic disk" [66, p. 355]—the cosmological implications of this shadow matter were discussed by Kolb et al. [43]; Schwarz later speculated that the shadow matter, in potentially accounting for half the mass of the universe, might constitute an "important ingredient in the solution of the dark-matter problem" [57, p. 275]. Much has been made about the fact that Witten emphasized the importance of the result to the rest of the particle physics community, almost as if that community had no decision-making power of its own—see, for example, Lee Smolin's comments in his book *The Trouble with Physics* [59, pp. 274–275] and Peter Woit's comments in his book *Not Even Wrong* [68, pp. 150–151]. However, the fact, as I've indicated above, that the cancellation of gauge and gravitational anomalies was being pursued *prior* to Green and Schwarz's announcement (and Witten's communication of the result in Princeton), is evidence enough that many particle physicists were already on the anomaly cancellation 'bandwagon' if not yet the superstring bandwagon of Green and Schwarz. (It is also worth mentioning that David Gross was at Princeton at the time too, and must also generated additional enthusiasm for the theory.) Anomaly cancellation is strong constraint on theory-building, and given that it was resolved in a way that pointed towards gauge groups that were know to have desirable properties, then there was a *very good reason* for physicists to suddenly focus on superstring theory.

phenomenologically resourceful compact spaces.[36] Dealing with these challenges, and the new challenges they throw up in their wake, occupied string theorists for the rest of the decade, which we turn to in the next chapter.

8.5 Summary

We have shown that increased interest in string theory had already begun prior to the anomaly cancellation results often credited with igniting the first superstring revolution. A tandem increase of interest in quantum gravity and grand unified theories (especially supergravity) was partly responsible for this, which, I argued, prepared the ground, leaving it fertile for the anomaly cancellation result to have the impact it did (or, at least, massively amplifying it). Not to mention the ongoing programme of investigation of anomalies at the same time. However, anomaly cancellation was a vitally important constraint on theory construction, and served to severely reduce the number of possible theories consistent with it. That string theories were by this stage understood to include gravitational and gauge theories, meant that there was the potential for mixed anomalies, which are even more highly restrictive. Finding theories involving cancellations given such severe restrictions is an obvious cause for celebration. The period following the news of these anomaly-free theories therefore saw a dramatic rise in publications relating to string theory, as the details were worked out.

References

1. Alvarez-Gaumé, L., & Witten, E. (1984). Gravitational anomalies. *Nuclear Physics, B234*(2), 269–330.
2. Adler, S. (1969). Axial-vector vertex in spinor electrodynamics. *Physical Review, 177*, 2426–2438.
3. Ashrafi, B., Hall, K., & Schweber, S. (2003). Interview with Alexander Polyakov, 6 February 2003. Retrieved from http://authors.library.caltech.edu/5456/1/hrst.mit.edu/hrs/renormalization/Wilson/Wilson3.htm.
4. Bell, J. S., & Jackiw, R. (1969). A PCAC puzzle: $\pi^0 \to \gamma\gamma$ in the σ-model. *Il Nuovo Cimento, 51*, 47–61.
5. Becker, K., Becker, M., & Schwarz, J. (2007). *String theory and M-theory: A modern introduction*. Cambridge: Cambridge University Press.
6. Brink, L., di Vecchia, P., & Howe, P. (1976). A locally supersymmetric and reparametrization invariant action for the spinning string. *Physics Letters B, 65*(5), 471–474.
7. Brink, L., Green, M., & Schwarz, J. (1982). $N = 4$ Yang-Mills and $N = 8$ supergravity as limits of string theories. *Nuclear Physics, B198*, 474–492.

---

[36] A further task pursued by Witten was to test the perturbatively non-anomalous string theories for the existence of *global* anomalies—that is, those not continuously connected to the identity. He was able to prove that the $SO(32)$ and $E_8 \otimes E_8$ theories are both free from such anomalies (and likewise for the chiral $N = 2$ theory)—see [67].

8. Brink, L. (2012). From the Nambu-Goto to the σ-model action. In A. Capelli et al. (Eds.), *The birth of string theory* (pp. 474–483). Cambridge: Cambridge University Press.
9. Candelas, P., & Raine, D. J. (1984). Compactification and supersymmetry in $D = 10$ supergravity. *Nuclear Physics, B248*, 415–422.
10. Chamseddine, A. H. (1981). $N = 4$ supergravity coupled to $N = 4$ matter and hidden symmetries. *Nuclear Physics, B185*, 403–415.
11. Chew, G. F. (1970). Hadron bootstrap: Triumph or frustration? *Physics Today, 23*(10), 23–28.
12. Clavelli, L., & Halprin, A. (Eds.). (1986). *Lewes string theory workshop*. Singapore: World Scientific.
13. Cremmer, E., Julia, B., & Scherk, J. (1978). Supergravity theory in 11 dimensions. *Physics Letters, 76B*(4), 409–412.
14. Cremmer, E., & Julia, B. (1978). The $N = 8$ supergravity. I. The lagrangian. *Physics Letters, 80B*, 48–51.
15. De Rújula, A. (1986). Superstrings and supersymmetry. *Nature, 320*, 678.
16. Deser, S., & Zumino, B. (1976). Complete action for spinning strings. *Physics Letters B, 65*(4), 369–373.
17. Duff, M. J. (1988). Supermembranes: The first fifteen years. *Classical and Quantum Gravity, 5*, 189–205.
18. Ellis, J. (1986). The superstring: Theory of everything, or of nothing? *Nature, 323*, 595–598.
19. Ferrara, S., Freedman, D., & van Nieuwenhuizen, P. (1976). Progress toward a theory of supergravity. *Physical Review D, 13*(12), 3214–3218.
20. Fine, D., & Fine, A. (1997). Gauge theory, anomalies and global geometry: The interplay of physics and aathematics. *Studies in the History and Philosophy of Modern Physics, 28*(3), 307–323.
21. Frampton, P., & Kephart, T. W. (1983). Cancelling the hexagon anomaly. *Physics Letters, 131B*, 80–82.
22. Frampton, P., & Kephart, T. W. (1983). Explicit evaluation of anomalies in higher dimensions. *Physical Review Letters, 50*, 1343–1346.
23. Frampton, P. (1986). *Dual resonance models and superstrings*. Singapore: World Scientific.
24. Freund, P. G. O., & Rubin, M. A. (1980). Dynamics of dimensional reduction. *Physics Letters, 97B*, 233–235.
25. Freund, P. G. O., Oh, P., & Wheeler, J. T. (1984). String-induced space compactification. *Nuclear Physics, B246*, 371–380.
26. Friedan, D., & Shenker, S. (1986). The integrable analytic geometry of quantum string. *Physics Letters B, 175*(3), 287–296.
27. Galison, P. (1995). Theory bound and unbound: Superstrings and experiments. In F. Weinert (Ed.), *Laws of nature: Essays on the philosophical, scientific, and historical dimensions* (pp. 369–408). Berlin: Walter de Gruyter.
28. Gell-Mann, M. (1985). From renormalizability to calculability? In R. Jackiw et al. (Eds.), *Shelter island II* (pp. 3–23). Cambridge: MIT Press.
29. Goddard, P. (1989). Gauge symmetry in string theory. *Philosophical Transactions of the Royal Society of London A, 329*(1605), 329–342.
30. Goddard, P., & Olive, D. I. (1985). Algebras, lattices, and strings. In J. Lepowsky, S. Mandelstam, & I. Singer (Eds.), *Vertex operators in mathematics and physics* (pp. 51–96). New York: MRSI, Publication No. 3. Springer-Verlag.
31. Goddard, P., & Olive, D. I. (1988). *Kac-Moody and Virasoro algebras: A reprint volume for physicists*. Singapore: World Scientific.
32. Gomez, C., & Ruiz-Altaba, M. (1992). From dual amplitudes to non-critical strings: A brief review. *Rivista del Nuovo Cimento, 16*(1), 1–124.
33. Green, M. B., & Schwarz, J. H. (1981). Supersymmetrical dual string theories and their field theory limits: A review. *Surveys in High Energy Physics, 3*(3), 127–160.
34. Green, M. B., & Schwarz, J. H. (1981). Supersymmetrical dual string theory. *Nuclear Physics, B181*, 502–530.

35. Green, M. B., & Schwarz, J. H. (1982). Supersymmetrical dual string theory. 2. Vertices and trees. *Nuclear Physics, B198*, 252–268.
36. Green, M. B., & Schwarz, J. H. (1982). Supersymmetrical dual string theory. 3. Loops and renormalization. *Nuclear Physics, B198*, 441–460.
37. Green, M. B., & Schwarz, J. H. (1982). Supersymmetrical string theories. *Physics Letters, B109*, 444–448.
38. Green, M. B., & Schwarz, J. (1984). Anomaly cancellations in supersymmetric $D = 10$ gauge theory and superstring theory. *Physics Letters, 149B*(1,2,3), 117–122.
39. Green, M. B., & Schwarz, J. (1985). The hexagon gauge anomaly In type 1 superstring theory. *Nuclear Physics, B255*(1), 93–114.
40. Green, M. B., & Schwarz, J. (1985). Infinity cancellations in $SO(32)$ superstring theory. *Physics Letters, 151B*(1), 21–25.
41. Green, M. B. (1986). Superstrings and the unification of forces and particles. In H. J. de Vega & N. Sánchez (Eds.), *Field theory, quantum gravity, and strings: Proceedings, Meudon and Paris VI, France 1984/5* (pp. 134–155). Berlin: Springer.
42. Hatfield, B. F. (1987). Introduction to quantum field theory, path integrals, and strings. In S. -T. Yau (Ed.), *Mathematical aspects of string theory* (pp. 1–12). Singapore: World Scientific.
43. Kolb, E. W., Seckel, D., & Turner, M. S. (1985). The shadow world of superstring theories. *Nature, 314*, 414–419.
44. Kuhn, T. S. (1996). *The structure of scientific revolutions*. London: University of Chicago Press.
45. Lerche, W., Lüst, D., & Schellekens, A. N. (1987). Chiral four-dimensional heterotic strings from self-dual lattices. *Nuclear Physics, B287*, 477–507.
46. Manin, Y. (1989). Strings. *Mathematical Intelligencer, 11*(2), 59–65.
47. Nahm, W. (1978). Supersymmetries and their representations. *Nuclear Physics, B135*(1), 149–166.
48. Polyakov, A. M. (1981). Quantum geometry of bosonic strings. *Physics Letters, 103B*(3), 207–210.
49. Polyakov, A. M. (1981). Quantum geometry of fermionic strings. *Physics Letters, 103B*(3), 211–213.
50. Polyakov, A. M. (1997). A view from the island. In L. Hoddeson et al. (Eds.), *The rise of the standard model: Particle physics in the 1960s and 1970s* (pp. 243–249). New York: Cambridge University Press.
51. Polyakov, A. M. (2005). Confinement and liberation. In G. 't Hooft (Ed.), *50 years of Yang-Mills theory* (pp. 311–329). Singapore: World Scientific.
52. Salam, A., & Strathdee, J. (1974). Super-symmetry and non-abelian gauges. *Physics Letters, B51*(4), 353–355.
53. Salam, A. (1982). Gauge interactions, elementarity and superunification. *Philosophical Transactions of the Royal Society of London A, 304*(1482), 135–153.
54. Salam, A., & Sezgin, E. (Eds.). (1985). *Supergravities in diverse dimensions* (Vol. 2). North Holland: World Scientific.
55. Schwarz, J. H. (1982). Superstring theory. *Physics Reports, 89*(3), 223–322.
56. Schwarz, J. H. (1983). A brief survey of superstring theory. In R. Jackiw et al. (Eds.), *Shelter Island II* (pp. 220–226). Cambridge: MIT Press.
57. Schwarz, J. H. (1987). Superstrings–An overview. In L. Durand (Ed.), *Second Aspen winter school on physics* (pp. 269–276). New York: New York Academy of Sciences.
58. Schweber, S. S. (1986). Shelter island, Pocono, and Oldstone: The emergence of American quantum electrodynamics after World War II. *Osiris, 2*, 265–302.
59. Smolin, L. (2006). *The trouble with physics: The rise of string theory, the fall of science, and what comes next*. New Yrok: Houghton Mifflin Books.
60. Tataru-Mihai, P. (1982). Polyakov's string theory and Teichmüller spaces. *Il Nuovo Cimento, 72*(1), 80–86.
61. Thierry-Mieg, J. (1984). Remarks concerning the $E_8 \times E_8$ and $D16$ string theories. *Physics Letters B, 156*, 199–202.
62. 't Hooft, G. (1993). *Under the spell of the gauge principle*. Singapore: World Scientific.

63. Townsend, P. K. (1984). A new anomaly free chiral supergravity theory from compactification on $K3$. *Physics Letters*, *139B*(4), 283–287.
64. Turner, M. S. (2012). History: Aspen physics turns 50. *Nature*, *486*, 315–317.
65. Witten, E. (1985). Fermion quantum numbers in Kaluza-Klein theory. In R. Jackiw et al. (Eds.), *Shelter Island II* (pp. 227–277). Cambridge: MIT Press.
66. Witten, E. (1985). Some properties of $O(32)$ superstrings. *Physics Letters*, *149B*(4,5), 351–356.
67. Witten, E. (1985). Global gravitational anomalies. *Communications in Mathematical Physics*, *100*, 197–229.
68. Woit, P. (2006). *Not even wrong: The failure of string theory and the search for unity in physical law*. New York: Basic Books.
69. Zumino, B. (1974). *Fermi-Bose supersymmetry (supergauge symmetry in four dimensions)*. CERN preprint TH.1901 Retrieved from http://cds.cern.ch/record/417139/files/CM-P00060488.pdf.
70. Zumino, B. (1975). Supersymmetry and the vacuum. *Nuclear Physics*, *B89*, 535–546.

Part III
String Theory Becomes Super: 1985–1995

Chapter 9
Superstring Theory and the Real World

> *There are certainly some indications that our colleagues may*
> *have found the "Holy Grail" of fundamental physics.*
> Murray Gell-Mann, 1988

The demonstration of anomaly cancellation itself was a significant event, but as the previous chapter argued, perhaps of more significance was the *manner* in which the cancellation occurred. The gauge group (and the particle content) of the standard model of particle physics are encoded (albeit with much surplus mathematical lumber) in the group E_8. That the anomalies canceled for $E_8 \otimes E_8$ (ignoring the 'shadow' E_8 factor) points to the possibility of constructing physically realistic string models of experimentally accessible low energy physics.[1] In other words, the results promised a genuine theory of all interactions (including quantum gravity) that is (internally) self-consistent and (externally) consistent with experiments and observations: the holy grail of physics, as Gell-Mann puts it.

If the 1984 paper by Green and Schwarz could be said to have triggered a string revolution, as many claim, then, on the basis of impact, so too could the paper, introducing the heterotic string ("a new type of quantum string theory"), by the so-called 'Princeton string quartet' (a term coined by John Schwarz to denote its authors: David Gross, Jeffrey Harvey, Emil Martinec, and Ryan Rohm): [74].[2]

---

[1] Of course the aim is to break this symmetry at lower energy scales (and dimensions), jettisoning much of the apparent surplus structure of $E_8 \otimes E_8$, leading to an effective four-dimensional theory with a gauge group structure matching the standard model of particle physics: $U(1) \otimes SU(2) \otimes SU(3)$.

[2] I'm not entirely convinced by the utility of speaking in terms of *revolutions* in cases like this. If anything really warrants the title of 'revolution' within string theory, then it is the central idea of string theory that physics can be done consistently with non-pointlike fundamental objects. That represents a *genuine* break with the physics of the past, a discrete jump. However, I will continue to use the terminology if only to map onto the common usage found in the string theory literature. Note, however, that I only intend the weaker notion of a 'high impact event' (e.g. as measured by direct citations and, indirectly, by the number of new outputs generated). That the citation

D. Rickles, *A Brief History of String Theory*, The Frontiers Collection,
DOI: 10.1007/978-3-642-45128-7_9, © Springer-Verlag Berlin Heidelberg 2014

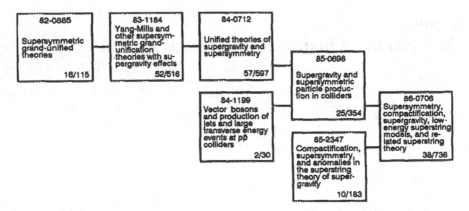

Fig. 9.1 Eugene Garfield's historiograph of the development of superstring theory and related areas of research (with lower numbers indicating number of cited/citing papers) [54, p. 331]

A similar claim could be made of the paper by Philip Candelas, Gary Horowitz, Andrew Strominger, and Edward Witten [20], that introduced Calabi-Yau compactifications into the string theory literature. As with the heterotic string, these compactifications were vital in demonstrating that realistic physics in the effective four-dimensional theory could be recovered from string theory.[3] Using them together, we can find structures with more than a passing resemblance to our own world.[4]

(Footnote 2 continued)

rates were heavily impacted can easily be seen by looking at SPIRES' top cited articles for the years 1984, 1985, 1986, and 1987 which feature a cluster of string theory papers at the top in each year following 1984, disturbed only by the Particle Data Group's Review of Particle Physics: https://inspirehep.net/info/hep/stats/topcites/index. (Interestingly, 1984 features several highly-cited papers on Kaluza-Klein compactification, which supports the claim in the previous chapter that the research landscape had been suitably transformed *prior* to the anomaly cancellation results.) See also Eugene Garfield's analysis of the most cited physical-sciences articles from 1984 for further evidence that a transformation took place in superstring theory ([54], see also Fig. 9.1).

[3] Gross et al. only considered the compactification of 16 (of 26) 'internal' dimensions in their paper on the heterotic string. This left open the task of wrapping up the remaining six (to leave the four standard spacetime dimensions), that was taken up by Candelas et al.—the preprint of which is briefly mentioned in the closing passage of Gross et al. Taken together with this compactification of the remaining six dimensions, Gross et al. conclude with the claim that "there seem to be no insuperable obstacles to deriving all of known physics from the $E_8 \times E_8$ heterotic string" [73, p. 283]. An obstacle (one of several) would, however, emerge not long after these words were written, in the form of an explosion in the number of possible vacuum configurations for the superstring, which dramatically reduced the prospects for deriving the *correct* one from the theory's equations.

[4] The idea underlying the search for such ground states (vacuum configurations) is that in order to have a computationally feasible string theory it was necessary to do that in the framework of string perturbation theory. But of course, that requires a background about which one expands, quantizing small fluctuations. Since this involves string theory at weak coupling, and the weak coupling limit is *classical*, this ground state had better be a match for the world we see around us.

The turning tide was also linked to a new set of questions stemming from perceived inadequacies with the standard model, as well as the dawning recognition that $N = 8$ supergravity would not be able to supply a satisfactory theory of quantum gravity. While it was admitted that (combined with general relativity) there were no experimental anomalies facing the standard model, there was something rather 'arbitrary' about its structure. There were too many unexplained parameters and inexplicable patterns in the organisation of particles and their properties (such as the partitioning of the elementary particles into 'generations'). The electroweak, strong, and gravitational interactions were annoyingly disunified. As it became clear that string theory might also be able to plug these kinds of gaps in the edifice of physics, it began to be taken far more seriously. Not only could it model gravity and aspects of particle physics, it seemingly promised to do so in a way that, true to its bootstrapping ancestry, left nothing arbitrary.

Interest in quantum gravity as a serious problem itself spiked at the same time and additional impetus must surely have come from the simultaneous rise of plausible alternative approaches to quantum gravity.[5] Chief amongst the central issues was the problem of renormalization and the UV behaviour of gravitational interactions. That string theory promised not only a renormalizable theory, but also a *finite* one, was a major source of motivation for those interested in quantum gravity. String theory was not quite able to convince the particle physicists (the 'tyranny of experimental distance' remained), but as a potentially finite, renormalizable, consistent theory of quantum gravity it had a strong case.[6]

Not only did string theory's fortunes change with the events of 1984, but, naturally, the status of string theorists changed too. In 1986, partly as a result of the anomaly cancellation result, but also because of this much improved state of superstring theory, John Schwarz was nominated for California scientist of the year, with the likes of Steven Weinberg offering support (see Fig. 9.2). However, just a few months later, the public controversy over string theory began, arguably with their now famous article in *Physics Today* by Paul Ginsparg and Sheldon Glashow, entitled "Desperately Seeking Superstrings?" (May, 1986).[7] The *Chronicle of Higher Education* stirred up

---

[5] For example, Gino Segrè wrote, 1.5 years after Green and Schwarz's result was announced, that the "delicate state of affairs ... in which a fair share of the high energy theory community is engaged in working on such a speculative theory ... seems at least logically natural as the great problem of incorporating gravity into our lore of knowledge of quantum field theory advances" [114, p. 123]. Again, reiterating an earlier point, it took some time for the particle physics community to *care* about the incorporation of gravitational interactions with the framework of quantum theory; but once they did, then string theory's killer app was given the credit it had been so patiently awaiting.

[6] For this reason, those working on supergravity will have been naturally interested (and, indeed, we have already the considerable professional overlap between supergravity and superstrings), since although renormalizability of $N = 8$ supergravity is arguably *less* of a problem than with the standard perturbative quantization of Einstein gravity, the problem is not entirely disposed of. Rather it is pushed into higher orders of the perturbation expansion for which supersymmetry fails to work its magic.

[7] Of course, Glashow, together with Howard Georgi, had proposed his own grand unification scheme involving $SU(5)$ in 1974 [57]. Glashow viewed this to be the simplest possible unification model for the strong and electroweak interactions (gravity was not included).

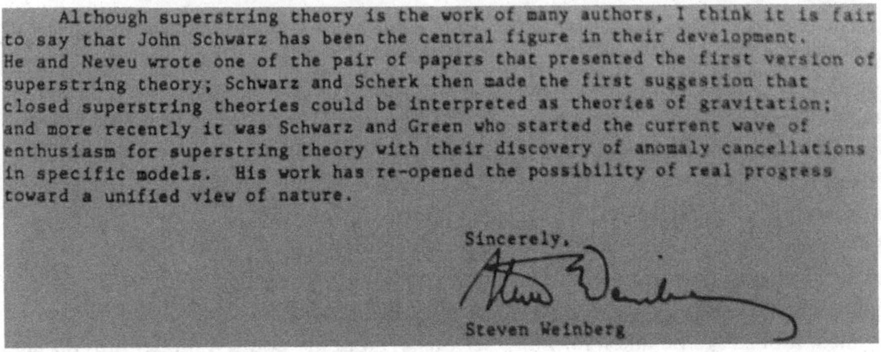

Although superstring theory is the work of many authors, I think it is fair
to say that John Schwarz has been the central figure in their development.
He and Neveu wrote one of the pair of papers that presented the first version of
superstring theory; Schwarz and Scherk then made the first suggestion that
closed superstring theories could be interpreted as theories of gravitation;
and more recently it was Schwarz and Green who started the current wave of
enthusiasm for superstring theory with their discovery of anomaly cancellations
in specific models. His work has re-opened the possibility of real progress
toward a unified view of nature.

Sincerely,

Steven Weinberg

Fig. 9.2 Extract from Weinberg's letter of support for John Schwarz's nomination for California Scientist of the Year. *Image source* Caltech archives [letter dated Feb. 13, 1986; Gell-Mann papers: Box 5, Folder 10]

the controversy some more with an article suggesting that while very imaginative, superstring theory might be impossible to verify (July, 23 1986). The same year, the *New York Times Magazine* had a front page article on string theory, in which the terminology of "theory of everything" is used (April, 1986: p. 24). In fact, it was earlier in July 1985 that the expression "superstring bandwagon" was coined [113]. These articles are direct ancestors of the recent books of Lee Smolin and Peter Woit, and make many of the same points (occurring in something like the same context as the debate over the string theory 'Landscape'). We return to the controversy briefly in the next chapter. In this chapter we focus on the surge of research that took place during the decade (roughly) 1984–1994.[8]

Before we turn to this (compactification, heterotic strings, and related themes), we should also mention another important event, from the point of view of the *establishment* of string theory; namely, the publication of the first textbook on superstring theory, which appeared in 1987 as the two volume set *Superstring Theory*, by Green, Schwarz, and Witten.[9] Textbooks have some historical importance in the development of a subject, signalling that a certain degree of maturity and consensus has

---

(Footnote 7 continued)
The idea is that $U(1) \otimes SU(2) \otimes SU(3)$ is a spontaneously broken subgroup of the larger group, $SU(5)$, so that there exists 'higher symmetry' (broken at energies beyond current means). Hence, Glashow was not writing from an entirely impartial perspective. Ginsparg, as we see later in this chapter, was 'pro-string theory' and had already done important work in the field.

[8] Something sociologically interesting happens with the renaissance of string theory, in terms of those physicists that had worked on string theory in its hadronic version. Many such physicists began contributing once again, returning to old research topics, and often covering similar ground a second time around (many after a gap of ten years). Alan Chodos puts it like this: "Older physicists who had contributed to the first wave of string theory dusted off their notebooks and reemerged from the woodwork. Suddenly reprints that had been yellowing undisturbed for 15 years were in demand again. The rumble of an unstoppable bandwagon, at first faint and far-off, grew quickly to a roar" [26, p. 253]. I wonder if there is any other episode like this in the history of physics?

[9] It would be more precise to say that this was the first textbook on superstring theory in its modern guise as a theory of quantum gravity and other interactions. Paul Frampton deserves

been reached.[10] They also solidify what the important tools, concepts, and key open problems are, which in itself can go some way towards forging a community of scholars.

9.1 Compactification

Describing compactification, in his diatribe on superstring theory, De Rújula writes that the "geometry of the 'lost' six-dimensional space must be gruesomely contrived" [15, p. 678].[11] Compactification was, as we have seen, old news by the time superstring theory was being more actively pursued, in the early to mid-1980s.[12] What was demanded were spaces that could be represented as a product space of our own low-energy world of four-dimensional Minkowski spacetime with some other six-dimensional space capable of going unnoticed ('lost') in low-energy physics, yet ultimately able to conspire with superstrings so as to both remain consistent with and generate features of such low-energy physics—these structures would provide classical (weakly coupled) solutions of the equations of string theory. This had originally been proposed in the string theory context as far back as 1975, by Scherk and Schwarz [31]—later pursued by Cremmer and Scherk the following year.[13]

In this work, Cremmer and Scherk studied "spontaneous compactification" solutions of general relativity, which was itself the outcome of an analysis of classical general relativity coupled to gauge and scalar fields [30]:

> These solutions represent a state where some of the matter fields have acquired position-dependent vacuum expectation values such that, in certain directions, space is so strongly curved that it closes upon itself. In other directions, where fields have constant vacuum expectation values, space-time does not close and is asymptotically flat. The non-compact directions can be thought of as the ordinary four-dimensional space-time, while the other, compact, directions are like an internal space [31, p. 61].

(Footnote 9 continued)

the credit for the first textbook on string theory, with his *Dual Resonance Models* [43] published in its first edition in 1974 (where is only a brief mention of "the rubber string model", with an expanded edition appearing with the title *Dual Resonance Models and Superstrings* [44] in 1986 (containing a supplementary chapter on superstring theory 'post-exaptation').

[10] See the chapters in Part III of [82] for a discussion of the role of textbooks in scientific fields.

[11] The contrived structure had, fortunately, already been constructed by Shing-Tung Yau: Ricci-flat Kähler manifolds having three complex dimensions [133]. See [136] for a very readable account of Yau's proof of Calabi's conjecture that led to the spaces now so ubiquitous in string theory, and their subsequent journey into physics (together with the subsequent feedback of ideas back into mathematics).

[12] Though I've not seen it mentioned as one of the many spin-offs of string theory research (and supergravity, to be fair), the work on compactification provided an enormously powerful set of tools and concepts for studying aspects of purely classical general relativity, and the more general notion of stable vacua therein.

[13] The idea of extending the Kaluza-Klein idea to more general, arbitrary non-Abelian gauge groups, can be traced at least as far back as 1968, to Ryszard Kerner: [84]. Peter Goddard mentions working to quantize the motion of strings on group manifolds to get out non-Abelian symmetries, while at the IAS in 1975 [63, p. 256].

This scheme was envisaged as a way of overcoming the problem of the extra dimensions in the original dual model context. Scherk's earlier work on zero slope limits is also being employed, since they explicitly draw attention to the fact that the dual models have low-energy limits of Einstein gravity in interaction with gauge fields (and might well constitute a renormalizable theory of gravity). The pheneomenon of spontaneous compactification "justifies the idea that in such models certain of the directions could be compact, and others not" (ibid, p. 62) so that they are neither unphysical nor *ad hoc*, nor is the compactification imposed as an "arbitrary condition" [30, p. 415]:

> If this mechanism, as is likely, can be generalized to an $SU(3)$ group, it could provide an alternative answer to the problem of quark confinement, as fields would have a tendency to bind into singlets to lower the energy. Another application is for dual models which are candidates for a renormalizable theory of gravity unified with other fields. In these models there are Yang-Mills massless bosons, a massless graviton, a massless fermion and some Higgs scalars in presence of other fields as well. An embarrassment was that the dimension of space-time had to be 10 for the model to be consistent. It has been proposed to compactify the extra spacetime dimensions and to use them to generate internal symmetries. An explicit example of a compactified dual model has been built in the simple case where the internal space is a flat hypertorus. Nevertheless, compactification seemed an arbitrary condition imposed on the model. Now we see that this compactification of unwanted spatial dimensions can spontaneously happen in a very simple model which has some of the salient features of a dual model [30, p. 415].

However, in the solutions studied the product space consists of Minkowski space and a real compact space of constant curvature.[14]

Recall that the critical dimension for bosonic strings is 26, while it is 10 for superstrings. There is an initial set of dimensions to deal with, then, in the form of the $26 - 10 = 16$ dimensional difference between these cases. These dimensions are understood to be non-spatiotemporal, but instead 'house' the internal degrees of freedom of the particles of the kind found in the standard model—16 is precisely the rank of the gauge groups in which the chiral anomaly cancels.[15] The remaining 10 dimensions are given a spacetime interpretation, and it is these which demand compactification into a product space of $K^6 \times \mathcal{M}^4$ of the kind introduced by Cremmer and Scherk. However, as we see below, later work would restrict a spacetime interpretation to just those four dimensions that we observe.

---

[14] Not quite gruesomely contrived enough!

[15] The initial compactification process from 26 dimensions, bringing in fermions, requires the lattices with the gauge groups identified by Green and Schwarz, namely the only 16 dimensional even, self-dual (unimodular) lattices $E_8 + E_8$ and D_{16}^+ (with automorphism groups $E_8 \otimes E_8$ and $SO(32)$ respectively). In the chapter introducing the heterotic string, Gross et al. [74, p. 254] made use of what Peter Goddard [62, p. 329] has called the 'Frenkel-Kac-Segal mechanism,' to generate, within a theory of closed strings, the desired gauge group by compactification (rather than by the usual Paton-Chan, open string, procedure of placing appropriate objects at string endpoints: delivering only the gauge groups $SO(N)$ and $Sp(2N)$)—Frenkel and Kac [46] had been thinking in terms of string vertex operator representations of Kac-Moody algebras. In the heterotic string theoretic context one views the initial 26 dimensional 'spacetime' as a quotient structure $\mathbb{R}^{25,1}/\Lambda$ (where Λ is one of the relevant lattices from above.) This amounts to compactification on a 16-dimensional torus in which points of the lattice, representing the space of interest, are identified.

Fig. 9.3 Flow diagram showing the superstring theorist's strategy for "making contact with reality". Image source: [19, p. 596]

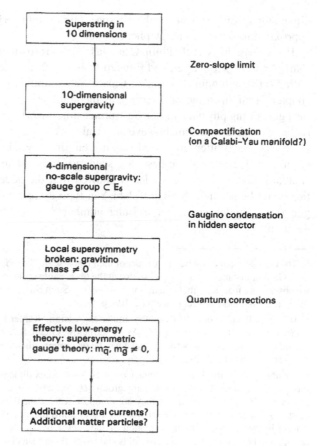

John Ellis[16] provides a useful schematic for the process of going from superstrings in 10 dimensions to the real world (see Fig. 9.3). He sees as essential the satisfaction of the following three conditions for the compact manifold [40, p. 596]:

- The four dimensional spacetime should be maximally symmetric (to stand any chance of describing the universe as we observe it at low energies).
- Supersymmetry should be preserved.
- The effective four-dimensional theory should be parity-violating (chiral).

At the IAS, Edward Witten and Andrew Strominger [121] lost no time in showing how, given appropriate compactifications of the kind Ellis circumscribes (though Ellis was writing after the fact), the Type-II theory could generate realistic-looking particle physics at low energies. More generally, there began a search for appropriate 6-real

---

[16] Note that the paper from which this is drawn is the original source of the phrase "theory of everything".

dimensional compact spaces (the K^6 space) that might serve to generate reasonable approximations to low energy phenomenology.[17]

Early on in this search, Philip Candelas, Gary Horowitz (a postdoc of Shing-Tung Yau), Andrew Strominger, and Edward Witten [20] argued that compact manifolds with $SU(3)$ holonomy (so-called Calabi-Yau manifolds—a phrase coined in this chapter[18]) might fit the bill, serving as the appropriate vacua for superstring theories, generating physically realistic 4-dimensional worlds.[19] In terms of generating a realistic particle spectrum, however, the Calabi-Yau manifold Strominger, Candelas, Horowitz, and Witten employed was not empirically adequate, yielding four generations of particles in the best case scenario, rather than the desired three, of the standard model. However, it clearly showed that the necessary replication properties could be secured through Calabi-Yau-type compactification schemes. A three generation model demanded an Euler number[20] χ of ± 6, in accordance with the formula:

---

[17] The compact space will be of the order of the Planck scale. The Planck scale involves energies of 10^{19} GeV, or distances of 10^{-33} cm, so that at the observable scale of real experiments the particles we observe are massless (that is, extreme low energy). Such particles are represented by zero modes of operators on the compact space (cf., [105, p. 55].

[18] However, Ivan Todorov had used the similar expression "Kähler-Einstein-Calabi-Yau metric" in his earlier study of $K3$ surfaces [125]. As we see below, Green, Schwarz, and West initially focused on such $K3$ surfaces as they tried to get a handle on compactifications onto spaces that might give physically realistic effective theories.

[19] Calabi-Yau manifolds can be characterised as N-complex dimensional manifolds, with unitary motions in the space defining a holonomy group $SU(N)$, which in the case of interest in superstring theory is $SU(3)$ (more precisely, they are compact Kähler manifolds with vanishing first Chern class). (A space's having $SU(3)$ holonomy simply means that parallel transport of a vector around a loop in the space results in a rotation of the vector by an element of $SU(3)$.) Yau puts the connection between Calabi-Yau manifolds and string theory succinctly as follows: "if you want to satisfy the Einstein equations as well as the supersymmetry equations—and if you want to keep the extra dimensions hidden, while preserving supersymmetry in the observable world—Calabi-Yau manifolds are the unique solution" [136, p. 131]. Luis Alvarez-Gaumé and Daniel Freedman had established the link between supersymmetry σ models and Ricci-flat Kähler manifolds in 1980: [1, 2] (a result leaned on in [74]).

[20] In fact, the link between the Euler number and the number of fermion generations was figured out, in the context of a ten dimensional field theory, by Witten in his talk at the 2nd Shelter Island conference in 1983 [128, p. 265], though there the number was equal to the modulus of the Euler number, $|\chi|$, rather than half). This included the claim that the number of fermion generations in a ten dimensional model is always even—it is perhaps no surprise that Witten's first attempt (together with Candelas, Horowitz, and Strominger: [20]) at a realistic compactification for the ten-dimensional heterotic string resulted in a *four* generation model. Witten also invents in this 1983 paper the method of regarding a spin-connection as a gauge field on the compact space (a method we meet below the context of Calabi-Yau compactification of the heterotic string). Witten was clearly well-armed for dealing with the compactification of ten dimensional string theories. He even makes the claim, further on, that, in order to get around certain problems with compactifying ten to four dimensions, there might exist "inherently stringy" methods "to compactify the string theory to 4 dimensions" noting also that "with the present incomplete understanding of string theory, it is difficult to pursue this possibility" (ibid, p. 267). This reiterates one of the key points of the previous chapter, which is that the research landscape was ripe for the anomaly cancellation results, which might otherwise have arrived dead born, as with many other significant results in the early

$$\frac{|\chi|}{2} = (h^{1,1} - h^{2,1}) = \frac{|-\chi|}{2} = \text{No. gen.} \quad (9.1)$$

This makes use of the generalization of the Euler number from real (involving Betti numbers describing the properties of the real manifold) to complex manifolds (involving Hodge numbers $h^{p,q}$ describing the properties of the complex manifold).[21] Hence, one needs to find a manifold with $(h^{1,1} - h^{2,1}) = 3$. Yau found such a $\chi = 6$ space while *en route* to give a talk (at the Argonne symposium on Anomalies, Geometry and Topology in 1984), involving the reduction of a $\chi = 18$ manifold with 3-fold symmetry to an orbit manifold ([134]; see also [136, p. 144].[22] The method involves identifying any points related by some group action, which will lie in some equivalence class (the group's orbit), which can then be eliminated by taking the quotient of the original space by the orbit. Candelas et al. also considered the possibility of obtaining new spaces by quotienting by discrete symmetry groups, and note how this can reduce the Euler number, thereby reducing the number of fermion generations ([20, p. 65]—though it still does not give them the three generation model, again resulting in four (or rather, $\chi = -8$). As a side-effect, such quotienting results in multiply-connected spaces, which, as we see in the next section, can be used as a kind of symmetry-breaking device to achieve physically more realistic models from the heterotic superstring theory.

As Yau tells the story ([136], chapter 6), Strominger and Candelas had first together isolated Calabi-Yau manifolds as having the right properties for the compact spaces of string theory, and then subsequently joined forces with Gary Horowitz, who had been at the IAS with Strominger and later moved to Santa Barbara (where both Strominger and Candelas were in 1984). It seems that Witten had independently followed his own path to the same destination via conformal field theory, as Strominger discovered during discussions on a return trip to the IAS. The convergence of distinct routes onto Calabi-Yau manifolds obviously increased the level of confidence in their applicability.

However, as Yau also points out [136, pp. 152–156], a potentially serious problem arose in the quantized form of the theory. Considering the behaviour of worldsheet in the target space from a quantum mechanical point of view, one has to envisage summing over all worldsheets interpolating between a pair of worldsheet boundaries, in this case weighting highest those worldsheets that have a smaller area (along the lines of the Nambu-Gotō action). A question arises as to the preservation of

---

(Footnote 20 continued)
days of string theory—e.g. the GSO tachyon cancellation result, which was in many ways just as significant as the anomaly cancellations.

[21] For a real manifold, the Euler number is $\chi = \sum_n (-1)^n b_n$, while for a complex Kähler manifold it is $\chi = \sum_{p,q} (-1)^{p+q} h^{p,q}$. The Betti and Hodge numbers are related by $b_n = \sum_{p+q=n} h^{p,q}$.

[22] Candelas and Rhys Davies [24] recently found an analogue of Yau's manifold after a detailed search of the tip of the distribution (with both Hodge numbers small) of Calabi-Yau manifolds.

conformal invariance; and it turns out that the answer depends on the metric on the target space in which the worldsheet lives: conformal invariance demands specific compatible metrics. The vanishing of the so-called β function (different from Euler's beta function) determines whether a theory is conformally (and thus scale) invariant. By the same token, a non-vanishing β function implies that conformal invariance has been spoiled by quantization (i.e. a conformal anomaly). To calculate the β function one applies perturbation theory, and finds that the first few levels (loops) involve a vanishing β function; but, as was shown by Marcus Grisaru, Anton van de Ven and Daniela Zanon in 1986 [70], the four-loop β function does *not* vanish in the Ricci-flat case.[23]

In other words, an exact Ricci flat metric cannot yield a consistent solution of the classical equations of motion as derived from a theory of strings (Type II or heterotic): all of the vacuum configurations on which so much effort had been expended, appeared not to be solutions to the string theory equations after all. Several proposals emerged in a bid to resolve this issue. Witten and Gross [71] proposed a perturbation scheme in which the metric should diverge ever-so-slightly from Ricci flatness, so as to make the β function vanish. This involved a microscopic readjustment of the vacuum. Dennis Nemeschansky and Ashoke Sen [99] showed that one can construct a fully conformally-invariant model on a Calabi-Yau manifold that, while not Ricci-flat (because of its conformal invariance) can be related to a Ricci flat metric by a non-local (metric) field redefinition such that the scattering amplitude describing the massless particles in the string theory (from which the string effective action is obtained, leading to the equation of motion of the metric) remains unchanged under such a local redefinition. These rescue attempts appear to have been quickly accepted, and business was able to continue much as before the crisis—Doron Gepner provided a more solid mathematical case that clinched the acceptance of Calabi-Yau manifolds as acceptable solutions, as we see later in this chapter.

Initially, the space of Calabi-Yau manifolds was uncharted territory.[24] Yau estimated that there might be tens of thousands, during his aforementioned talk at

---

[23] This was quite against expectations: see, e.g. [3]. Note that Grisaru et al. were themselves following a trail left by Pope, Sohnius, and Stelle [100], in which they pointed out certain flaws with the arguments for the vanishing of beta functions.

[24] Yau claims that he knew of just two such (known, constructed) manifolds before his trip to the conference at Argonne National Labs: the quintic 3-fold and a certain kind of product manifold, formed by 'stitching' three 1-dimensional tori together with alterations made to the resulting structure. The former lies at the root of a famous piece of 'physical mathematics,' involving Candelas [23], together with Xenia de la Ossa, Paul Green, and Linda Parkes, in which a feature of a plot of Calabi-Yau manifolds they had generated (mirror symmetry) was used to inspire the computation of a near-intractable problem in enumerative geometry: calculating the number of curves of a given degree $d \geq 3$ intersecting a particular surface. Famously, the physicist's method, employing a curious connection between the number of curves and the instanton number, was able to outperform the method of the mathematicians—this is somewhat simplified: see [52] for a nice historical account of this episode ([104] uses the same example, with some of the formal details worked out, to argue for the methodological legitimacy of string theory).

Argonne National Labs. In 1988, Candelas (together with his students, Dale, Lutken, and Schimmrigk) produced a computer algorithm (based on a technique devised by Yau) to classify *all possible* complete intersection Calabi-Yau 3-folds (that is, with 3-complex dimensions and that are expressed as complete intersections in products of projective spaces) [21]. This generated a data set consisting of 7890 such manifolds, more or less matching the prediction made by Yau in 1984. This represents an interesting, and potentially novel, interaction between physicists and mathematicians. As Peter Galison puts it, unlike the mathematicians' stance on Calabi-Yau manifolds, the physicists were interested in "the grubby details of their internal geometric structure", and they wanted to provide a complete map of the space of such manifolds since "[s]omewhere in the panoply of manifolds might lie one solution to the theory of everything" ([52], p. 38).

John Schwarz, together with Michael Green and Peter West, had, around the same time as Candelas et al. were constructing their models of compactification, initially focused on (2-complex dimensional) $K3$ surfaces as toy models, since Yau had shown that such spaces could support a Ricci-flat metric (thus proving an earlier conjecture of Calabi).[25] $K3$ surfaces had been well studied for almost 100 years by the time they entered string theory, so much of the hard labour in figuring out their properties had been done. But as Paul Aspinwall and others' work on duality revealed, $K3$'s uses went beyond its provision of a nursery for realistic compactification to become "almost omnipresent in the study of string dualities" [8, p. 2].

The study of compactification brought with it various unexpected novelties of spacetime structure.[26] One of the more conceptually interesting of these is the fact that multiple formally distinct spacetime structures[27] may correspond to one and the same conformal field theory. This occurs at several levels of complexity, from compactification on a circle (a 1-torus) to more general tori to Calabi-Yau manifolds and orbifolds. The basic symmetry, now known as target-space duality or 'T-duality' (named by Font, Ibáñez, Lüst, and Quevedo in 1990, [50], pp. 39–40—

---

[25] As Yau puts it, the name 'K3' "alludes both the K2 mountain peak and to three mathematicians who explored the geometry of these spaces, Ernst Kummer ... Erich Kähler, and Kunihiko Kodaira" [136, p. 128]. He goes on to note that Green, Schwarz, and West had in fact been (mis)informed that the $K3$ surface represented the maximum, in terms of number of dimensions, for a space with the desired properties—namely Ricci-flatness, in order to cope with the value of the cosmological constant. (However, Schwarz claims to have no personal recollection of such an episode, noting only that they studied $K3$ because it was what they knew.)

[26] Although string theory is often ignored by philosophers of physics, it (since the mid-1980s at least) contained a steady stream of conceptually important work regarding spacetime. T-duality, and its generalisations, represent the tip of an iceberg yet to be properly explored by philosophers of physics. Fortunately, there are signs that this neglect is changing, with several recent articles on this and related issues in string theory (e.g., [25, 91, 102, 103]), a book [32], and PhD theses now appearing on philosophical aspects of duality in string theory (e.g. [92, 127]).

[27] What Aspinwall, Greene, and Morrison label "string equivalent spaces" [12, p. 5325]. These can differ topologically, as well as geometrically, and yet still generate the same physics at the level of the string theory. For a philosophical examination of this phenomenon, see [102].

Fig. 9.4 Graph of the effective potential V_c for a closed string against a scale a_i. One can see the minimum at $a_i = \sqrt{\alpha'}/R_i = 1$ and the invariance under $a_i \to a_i^{-1}$.
Image source [85, p. 359]

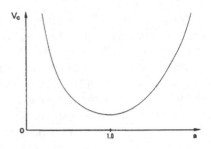

see also Schwarz and Sen: [112][28]), was first pointed out (for circles) in Kikkawa and Yamasaki's 1984 study of Casimir effects in string theory [85].[29] This chapter also contains the result of a kind of minimum length imposed by the invariance with respect to the interchange of $a_i \to a_i^{-1}$ (where a is a common scale: $a_i \equiv \kappa_i a$). The minimum must occur at $a_i = \sqrt{\alpha'}/R_i = 1$ (see Fig. 9.4).

The explanation given by Kikkawa and Yamasaki has remained the *de facto* one:

> Once a subspace happens to be compactified, the string modes wound around circumferences of the torus tend to squeeze the radii further to reduce the tension energy. On the other hand the kinetic energy increases as $R_i \to 0$ according to the uncertainty principle. There must be somewhere finite radii at which the space is balanced. If $\sqrt{\alpha'}$ is of the order of the Planck length, so must be R_i according to [$a_i = \sqrt{\alpha'}/R_i = 1$] [85, p. 359].

The following year, the toroidal (for an r-dimensional torus) case was also presented in Norisuke Sakai and Ikuo Senda's study of vacuum energies (and the avoidance of double counting) for a string compactified on torus: [107]—the key result is the modular invariance of the vacuum energy.[30] Here, one simply finds that r-dimensional

---

[28] There were earlier discussions (in the context of the effective string action: [41, 51]) in which the letter 'T' was introduced (as a complex scalar field that is transformed according to the duality mappings). Font et al. [50] is also is the source of the label 'S-duality' to stand for the nonperturbative $SL(2, \mathbb{Z})$ symmetry in string theory—Dieter Lüst notes that this work on S-duality was in part triggered by their work on T-duality (personal communication). Note that Schwarz, together with Brink and Green [66], carried out some important 'preparatory' work on the study of the behaviour of strings with respect to varying radii in 1981—in this same work they recover $N = 4$ Yang-Mills theory in four dimensions as a limit of an interacting theory of open and closed strings with simple ten-dimensional supersymmetry and $N = 8$ supergravity in four dimensions as a limit of an interacting theory of closed strings only with extended ten-dimensional supersymmetry.

[29] The basic idea is very simple: compactifying one of the dimensions onto a circle of radius r, one finds that the zero radius limit is identical to the infinite radius limit, given a switching of momentum and winding modes, m and n, of the strings—of course, compactifying onto a circle, in which points differing by $2\pi r$ are identified, leads to a discretisation of momenta. More formally, one finds that the mass spectrum $M^2 = \frac{m^2}{r^2} + \frac{n^2 r^2}{\alpha'^2} + \frac{2}{\alpha'}$ is invariant under the simultaneous transformation $\langle m, n, r \rangle \to \langle n, m, \frac{\alpha'}{r} \rangle$.

[30] In 1986, Paul Ginsparg [60] also looked at the toroidal compactification case, for heterotic strings, with a focus on the deepened understanding of the relationships between the different string theories that it offers. In particular, one can "continuously interpolate between compactified versions of the $E_8 \otimes E_8$ and $\mathrm{Spin}(32)/\mathbb{Z}_2$ theories by turning on appropriate background gauge fields and adjusting

discrete momentum and winding number vectors are introduced, but the same 'large-small' symmetry holds.

This opened up the general study of characteristically string theoretic phenomena concerning spacetime structure. Along these lines, Aspinwall et al. [11, pp. 414–415] distinguished two forking paths concerning string theory research: the first aims to "extract detailed and specific low energy models from string theory in an attempt to make contact with observable physics", while the second aims to examine "those properties of the theory which are generic to all models based on strings and which are difficult, if not impossible, to accommodate in a theory based on point particles". The former path, as we will see, meets with difficulties in the face of a profusion of possible string backgrounds. The latter path leads to a host of conceptually interesting features, including the idea that strings can propagate on kinds of spacetime that would simply be impossible for point-like particles, including those possessing singularities generated by quotienting procedures (by lattices).

In 1988, Amati, Ciafaloni, and Veneziano also made an important contribution to the understanding of the small-scale structure of spacetime according to string theory by looking at ultra-high energy string scattering, which they argued pointed to the existence of a new kind of 'quantum geometry'. The work is based on the premise that it is at "high energies—or short distances—i.e., around planckian scales, that strings should reveal their virtues, offering novel solutions to the long-standing problems of classical singularities and quantum infinities in general relativity" [5, p. 81]. In other words, they use string theory as a tool to figure out what space time is like at scales where quantum gravitational effects are expected to become non-negligible. In their paper "Can Spacetime be Probed Below the String Size?" they demonstrate that, due to the 'softness' of the string (in part, responsible for the failure to provide a model for the hard scattering events explained by QCD's point-like quarks), it is operationally impossible to resolve distances shorter than the string length $\lambda_s = \sqrt{\hbar \alpha'}$. In other words, string theory possesses a minimal observable length, corresponding to that identified by Kikkawa and Yamasaki.[31]

(Footnote 30 continued)

radii" [60, p. 648]. In a move that would parallel later work on dualities, Ginsparg argues that, despite the two theories having very different spacetime interpretations (differing with respect to their radii) from a "mathematical point of view, compactified versions of the $E_8 \otimes E_8$ and $Spin(32)/\mathbb{Z}_2$ theories, insofar as they are continuously related, may thus be regarded as different ground states of the same theory" but where "a physical observer would choose one or the other as the natural interpretation" ([60, p. 652]. We return to similar results below. Shapere and Wilczek also considered toroidal compactifications, as a feature of spacetime modular invariance—they draw parallels between such string theoretic dualities and what would be called self-S-duality (physical equivalence with respect to inversion of coupling constants), focusing on situations in which a theta term is added (in which case the duality is "extended to an invariance under an action of an infinite discrete modular group on the coupling parameter space" [118, p. 669].

[31] The idea that quantum gravitational considerations might lead to a fundamental length, which could serve to regulate problematic fields at high energies, has an old and venerable history: both Pauli and Landau had considered it. However, in this case, the minimum length is issuing entirely from the ability of closed strings to wind around compact dimensions and has nothing directly to do with gravitation.

David Gross ([75], §4) considered the same issue at around the same time, again focusing on the attempt to probe sub-string scale distances for a breakdown of standard spacetime continuum concepts, by using fixed-angle high-energy scattering experiments, developing in this case an interesting (operational[32]) 'string microscope' thought-experiment to make the issues more transparent.[33] Gross directly links the issue to other issues from quantum gravitational physics: spacetime foam (from a fluctuating metric) and gravitational regulators providing cutoffs. He argues that while these notions provide a nice qualitative description of what is expected to occur at the Planck scale, string theory is able to offer quantitative results. Again, as in the Amati et al. account, it is the non-local nature of the strings that is responsible for an inability to resolve sub-Planck distances, despite the fact that there is no physical cutoff in the theory. The result that we cannot probe limitlessly comes about from the fact that the size of the strings, defining the interaction region (that is the distances explored), grows with the energy (which, of course, itself grows as the scale one wishes to probe is reduced!). The expression Gross finds for the size of the collision area, X_{coll} is:

$$X_{coll} \approx \frac{\alpha' E}{N} \tag{9.2}$$

In other words: trying to measure smaller distances with strings (treating them as 'local probes'), causes them to increase in size. Thus, this way of viewing the behaviour of strings as the scale is adjusted can be seen as a more physical way of understanding the $r \rightarrow \alpha'/r$ T-duality. This is a purely string-theoretic effect. From this analysis, like Witten, Gross draws deep implications, writing that he expects "in the final formulation of string theory that space and time will emerge only as an approximate concept, valid or useful for certain approximations to the theory" [75, p. 413].

This increased awareness of the subtle nature of stringy spacetime posed problems for at least one otherwise natural candidate for a definition of string theory beyond perturbation theory: (nonperturbative) superstring field theory. Witten [130] first analysed the open string case; and Strominger [123] analysed the closed string case.[34]

---

[32] In other words, it is not presented as a demonstration of a discrete metric structure 'in spacetime itself,' but concerning our ability to *measure* distances. As Gross says: "I do not know of a direct way to tell whether string theory will truly require a modification of the notions of space and time at short distances" [75, p. 412].

[33] Such implications led Witten to remark: "What one might imagine would be a world in which at distances above $\sqrt{\alpha'}$, normality prevails, but at distances below $\sqrt{\alpha'}$, not just physics as we know it but local physics altogether has disappeared. There will be no distance, no times, no energies, no particles, no local signals—only differential topology, or its string theoretic successor" [132, p. 351].

[34] Strominger's lectures include a useful presentation of Witten's approach. Strominger lists three conditions for a good (closed) string field action: (1) "it should reproduce the Virasoro-Shapiro amplitudes when performing an expansion around a ground state; (2) it should be diffeomorphism invariant (or some stringy generalization thereof); (3) it should not feature a spacetime metric (or "background fields") since there is one "included as a component of the string field" [123, p. 311]. He offers the action: $S_C = \int_C \mathscr{A} \cdot (\mathscr{A} \cdot \mathscr{A})$ (where $\mathscr{A}$ is the string field). The equation of motion for the field is generated from: $\delta S_C = \int_C (\delta \mathscr{A} \cdot (\mathscr{A} \cdot \mathscr{A}) + \mathscr{A} \cdot (\delta \mathscr{A} \cdot \mathscr{A}) + \mathscr{A} \cdot (\mathscr{A} \cdot \delta \mathscr{A}))$.

By analogy with standard quantum field theories, this would involve string fields that create and annihilate strings (rather than particles)—hence, the new objects would be functionals $\Phi[X^{\mu}(\sigma, \tau)]$ of the old string coordinates (or, rather, the space of string configurations X^{μ}). But problems are faced with the spacetime picture involved in these approaches, since they can at best provide a treatment of a perturbation expansion around a *specific* classical background.[35]

An alternative approach to the standard form of 10 $\to$ 4 Calabi-Yau compactification, known as 'orbifold compactification,' also led to some conceptually curious results, that link back to some of those discussed above. These new approaches were also mined for their phenomenological minerals. This relates both to the quotienting method that leads to the multiply-connected spaces in [20] (discussed further in the next section) and also the Frenkel-Kac mechanism, discussed earlier in the context of eliminating those degrees of freedom that were not to receive a spacetime interpretation.

The first analysis of string propagation on such a quotient space[36] was carried out by Lance Dixon, Jeff Harvey, Cumrun Vafa, and Edward Witten [35]. They use a

---

[35] I take this brief survey (which barely skims the surface of a rich vein of similar literature) to point to a clear openness of string theorists to deal with conceptual and foundational issues having to do with spacetime and the notion of background independence—I mention this since string theory (as a quantum theory of gravity) is often castigated for not being sufficiently sensitive to such considerations (see, e.g., [119]).

[36] This is the *orbifold* (a contracted form of orbit manifold), coined, I believe, in the same paper. Strings on orbifolds have some interesting and unexpected historical links to finite group theory, most notably the so-called 'Monster sporadic group'—an exceptionally clear discussion of these developments can be found in [48]. In 1973, around the same time the dual resonance model was recognised to be equivalent to a theory of strings, mathematicians Robert Griess and Bernd Fischer had (independently) predicted the existence of a novel sporadic finite simple group, the largest such group—a group G is 'simple' just in case its normal subgroups are the group G itself and the trivial subgroup containing the identity element of G (simple groups are elementary or atomic: they have no nontrivial normal subgroups); finite groups are composed of simple groups; sporadic simple groups are amongst twenty six exceptions that do not fit into the twenty or so infinite families charted in the classification of finite simple groups [28] (the sporadic groups include the Leech lattice groups, related to the physical Hilbert space [24 transverse components] of the Veneziano model: [29]). Their group was conjectured to have as its smallest non-trivial representation a 196883 dimensional structure, and so was duly labeled the *Monster*—though Griess called it 'the friendly giant' [14]. Richard Borcherds estimates the number elements to be roughly equal to the number of elementary particles in the planet Jupiter [17, p. 1076]:

$$2^{46}.3^{20}.5^9.7^6.11^2.13^3.17.19.23.29.31.41.59.71 \tag{9.3}$$

In 1974, John McKay noticed a remarkable coincidence: the number 196883 differed by one from the linear term of $q = e^{2\pi i \tau}$ in the expansion of the elliptic modular function $j(\tau) = q^{-1} + 744 + 196884q + c_n q^n$. McKay took the fact that these numbers are so large and yet so close (and also so 'unusual') to point to a close relationship between the two apparently disconnected host fields—such 'large number' reasoning has a history in physics and cosmology, of course: Dirac and Eddington famously placed a lot of weight on the coincidence of large numbers appearing in physics. This relationship became known as the 'Moonshine conjecture' (named by John Horton

simple case to establish their results: compactification onto a torus.[37] However, their example is an orbifolded product of tori—it is, in fact, the same example used in [20], which they called the 'Z manifold'. Z involves a product of three two-dimensional tori, $T = T_1^2 \otimes T_2^2 \otimes T_3^2$, each defined by $z_i \approx z_i + 1 \approx z_i + e^{i\pi}/3$, with a discrete symmetry group G (isomorphic to $\mathbb{Z}_3$) whose generators are related to $\frac{2}{3}\pi$ rotations about an origin. This results in twenty seven fixed points, rendering the space $Z = T/G$ an orbifold, rather than a genuine smooth Calabi-Yau manifold.[38]

What is surprising about this scenario is that, despite the existence of such singularities in the spacetime (which can be smoothed out or not, as I mentioned), the string propagation is perfectly consistent: it 'senses' no obstructions. Hence, here is an example of another distinctly 'stringy' phenomenon. As Dixon et al. point out, the orbifold constructions inherit additional so-called 'twisted sectors' of states ($x(\sigma + \pi) = gx(\sigma)$, where $g \in G$), forced by the preservation of modular invariance (see their second paper, [35], §4). These twisted boundary conditions provide the authors with an alternative approach to symmetry-breaking, and a new path to phenomenologically realistic theories—Aspinwall notes that we can view them as referring to "open strings in M [the unreduced space] whose ends are identified by the element $g \in G$" in which case the state is in "the g-twisted sector" [9, p. 367]. For example, Dixon, together with Daniel Friedan, Emil Martinec, and Stephen Shenker,

---

(Footnote 36 continued)

Conway and Simon Norton: [27]). Its evolution and proof involves a curious blend of algebra, the theory of modular forms, and physics (see [53]). Griess constructed the group in 1981. In 1984, Frenkel, Lepowsky, and Meurman [47] constructed the 'Moonshine module' V for the Monster by using tools from conformal field theory (via their chiral or vertex algebras). In doing so they had also written down something equivalent to the theory of bosonic string propagation on a $\mathbb{Z}_2$ asymmetric orbifold (see [48], §IV)—in a letter to Murray Gell-Mann (dated November 16, 1984), Lepowsky writes: "On the hunch that some of the enclosed [on the moonshine module] might be relevant to the recent string theory discussions, I'm enclosing you some material..." [Murray Gell-Mann Papers (Caltech): Box 11, Folder 43]—the paper enclosed was [47]. The link to strings on orbifolds was made by Dixon, Ginsparg, and Harvey in 1988: [39]. Richard Borcherds followed up the string connections (including the no-ghost theorem), using the vertex operator algebra techniques to prove that V satisfies the moonshine conjecture [14, 15] (see also: [16]). See [64] for a nice account of these ideas.

[37] Note that this is not phenomenologically feasible since it is unable to generate chiral fermions. The solution to the problem of chirality was to compactify onto a torus quotiented by a discrete group. The resulting manifold will have a conical singularity from the identifications (a fixed point that is mapped to itself under the action of the group)—these can be removed by a process called 'blowing up,' involving the smoothing out of such points, or they can be left as is. Given an interpretation of such an orbifold as a *spacetime* (i.e. a structure involved in physical compactification scenarios), the fixed points correspond to curvature singularities, since the group action (a $\frac{2}{3}\pi$ rotation) on a vector encircling such a point would be the identity mapping. Note that this imposes obvious conditions on the physical states; namely that they must commute with such discrete rotations (and also the translations of the underlying torus, which are also symmetries).

[38] As Paul Aspinwall points out, these orbifolds were initially viewed as "little more than a step in the construction of a smooth Calabi-Yau manifold" ([9, p. 355]. However, this work of Dixon et al. meant that orbifold compactifications could stand their ground; besides which, "the torus itself is a little too trivial whereas a general Calabi-Yau manifold can render many calculations very difficult" ([9, p. 356].

went on to compute Yukawa couplings in the effective field theory generated by the orbifold compactification of fermionic strings [37].[39]

Finally, at about the same time as these orbifold methods were used to reduce the theory from ten to four spacetime dimensions, K. S. Narain [96] introduced the idea of an intrinsically four-dimensional theory, in which there was no spacetime reduction as such, but a kind of Frenkel-Kac mechanism applied on all $d - 4$ dimensions. The alternative involves a kind of 2-step procedure to get to $d = 4$, of course: (1) compactify the twenty six-dimensional left moving sector down to ten; (2) compactify the ten-dimensions of the both the left- and right-moving sectors down to four. Narain proposed a direct compactification to tori, for both sectors: there is a lattice for the 22 (26-4) left-movers and another for the 6 (10-4) right-movers: he has a $(10 - d)(26 - d)$-parameter family of one-loop finite string theories. Rather worryingly, Narain argues that there are infinitely many heterotic string theories generated in this way. However, as he admits, the theories are $N = 4$, and so are not phenomenologically viable as they stand—he does suggest that his approach might be integrated with the orbifold methods of Dixon et al. in order to break this down to an $N = 1$ theory. Following up on this, in a paper co-authored with Sarmadi and Witten, it is argued that in fact Narain's earlier work pointed to new toroidal compactifications of the already existing heterotic superstring. Interestingly, too, it is argued that Narain's analysis opened up the idea that the two forms of heterotic superstring, $E_8 \otimes E_8$ and $SO(32)$, were really "two different vacuum states in the same theory" [97, p. 378]—as they note, this had been suggested earlier (not least in [35]), but such convergent results must have strongly increased the belief in this idea.[40]

Narain collaborated with Sarmadi and Vafa on the topic of 'Asymmetric orbifolds' [98]. The central idea is to demonstrate that the left- and right-movers of heterotic strings can be treated independently on different orbifolds, so that the remaining $d - 4$ 'spacetime' dimensions might not be "real" [98, p. 571].[41] In this work they too notice the profound impact that string theory has on our conception of spacetime, and the relationship between matter and spacetime. Indeed, they draw attention to a kind of snapping-point with respect to the kinds of point-particle/field theory analogies that have often worked out so well in the history of string theory:

> [T]here are a number of examples showing how our naive intuition fails. For instance, the field theory of particles gives us no clue as to why compactifying on special tori, such as is the case for heterotic strings and their toroidal compactifications to lower dimensions, should give rise to gauge invariances besides the ones expected from the Kaluza-Klein picture. These are "stringy" effects which cannot be explained from a point particle viewpoint. The

---

[39] They suggest that the orbifold technique might provide a "laboratory for dissecting the rich structure of conformal field theory" [37, p. 72], since it allows one to generate very many solutions of classical equations of motion of strings (given the correspondence between 2D conformal field theories and such solutions, that I describe below).

[40] An idea that grew, as new equivalences were found linking what were once thought to be distinct theories. Such works are clearly direct ancestors of the recent work on M-theory and the AdS/CFT duality (considered in the next chapter).

[41] This point was made somewhat earlier in [86].

winding of strings around the non-trivial loops of the tori is responsible for giving rise to gauge degrees of freedom. Since this involves extended states, it is not expected from a point particle point of view. ... Another example where our naive point particle intuition fails is for strings propagating on orbifolds. Orbifolds, which generally are not manifolds, would give problems with the unitarity of point particle theory (due to the points where orbifolds fail to be manifolds). However, the string propagation on them is not only consistent, but in cases where they are formed from tori, is natural and rather simple to describe [98, p. 551].

In other words, strings 'see' spacetime differently to point particles. It is precisely this feature that allows one to use the combination of strings and compact (non-trivial) spaces to generate all sorts of interesting (realistic) possibilities for the low-energy effective theory.

9.2 Heterotic String Theory

Basic citation analysis reveals that the paper on the heterotic string was a landmark event, amassing 194 (recorded) referring publications in 1986 alone—remembering, of course, that this was still the days before arXiv. The explosion of publications involved the search to extract something close to real world physics from the new string theory, using compactification (since a ten dimensional theory) as the tool to get at it. The package of heterotic strings plus Calabi-Yau compactification provided the basic framework for a large proportion of the work that took place in string theory between 1984 and 1994.

Heterotic string theory is a theory of closed strings which, upon compactification of sixteen (internal) dimensions (of the 'left-movers') delivers the anomaly-free gauge groups. Hence, there are in fact two kinds of heterotic string theory: one implementing the gauge group $E_8 \otimes E_8$ and the other implementing $SO(32)$—Green and Schwarz had considered $SO(32)$ Type I strings.[42] The reasoning behind the construction of heterotic string theories is obvious: given the anomaly cancellation results of the previous chapter, string theories that implement these gauge groups would be automatically anomaly free (and 1-loop finite) and, in the case of $E_8 \otimes E_8$ at least, phenomenologically promising.

The phenomenological promise increases with the Calabi-Yau compactification result, since the solution found in [20] involves a space with $SU(3)$ that breaks one of the E_8 factors to $E_6 \otimes SU(3)$. For multiply-connected spaces (i.e. those with one or more noncontractible closed curves, as generated through the process of modding out by a discrete group action, mentioned in the previous section), this E_6 can be further broken down into the standard model's $U(1) \otimes SU(2) \otimes SU(3)$ by invoking a procedure analogous to the Aharonov-Bohm effect, threading flux through the holes which leads to physical effects generated by the non-trivial topology (much

---

[42] As we see in the next chapter, the puzzle over the $SO(32)$ (heterotic and Type I) duo would be resolved by finding a duality linking them.

as a solenoid can act as a hole separating charges from the magnetic field).[43] The holonomies (or Wilson loops) completing circuits through the holes (and, therefore, possessing non-contractible paths) are non-vanishing and act like Higgs bosons, which provides the desired symmetry breaking mechanism, reducing E_6 down to the largest subgroup that commutes with all the Wilson loops (see [55, p. 333], and [110, p. 273] for more details).

The heterotic string is hard to gain an intuitive grasp of.[44] It is, as the name suggests, a hybrid of two string theories: one of left movers (bosonic strings, of the old dual model type), and the other of right movers (superstrings, of the Green-Schwarz type). What is so hard to grasp is the fact that (1) these motions are often taken to live in spacetimes of different critical dimensionality (26 and 10 respectively); (2) they involve one and the same string. However, as Schellekens pointed out in his talk[45] on four-dimensional string theories, one need not give a full spacetime interpretation here, and to do so courts mystery where there's none present:

> The old notion of a "critical dimension" follows if one assumes that all resulting bosons have to be paired into spacetime coordinates. (The known possibility of torus compactification gave no reason for drastically altering this notion.) This pairing was no longer possible for the heterotic string, and one was forced to find a different interpretation for 16 unpaired left-movers. Old notions die hard, and for a while the heterotic string was described as a mysterious object, whose left-movers lived in 26 dimensions (of which 16 were compactified), and whose right-movers lived in 10 dimensions. This soon gave way to a more sensible interpretation, emphasizing mathematical and physical consistency over a space-time interpretation for the 16 extra modes ([108, p. 1]).

Of course, as we saw in the previous section, the Frenkel-Kac-Segal mechanism was used for 16 of the dimensions. This leaves us with the left-movers living on a product space of ten-dimensional spacetime and a sixteen-dimensional torus (one of the two lattices of the Frenkel-Kac construction), with the right-movers on the *same* ten-dimensional spacetime as the left-movers. Schellekens goes on to argue that our knowledge of the expanse of possible four-dimensional string theories depended on this conceptual split between the critical dimension and spacetime dimensions.[46]

---

[43] This is first laid out in Candelas et al. [20, p. 65]. As they note, for holonomy U, so long as the path around which the vector is transported is non-contractible, one can have $U \neq 1$ even though the E_6 (ground-state) gauge field strength F vanishes. (For some subsequent discussions of this method of symmetry breaking, following soon after the appearance of [20], see: [19, 115, 129].)

[44] The nickname "heterotic" derives, as the authors point out, "[f]rom the Greek "heterosis": increased vigor displayed by crossbred animals or plants" [74, p. 505]. Characteristically, during his talk at the First Aspen Winter Physics Conference (January 6–19, 1985), as heterotic strings had just appeared, Murray Gell-Mann attempted, rather sensibly, to impose some nomenclatural order on the various string theories by calling the heterotic string theories "Type III/II" ([55, p. 332]. So far as I can tell, nobody followed suit.

[45] Note that this talk is essentially a summary of Schellekens' earlier, joint work with Lerche and Lüst: [86].

[46] In a related, earlier analysis, Freund, Oh, and Wheeler drew a similar lesson about a possible disconnection between the critical dimension and the spacetime dimensionality: "the moral of all this is that one should not be too dogmatic about the critical dimension of space-time, as it may change

Table 9.1 The three generations of fermions, split into two pairs of quarks and leptons per generation

1st Generation	2nd Generation	3rd Generation
Up Quark	Charm Quark	Top Quark
Down Quark	Strange Quark	Bottom Quark
Electron Neutrino	Muon Neutrino	Tau Neutrino
Electron	Muon	Tau

9.3 The Generation Game

The catalogue of fermionic particles in our Universe divides into two broad types: leptons and quarks. The quarks, as we know, are subject to the strong force, while the leptons are not. For every one of the particle types falling within these categories, there is an anti-particle possessing the same mass but the opposite charge. These particles fall into a curious pattern of (three) 'generations' (see Table 9.1). In each generation sit a pair of leptons and a pair of quarks. In the first generation we find up and down quarks and the electron and the electron neutrino. In the second generation we find the charm and strange quarks, and the muon and the muon neutrino. In the third generation there are the top and bottom quarks, and the tau and tau neutrino. The top quark was a prediction made on the basis of this pattern and was successfully detected at Fermilab's Tevatron in 1994.[47] Why the particles should fall into such a pattern is a mystery, and if it were to be derived from string theory, then that would constitute a major empirical success.

The features of the heterotic string combined with Calabi-Yau manifolds opened the door to string theory phenomenology, as we have seen. In particular, an explanation (of sorts) could be given for why there appeared to be this three generations of particles.[48] The topology of the compact space determines key features of the low energy physics, including the low-energy symmetry groups, symmetry-breaking scales, masses and lifetimes of particles, couplings, and how many species of parti-

---

(Footnote 46 continued)
when the world-manifold ceases to be flat" ([45, p. 374]). Peter Freund, together with Freydoon Mansouri, had also shown that increasing the dimensionality of the fundamental objects (from strings to membranes or "bags", and higher-order objects) itself can reduce the critical dimension. Unlike the case for point particles, which has no notion of a critical dimension, strings and other higher-order objects have specific critical dimensions dependent on the dimensions of the objects, due to the restriction of cancelling the conformal anomaly arising from the coordinates of the string points (*qua* two dimensional scalar field) relative to the anomaly arising from the two-dimensional metric of the string's worldsheet (cf. [49, p. 279]).

[47] The first evidence was found in 1994 (by the CDF [the Collision Detector] at Fermilab)—see Kent Staley's excellent book [120] for a historic-philosophical account of the discovery of the top quark.

[48] The existence of three generations is an experimental result without a current theoretical explanation. A 'generation' here refers to a family of particles united with respect to the kinds of interactions they display with respect to the electric and nuclear forces (which in turn further pins down the particles' properties). What we find is that the properties of particles in one generation are replicated in the other generations *with the exception of their masses*.

cle there are (including the number of fermion generations). It isn't surprising that a small industry built up around the search for such a compact space that was 'just so'. The search for such classical worlds from the equations of heterotic string theory accounts for a huge proportion of the string theory literature in the mid-to late 1980s.

In general, the problem is to figure out how to get from the high energy theory, with its exceptionally high degree of symmetry, and break the symmetries in such a way as to get out the kind of low energy physics we see in our experiments. Specifically, what was desired was a single ground state that latched on to this low energy physics and enabled predictions to be made. This kind of procedure was considered vital for the justification of string theory since it was (and is) clear that the string scale itself is out of bounds experimentally speaking. As Philip Candelas and Sunny Kalara put it, any "contacts as will be possible with direct observation will be through the effective low energy theory" ([22, p. 357]). As we saw, the solution has to do with the topological invariants of the compact spaces, and in particular the Euler number χ of the space specifying the number of holes which determines the number of generations of particles. Hence, the compact topology, once something of an *ad hoc* device, to be hidden away, becomes a powerful *resource* for generating the low energy physics: it is part of the machinery providing explanations of particle physics, rather than an *explanandum*.[49]

Following an extensive search, a Calabi-Yau manifold was discovered that would allow the three generations of elementary particles to emerge when selected as the classical vacuum configurations of $E_8 \otimes E_8$ heterotic superstrings. It happened to be the very first such manifold discovered by Yau (see [10, p. 193]). Such systematic searches through mathematical objects, for the desired kind, became almost commonplace with respect to Calabi-Yau manifolds.[50] The Calabi-Yau compactifications were chosen to preserve ($N = 1$) supersymmetry at the Planck scale, which was expected to be able to go some way towards solving the hierarchy problem.[51]

Recall that according to the compactification scheme, four dimensions remain non-compact, with the string freely propagating in these, but constrained in the remaining compact dimensions. The tiny scale of the compact space means that the effective theory thus generated will appear, from the energy scale of our experiments (and considerably higher: until they are probing near the radius of the compact space) to be four-dimensional. The specific kind of compactification will generate a low-

---

[49] This is, of course, closer to Oskar Klein's usage of a compact fifth coordinate in order to explain charge quantization. However, it differs in an important way from Kaluza-Klein compactification in that the procedure begins with fields living in the pre-compactified theory that are 'funnelled' through the compactification procedure to get the right kind of effective structure out (namely chiral fermions)—this notion was introduced by Witten at the Shelter Island II conference [128].

[50] Rolf Schimmrig found another, with $\chi = -6$ in 1987, following a classification of all possible Calabi-Yau manifolds embedded in $\mathbb{P}_2 \otimes \mathbb{P}_3$ (where $\mathbb{P}_n$ is the n-complex-dimensional complex projective space): [116].

[51] That is, the problem of explaining the gap separating the strength of the extremely weak gravitational interaction and the weak nuclear force (a difference of 10^{32} orders of magnitude). Again, not explained by the standard model of particle physics.

energy physics that can be tested against experiments; though this amounts to more of a 'calibration' with existing results, than novel predictions.

Recall also that each E_8 factor associated with the $E_8 \otimes E_8$ heterotic string has dimension 248. Given the decomposition $E_8 \supset E_6 \otimes SU(3)$ (the maximal subgroup of E_8, induced by the Calabi-Yau compactification[52] of Candelas et al., where the $SU(3)$ lives in a single copy of E_8), we get the following further decomposition of the adjoint representation of E_8:

$$(248) = (78, 1) \oplus (27, 3) \oplus (\overline{27}, \overline{3}) \oplus (1, 8) \tag{9.4}$$

This results in an E_6 gauge theory closely related to E_6 four-dimensional grand unification models, where standard model-like chiral fermions are possible, thanks to the group's complex representations[53]: a realistic generation of quarks and leptons can be accommodated by the 27 fundamental representation of E_6 (see [20, pp. 61–62]).[54] In other words, the gauge group here is capable of accommodating the kinds of particle found in the standard model. That is to say, one can, given a suitable compactification and decomposition schemes, 'recover' the standard model (or something close) as an output. The problem, obvious from the beginning of this enterprise, was that in order to stand a chance of pinning down a space and decomposition that could match up to the results of experiment and observation, one had to do a certain amount of 'fixing by hand' ("invoking phenomenological criteria," as Ross puts it: [105, p. 382]) to get a reasonable vacuum solution, which is of course against the *ethos* of the theory of leaving nothing *arbitrary*.[55]

---

[52] More precisely, given such a compactification it is necessary to identify the $SU(3)$ spin connection (i.e. the $SU(3)$ gauge field) on the Calabi-Yau manifold with an $SU(3)$ subgroup of the $E_8 \otimes E_8$ gauge potentials (for just one factor, so that the second factor has vanishing spin connection), in order to satisfy the classical equations (cf. [110, p. 273]).

[53] Recall that unbroken E_8 admits only real representations and therefore is not capable of supporting chiral states.

[54] This includes the electron, electron neutrino, up and down quarks. A right-handed neutrino and two Higgs doublets can also be accommodated. Note that E_6 contains as a subgroup $SU(5)$, as utilised in Georgi and Glashow's [57] grand unification scheme.

[55] Of course, if certain features of the world are going into the construction of a theory, then the ability of that theory to accommodate those features is not going to be taken as strong evidence for the theory. As Deborah Mayo puts it: "evidence predicted by a hypothesis counts more in its support than evidence that accords with a hypothesis constructed after the fact" ([93, p. 251]). Inasmuch as the evidence (of the generations, etc.) support the specific string theory (suitably compactified), it apparently does so only trivially, for this reason. However, Wayne Myrvold [95] has described a Bayesian account of theory-confirmation whereby simplicity and unification can be regarded as contributing to the level of empirical support afforded some theory—where by "unification" he means the establishing of informational relevance links between apparently independent phenomena. As he puts it: "the ability of a hypothesis to unify a body of evidence contributes in a direct way to the support provided to h by the body of evidence" (p. 412). Bayesian string theorists can then help themselves to the theoretical virtue of unification as offering confirmational support to their theory. Of course, such schemes generally depend on there being a family of *rival* theories between which one would like to choose. Yet in the case of string theory *qua* unified theory of the interactions, it has

Given the obvious (and well-vocalised) disconnection between superstring theory and experimental physics, it was of vital importance to the research programme that a uniqueness claim could be upheld at this sensitive stage. However, already in the late 1980s, soon after apparent progress had been made on the phenomenological side of the theory, a serious problem emerged involving the number of possible vacuum states which have a very high degree of degeneracy. This degeneracy ruins the uniqueness of the theory, and in so doing also damages the predictive power of the theory.[56] Strominger's paper on "Superstrings with Torsion" [122], from 1986,[57] brought into focus the full impact of this plurality, in a way closely related to what would later became known as 'the string landscape' (while also pointing towards several other future paths that would be taken, including anthropic reasoning, nonperturbative effects, and dualities linking apparently different theories). A little later, Lerche, Schellekens, and Warner developed these further (based on earlier ideas developed in [86]):

> [S]tring theory appears to have succeeded too well. One can debate whether this abundance of consistent theories is good or bad in principle, but it certainly poses horrendous practical problems. With regard to matters of principle, several schools of thought exist. Some people hope to find a criterion which favors one four-dimensional theory, hopefully containing the standard model, over all others. The most obvious place to look for such a criterion is in nonperturbative effects, but unfortunately next to nothing is known about such effects. Others hope for some universality theorem stating that if one considers some ensemble of all vacua, the dominant contribution will come from those theories that lead to the standard model. Still others speculate that nothing but our own existence selects the vacuum we live in (this is also known as the anthropic principle). If one takes this last point of view, the main worry is that the number of consistent string theories is too small rather than too large.[58]

(Footnote 55 continued)

no rivals—though, of course, it does have rivals for specific 'sub-problems' (such as the problem of quantum gravity).

[56] Another level of arbitrariness was the fact that, as John Schwarz puts it, "there is no compelling theoretical reason to separate off four-dimensional spacetime or to require that it be a Minkowksi space" ([111, p. 359]). This was essentially the feature that so irked Richard Feynman: "maybe there's a way of wrapping up six of the dimensions. Yes that's possible mathematically, but why not seven? When they write their equation, the *equation* should decide how many of these get wrapped up, *not* the desire to agree with experiment" ([42, p. 194]). Of course, this was a known problem: Feynman was simply repeating what many string theorists were also saying. For example, in 1986, Ramond explicitly writes that a good string theory (or rather a good string vacuum) "must tell us why theories which are apparently perturbatively healthy in higher dimensions feel the need to compactify" ([101, p. 104]). However, it highlights the emergence of a slightly different unificatory approach in which known (yet often inexplicable and disparate) evidence goes into the construction of the theory, reducing the number of possible theories or solutions. Relating to the previous footnote, the fact that there is *no other known way* of bringing together quantum gravity and other forces in a unified scheme is, of course, a key part of string theory's case to the present day (and, indeed, this 'no alternatives' claim forms the basis of a recent defence of string theory by Richard Dawid: [32]).

[57] Dieter Lüst also discussed superstring compactification with torsion, independently, at around the same time, employing homogeneous coset spaces (see [89]).

[58] As we see in the next chapter, this worry would be eliminated by later work, which revealed an even greater abundance of vacua.

These points of view can make sense only if the four-dimensional string theories are not really different theories, but different vacua of the same theory. This is generally believed to be likely (circumstantial evidence exists in the form of non-trivial connections between different theories, for example via torus compactification), but it would be interesting to know of what exactly they are ground states. The recent progress in string theory has taught us a lot about possible ground states, and perturbations to arbitrary order about them, but we still know disappointingly little about effects that go beyond perturbation theory. Such effects must certainly be there: string theory contains Yang-Mills theory, which does have non-perturbative effects that are unlikely to result from string perturbation theory ([87, pp. 104–5]).

At this stage, then, as Schellekens expresses it, "string theory had become a victim of its own success" ([109, p. v]). The search for phenomenologically acceptable theories led to a profusion of theories to choose between, yet with no principle of selection. However, Lerche et al. don't see this as necessarily problematic: if the features of the standard model can indeed be located somewhere in the space of possible superstring theories, then that warrants further study, for it leaves open the possibility of schemes that could somehow eliminate those examples differing from the solutions containing the standard model:

Although at first it seemed that the theory would fail its first confrontation with experiment by predicting the wrong space-time dimension, that too changed very quickly after its birth. Indeed, what once may have seemed an insurmountable obstacle is now a gaping hole: not only can we construct string theories in four dimensions, but it is also so easy that we can make them in embarrassing abundance. Meanwhile the miracles that started the excitement have survived, although we are now so used to them that we tend to forget them: chiral string theories in four dimensions exist and their spectra have the same general features (gauge groups, representations and family replication) as the standard model. If all this is realized in a class of theories that holds the promise of offering a finite and consistent theory of quantum gravity, then there is certainly reason for excitement ([87, p. 3]).

In 1986, Schellekens went on to dismiss the goal of uniqueness as "philosophy" (a throwback to Chew's, by this stage largely discredited, approach to physics perhaps?):

The prevailing attitude seems to be that "non-perturbative string effects" will somehow select a unique vacuum. This is unreasonable and unnecessary wishful thinking. We do not know at present how to discuss such effects, and have no idea whether they impose any restrictions at all. One cannot reasonably expect that a mathematical condition will have a unique solution corresponding to the standard model with three generations and a bizarre mass matrix. It is important to realize that this quest for uniqueness is based on philosophy, not on physics. There is no logical reason why the "theory of everything" should have a unique vacuum. All we can reasonably demand is that there exists a consistent and stable ground state which describes all physics correctly. The recent "ground state explosion" (which may well turn out to be just the tip of an iceberg), certainly enhances the chances that such a ground state does indeed exist, although pessimists will probably take a dimmer view of the recent developments ([108, p. 5]).

One such suggestion for a non-perturbative mechanism selecting a unique vacuum comes out of the analysis of string perturbation theory for bosonic strings by David Gross and Vipul Periwal, in 1988 [72]—Gross was (and still is) a strong supporter of the uniqueness position. They argue that the theory diverges (and is not Borel summable, similarly to QCD[59]) for arbitrary values of the coupling constant. This is taken to point to a non-perturbative instability of the vacuum, by analogy with the situation in ordinary quantum field theories. They suggest that "the enormous multitude of classical heterotic solutions might all be unstable, and that the truly stable (and perhaps unique) ground state is picked out by the nonperturbative dynamics that destabilizes them ([72, p. 2107]; see also [111, p. 370]).[60] Note that this implies that perturbative string theory is incomplete: non-perturbative factors[61] must come into play in order to have a proper grasp of the theory. This is the topic of the next chapter.

The most common strategy for dealing with this degeneracy in the framework of string theory—thus restoring a kind of theoretical uniqueness—one that has continued to this day (though with the nature of the degeneracy altered by later developments in string theory), has been to view the space of string vacua as a space of solutions of one and the same theory—this is mentioned by Strominger above. This same kind of strategy applies to the multitude of *types* of consistent superstring theory,[62] though initially the evidence for the connections (dualities) wasn't as strong as it is today:

> [I]t was puzzling to us why there seemed to be five consistent superstring theories. We only needed one theory to describe the world, we felt, at least if it was the right one. There were much fewer than in quantum field theory. In quantum field theory, there are an infinite number of quantum field theories, with all sorts of freedom to do things differently, although none of them contains gravity in a consistent way. Here we had five theories, each of which contained gravity, but we only wanted one. The story really changed at the end of '94 or beginning of '95. I think the climax was a lecture that Ed Witten gave at a conference at USC [University of Southern California] in the spring of '95, which was a conference called "Strings '95," the annual strings conference. There had been some related work earlier; I won't say that

---

[59] Though, of course, each term in the series is nonetheless ultraviolet finite.

[60] John Moffat [94] had argued in 1986 that the case for N-loop finiteness had yet to be proven, despite the existing 1-loop proofs and expectations that all was well at other orders. He believed that the 'finiteness' position was not really any better than supergravity theory in that cancellation had to be demonstrated for all N in the N-loop amplitude, which he expected to hold only for a very special kind of superstring theory. However, according to John Schwarz, Moffat's claims were understood to be wrong even at the time of writing (private communication).

[61] More specifically, one expects the appearance of 'instantons,' behaving as a power of e^{-1/g_s^2}. Though most of the pieces of the puzzle were available from the late-1980s onwards, these were not explicitly identified with D-branes until 1994/5, when string theory becomes part of the larger M-theory.

[62] I propose calling this profusion in the number of types of string theory (Type I, Type IIA, Type IIB, Heterotic $SO(32)$, Heterotic $E_8 \otimes E_8$, etc.) a *Plurality of Type 1* (with arabic numerals distinguishing this classification from that involving the various types of string theory, with its associated roman numerals). This can be contrasted with a *Plurality of Type 2* describing the degeneracy in the ground states (Schellekens' "ground state explosion") of some particular string theory (selected from the elements in the type 1 plurality). Once D-branes are introduced as central elements of string theory, another level is introduced, partly as a consequence of dealing with issues caused within level 2 (namely fixing the arbitrariness of the moduli describing the shape and size of particular Calabi-Yau manifolds by stabilising them with flux: "compactifications with flux" [81]). This level 3 plurality is often called 'the Landscape' (or, alternatively, 'the discretum': i.e. almost enough elements to seem like a continuum, but not quite [18]). We briefly return to such issues in §10.3. For now, it is enough to notice that at we 'zoom in' on one kind of plurality, another larger one takes its place—however, thus far, the type 3 plurality appears not to degenerate any further into a fourth category.

every new idea was in that one lecture, but a remarkable number were. People were kind of blown away by it. So what became clear, with that lecture and subsequent developments, was that in fact all of these string theories are related to one another in different ways by what we call dualities. So we were able to recognize them as just five different special cases of a single underlying theory; it became clear that there really is just one theory. And this non-uniqueness has to do with the fact that this single theory has many different solutions, or quantum vacua. And that's what we had actually been counting; we hadn't been counting different theories. So there seems to be a unique, underlying theory.[63]

However, these varieties of plurality led to a bifurcation point in approaches: on the one hand, there were those who stuck with the plurality, and sought to tame it somehow, perhaps by finding links between the apparently diverse theories, as Schwarz describes.[64] On the other hand there were those who looked elsewhere in string theory. For example, along the latter lines, the explosion in the number of phenomenologically-realistic models, combined with the absences of a dynamical selection principle, led directly (and explicitly) to an attempt to resuscitate the older $N = 2$ Type II theories (containing only closed strings) in a bid to find a more predictive scheme—the oriented variety, IIB, contains gravitation but no gauge symmetry; the unoriented version has both gravitation and a $U(1)$ gauge symmetry. These are tachyon-free, but went largely ignored on the grounds of their poor phenomenological prospects (only generating the $U(1)$ gauge group)—Witten had shown that one can't get chiral fermions from such theories using the Kaluza-Klein mechanism alone: [128].[65] However, by using some of the new purely string-theoretic tricks developed in heterotic string theory, some four-dimensional ground states were in fact discovered, by three groups [13, 83, 86][66], that led to quasi-realistic physics, including the gauge groups of the standard model. Even chiral fermions were found in later work, by Lance Dixon, Vadim Kaplunovsky, and Cumrun Vafa [38]. However, the problem was, as Dixon et al. also demonstrated, that *both* the physically correct gauge groups *and* chiral fermions could not be implemented simultaneously in the same model: "*no models based on the type II superstring can possibly contain the standard model gauge group and its fermion content!*" ([38, p. 45]). Given this

---

[63] Interview with John H. Schwarz, by Sara Lippincott. Pasadena, California, July 21 and 26, 2000. Oral History Project, California Institute of Technology Archives. Retrieved [2nd jan, 2012] from the World Wide Web: http://resolver.caltech.edu/CaltechOH:OH_Schwarz_J.

[64] However, as Schellekens points out, while the position that all of the different theories are really different vacua of one and the same theory (of "one generic heterotic string"), is "attractive from a philosophical point of view" (i.e. it restores a form of uniqueness), it doesn't do much to help with the phenomenological project since "[o]ne still has to understand the gigantic space of ground states to be able to make progress" ([109, p. 171]). Calling them theories or solutions doesn't reduce the number of specific entities, whatever they might be. However, a point we return to is that the philosophically attractive strategy does at least recommend *other* strategies for trying to accomplish a genuine reduction in the number, by finding equivalences between various of the solutions, which doesn't seem as well motivated if we suppose that they are distinct theories.

[65] This, no doubt, was one of the contributing factors behind the slow uptake of Green and Schwarz's new superstring theories in the very early 1980s.

[66] Kawai, Lewellen, and Tye proposed calling their strings "Type III" (perhaps harking back to Gell-Mann's earlier attempt?) since they shared properties of heterotic strings, namely the asymmetric treatment of left- and right-movers by the spin structures and worldsheet supercurrents ([83, p. 63]).

sorry state of affairs, the recommendation of Dixon et al. is to "give up on the type II superstring itself" ([38, p. 79])—they reach this conclusion after dismissing several proposals to rescue the approach.[67]

Doron Gepner notes the jubilation followed by disappointment that resulted from the discovery that quantum superstring theories were consistent (anomaly free) coupled with the potentially realistic discovery of the heterotic string, followed by the non-uniqueness that followed when the inevitable compactification took place:

> Unfortunately, in the process of compactification down to four dimensions this uniqueness is replaced by a vast number of possibilities, each leading to different physical predictions. Moreover, the geometrical interpretation of strings was lost due to the constructions of apparently non-geometrical compactifications based on free fermions or bosons [58].

In the paper from which this passage is drawn, Gepner attempts to try and restore some uniqueness (thus following the second approach to the plurality of theories, alluded to above) by identifying "enhanced gauge symmetries" holding between various of the string backgrounds, rendering what look like distinct string theories physically equivalent—we saw this idea in operation in [97] above.

Gepner [58] introduced a 'minimal model' framework—I'll call it the 'CFT-CY Correspondence' (where 'CFT' stands for 'conformal field theory')—into the study of string theory that transformed the way compactified string theories are understood. The programme of seeking classical solutions to the string equations of motion is transfigured into the seemingly unrelated task of analysing 2D conformal field theories (satisfying certain constraints)—of course, the two dimensions (one space; one time) correspond to those of a string worldsheet. As Gepner puts it himself: "any supersymmetric heterotic compactification with an arbitrary conformal field theory with $N = 2$ superconformal symmetry is equivalent to string propagation on a Calabi-Yau manifold" ([58, p. 380]). Hence, a string theory compactified along the lines of a $\mathcal{M}^4 \times \mathcal{K}^6$ Calabi-Yau manifold approach corresponds to a two-dimensional free conformal field theory in four-dimensions and a two-dimensional field theory in six-dimensions: *the conformal field theory determines the physics of strings.*[68] A crucial feature, relating back to the discussion on p. XX—over the validity of treating Calabi-Yau compactifications as providing consistent solutions of the classical equations of motion as derived from a theory of strings—is that Gepner's models constitute exact solutions to the string equations "despite the breakdown of conformal invariance at

---

[67] A consensus appears to have been reached here that the conclusion holds good—Schellekens calls this episode in the life of Type II strings "a rather short-lived revival" ([109, p. 413]). Type II strings dropped out of favour once again. Curiously, however, the other path leading from the bifurcation point mentioned above, would lead (via non-perturbative explorations), in a matter of several years, to the re-establishment of Type II theories as worthy objects of attention. Interestingly, in his assessment of the results of Dixon et al., Bert Schellekens presciently muses: "Time will tell whether the negative result of [DKV] is the final nail in the coffin of Type-II strings, or just another no-go theorem that can be evaded because one of the assumptions was too strong in an unforeseen way" ([109, p. 414]).

[68] Hull and Witten [77] developed a simple (p, q) notation to describe the number of left-moving (p left-handed Majorana-Weyl supercharges) and right-moving (q right-handed Majorana-Weyl supercharges) supersymmetries on the worldsheet.

the four-loop level" ([58, p. 387]) so that Calabi-Yau manifold compactifications too constitute exact solutions (given the correspondence).

A variety of people and groups were converging on similar results, in each case pointing to strange new kind of 'stringy quantum geometry' in which one derived the same physical results (i.e. the observable correlation functions, or Yukawa couplings) despite switching between apparently quite distinct Calabi-Yau manifold backgrounds. In other words, the mapping from the structure ⟨Calabi-Yau manifold, Strings⟩ to an observable structure was many-to-one, implying that some kind of symmetry was generating unphysical, surplus structure.[69] Explicit computations of Yukawa couplings were carried out by Jacques Distler and Brian Greene, in 1988 [34], and shortly after by Gepner: [59].

This brings us back to the mirror symmetry briefly mentioned in fn., 24. Mirror pairs are very specific examples of Greene's string equivalent spaces which differ by an exchange of certain Hodge numbers characterising the spaces (which can cause the Euler number to change sign). Naturally, this resolves the above ambiguity inherent in the many-to-one map between Calabi-Yau manifolds and CFT (cf., [135, p. xi]). Yau ([136, p. 157]) credits Lerche, Vafa, and Warner with the first statement of mirror symmetry in [88], from 1989.[70] Though the terminology is somewhat different to standard treatments of mirror symmetry, Yau's claim seems to be borne out by the following passage ([88, p. 442]):

> [F]or superconformal models coming from compactification on Calabi-Yau manifold, the
> (c, c) ring becomes isomorphic to the structure of the cohomology ring of the manifold in the

---

[69] Brian Greene sums the idea up concisely as follows: "[t]wo distinct spacetime M_1 and M_2 (distinct in the classical mathematical sense of not being isomorphic as (complex) manifolds) are said to be string equivalent if they yield *isomorphic* physical theories when taken as backgrounds for string propagation" ([69, p. 30]). In plainer words: "a classical mathematician would describe M_1 and M_2 as being distinct while a string theorist would say they are the same in the sense that absolutely no experiment can distinguish between them" (ibid.).

[70] Yau claims that Dixon and Gepner had talked about something very close to this idea a little earlier (though without publishing their results), though they used $K3$ surfaces, rather than Calabi-Yau manifolds, which are trivial by comparison (since all $K3$'s are homeomorphic)—Candelas et al. ([23, p. 119]) also make the same priority claim. However, Rolf Schimmrigk seems to have been on a similar track in his 1987 paper describing a novel three-generation Calabi-Yau manifold. Thus, he writes: "It is interesting to note that this manifold has the same Hodge diamond and fundamental group as the two non-simply connected manifolds with Euler number -6. This makes it conceivable that all three manifolds are in fact diffeomorphic although they are constructed in different ways starting from different ambient manifolds" ([116, p. 179])—this equivalence was proven shortly afterwards by Brian Greene and Kelley Kirklin [67]. However, Schimmrigk assumes, as seems *prima facie* sensible, that physical differences would result: "from a physical point of view these examples should behave quite differently, since their complex and Kahler structures and therefore the coupling between the various multiplets are expected to be different because these are determined by the choice of the embedding space and the particular restriction of the moduli" (ibid.). It is, of course, precisely the roles of the complex structure $h^{2,1}$ and Kähler moduli $h^{1,1}$ that are interchanged by the mirror duality. However, testing for equivalence demanded a detailed analysis of the conformal field theories on the two manifolds (which itself called for Gepner's framework in [58]). Yau himself wasn't initially convinced by the mirror conjecture, since most of the manifolds found had been of negative χ—this changed later as more systematic means of generating new examples emerged.

large radius limit. Of course it should be clear in general that it is a matter of convention as to which ring we call (a, c) and which ring we call (c, c) because we can always flip the relative sign of the J and $\bar{J}$, and change our conventions. So more precisely we should say that one of the two rings (c, c) or (a, c) is a deformation of the cohomology ring of the manifold. One of them gives the Poincaré polynomial of the manifold $[P_{(a,c)}]$ and the other gives a Poincaré polynomial of the form $[P_{(c,c)}]$, which in general differs from the Poincaré polynomial of the manifold. One would clearly like to have a geometric interpretation of the other Poincaré polynomial. One possibility might be that this polynomial is the Poincaré series for (a deformation of) the cohomology ring of another manifold. This is quite possible in light of the fact that string propagation on topologically distinct manifolds can be isomorphic. This happens, for example, for certain orbifolds. If so, there must be another manifold $\tilde{M}$ which the betti numbers satisfy

$$b^{\tilde{M}}_{p,q} = b^{M}_{d-p,q}$$

This latter expression (where I assume by "betti numbers" they mean "Hodge numbers") corresponds to what would later be called 'the mirror map,' producing a topologically distinct manifold, yet preserving the observable features of string propagation, as encoded in Gepner's minimal models (the conformal field theories associated with the compactifications). The "conventional" aspect described by Lerche et al. quite clearly corresponds to a duality in the theory: $h^{p,q} \leftrightarrow h^{q,d-p}$. This "Calabi-Yau manifold duality" was made more precise, and the terminology of 'mirror pairs' introduced, by Greene and Ronen Plesser in 1990 [68]—where the authors also link mirror duality to other dualities, showing how it fits in a continuing saga stretching back to the large/small radius duality of Kikkawa and Yamasaki.[71]

This is, of course, part of a pattern that runs through the work on compactification from 1984 onwards: strings and spacetime conspire together in a way that is quite distinct from a physics of particles. These converging lines, on the idea that the spacetime picture of string theory must be radically different from those encountered previously, paved the way for the non-perturbative work that was to follow, in which even the notion of 'dimension' is deemed conventional. Seen in this light, however, the apparently radical shifts that we see in the next and final chapter can be seen as part of a more continuous historical lineage.

9.4 Quantum Gravity Elsewhere

In his talk at the 1986 Nobel Symposium on the Unification of Fundamental Interactions, Stanley Deser opened with the statement: "Taming the problems of quantum gravity is one of the most dramatic promises of string theory" ([33, p. 138]).[72] It was only around the mid-1980s that string theory began to be taken seriously as an

---

[71] This link was made exact in a later paper by Strominger, Yau, and Zaslow unambiguously entitled "Mirror Symmetry is T-Duality" [124].

[72] Deser, of course, was no stranger to quantum gravity, having been immortalised as an initial in the ADM [Arnowitt, Deser, Misner] collaboration. Another initial, 'A' = Richard Arnowitt, also made contributions to string theory, primarily in the area of string phenomenology.

approach to quantum gravity. In both 'Oxford Symposia' on quantum gravity, string theory was nowhere to be seen—though supergravity was.[73]

There was an interesting conference on the Conceptual Problems of Quantum Gravity in May of 1988, co-organised by Abhay Ashtekar (an important figure in the non-perturbative approach to canonical quantum gravity) and John Stachel. One of general relativist Robert Geroch's's former students, Gary Horowitz (of the "Vacuum Configurations" paper that launched Calabi-Yau compactifications), delivered a talk on spacetime (or rather its *absence* in string theory. This talk tackles head-on the problem of background independence (the idea that space-time geometry is something that one solves for, rather than putting it in by hand). As canonical quantum gravity researchers are fond of pointing out, the machinery of perturbative quantization methods, with their presuppositions of smooth, fixed, spacetime backgrounds, are bound to be inadequate when it comes to quantizing gravity, since gravity is inextricably entangled with spacetime geometry. The basic dynamical variable in general relativity is the metric, which serves a dual purpose in this theory: it both determines the geometry of spacetime (and so the kinematic structure against which physical processes are defined) and acts as a (pre-) potential for the gravitational field. Since it is a dynamical variable that also is responsible for spacetime geometry, it follows that geometry itself is dynamical: one has to solve the dynamics in order to get to the kinematics. This feature is a central part of 'background independence'—in some ways it is similar to the boot-strapper's expulsion of arbitrary elements. Horowitz makes the same point in the string theoretic case:

> If space-time is to be a derived concept, the theory must be formulated in terms of something more fundamental. Since string theory has the potential for being our first consistent quantum theory of gravity, it is important to ask what is its prediction for this more fundamental structure. In many discussions of string theory, one has the impression that there isn't any. One considers strings moving in a fixed background space-time. One mode of the string corresponds to a massless spin-two particle called the graviton, and the scattering amplitudes for these gravitons are calculated. In effect, the metric has been divided into a kinematical background space-time and a dynamical gravitational field described by the string. If this was the complete description of the theory, one might conclude that string theory represented a real step backward rather than forward in our understanding of space-time. Of course, this is not the complete theory. The graviton-scattering amplitudes mentioned above are calculated perturbatively, and the background space-time represents the solution one is perturbing about. In a more fundamental, nonperturbative formulation of string theory, both the dynamical and the kinematical aspects of space-time should be derived concepts ([76, pp. 299–300]).

There was much that was of a promissory nature in the talk, based on the purely cubic action defining Witten's string field theory [131], but it shows quite clearly that the problem of background independence was understood early on by the string theory

---

[73] It is hardly surprising that strings didn't appear in the first of these [78], in February 1974 (when string theory's potential to describe gravity had just emerged), but the second symposium [79], was held in April 1980, when superstring theory's potential was certainly known. The supergravity papers in the second volume did, however, mention the dimensional reduction models from some of the superstrings papers, but failed to mention their stringy heritage.

community in a way that matches the way it is understood in the canonical quantum gravity community and imbued with just as much importance.[74]

One should not neglect the role of other approaches to quantum gravity in solidifying and spreading string theory. If string theory was shown to provide a finite quantum theory of gravity that would, of course, be an extremely significant achievement quite regardless of its ability to predict the features of the standard model of particle physics. It would itself constitute the only known mathematically consistent theory of quantum gravity. However, strides were being made elsewhere in quantum gravity. Two developments, pulling in different directions, are of particular significance:

1. In 1986, Marc Goroff and Augusto Sagnotti [65] proved the non-renormalizability of Einstein gravity, even for pure gravity. The focus on non-renormalizable ultraviolet divergences cannot failed to have directed attention at string theory's elimination of such. As Gell-Mann expressed it in 1988: "no Band-aid will fix the quantum version of Einsteinian gravitation" ([56, p. 138]). This suggested that perhaps a radical approach was necessary, of which string theory was an example, with its cure provided by the absence of point-particles. It seemed, at the time, that it was the only example of an approach to quantum gravity free of unrenormalizable infinities identified by Goroff and Sagnotti.[75]

2. Competition began to arise in the form of an advance within the canonical approach to gravity. Though perturbatively non-renormalizable, this does not rule out the existence of non-perturbative quantizations of general relativity. New variables (implying that geometrical information is contained in self-dual spin connections, or fluxes) had been found for general relativity by Amitabha Sen in 1982 [117], resulting in a considerable simplification of Einstein's equations. In 1986 Abhay Ashtekar [6] figured out how to employ this in canonical quantum gravity, replacing the 3-metric and its conjugate with (complex) SU(2) connections and SU(2) soldering forms (rendering the metric a derived quantity). This method established the beginnings of a link to Yang-Mills theory and its associated techniques, and launched non-perturbative quantum gravity onto the scene.[76] The 'loop representation' of gravity, developed by Ted Jacobson and Lee Smolin

---

[74] In the discussion period ([76, p. 314]), Ashtekar didn't quite agree that the string position matched the canonical quantization position on this issue. The disagreement boiled down to the fact that string theory does not *quantize* the classical gravitational field, to get a microscopic structure of spacetime (with the spacetime picture coming instead from a classical solution to the string field equations), whereas that it precisely what occurs in the canonical approaches. Horowitz responds by pointing to the kinds of 'quantum geometrical' features that we saw in Amati, Ciafaloni, and Veneziano, and Gross' work on high-energy string scattering. Unfortunately, common ground could not be found.

[75] In fact, the proof of string theory's finiteness was some time coming. There was word circulating that Mandelstam had discovered a proof in the mid- to late 1980s. He finally supplied an explicit proof in 1991 [90] (supplying formulas for the n-loop amplitude, for Bose strings and superstrings, that "could be put on a computer (but may require an unreasonable amount of computer time)" ([90, p. 82]).

[76] [7] gives a good overview of the state of play in the aftermath of the new canonical variables. Interestingly, canonical quantum gravity had undergone its own 'dark ages' in which seemingly intractable problems (involving products of operators sitting at the same spacetime point, amongst other things) stalled the programme.

[80], was able to construct a class of exact solutions to the Wheeler-DeWitt equation based on Wilson loops. Carlo Rovelli then joined this collaboration, in which the Wilson loops were viewed as the fundamental variables [106].[77]

Competition can, of course, be a good thing, and in having new targets (non-perturbative formulation, background independence, etc.) also suggested by alternative approaches, string theory was forced to develop and advance in ways it might not otherwise have done. The availability of a non-perturbative formulation of quantum gravity was the big selling point of the new approach to canonical quantum gravity emerging in the mid-1980s. It is possible that some of the research on the importance of having such a formulation in the context of gravitational theories may have seeped into string theoretic research.[78] It is difficult to test this idea, from a historical point of view, since both the superstring renaissance and the renaissance of canonical quantum gravity occurred at around the same time.

9.5 Summary

By 1987 superstring theory was certainly secure. It was often featured in news stories in the popular press, with writers naturally jumping on the title "theory of everything". Schwarz was one of four string theorists to receive a MacArthur Foundation fellowship in 1987—in subsequent years they were also awarded to three other string theorists (Eva Silverstein, Nathan Seiberg, and Juan Maldacena).[79] In his own summary of the state of superstring theory, at a talk he gave at the birthday celebrations for Valentine Telgedi in July 1987, Murray Gell-Mann lists the following features that the theory offered ([56, p. 140]):

- an elegant, self-consistent quantum field theory,
- generalising Einstein's general-relativistic theory of gravitation treated quantum-mechanically,
- in the only known way that does not produce infinities,
- parameter-free,
- based on a single string field,
- but yielding an infinite number of elementary particles,

---

[77] Following the foundational work, a considerable amount of formal work was carried out to make the theory consistent: regularization, defining the inner-product structure, and so on. A change of basis (to so-called 'spin-networks') helped resolve many of these. See Rovelli's entry on "Loop Quantum Gravity" in the *Library of Living Reviews* for a more detailed breakdown of key events: http://relativity.livingreviews.org/Articles/lrr-1998-1.

[78] Gary Horowitz, for example, would, for a time, have been a natural interface between the two approaches.

[79] A spokesman for the MacArthur Foundation noted that the award was "an indication of the excitement the theory is causing in physics" (http://articles.latimes.com/1987-06-16/news/mn-7719_1_strings-super-studied).

- some hundreds of which would have low mass (although we don't know why they would be so very low!),
- including particles with properties like those of electrons, quarks , photons, gluons, etc.,
- with the underlying symmetry system essentially determined,
- and with the symmetry breaking connected with the behaviour of some extra, but perhaps formal dimensions.

Hence, the theory was characterised, as it had been throughout its life, by a beautiful mathematical structure, now with consistency proven beyond doubt, but with stubborn problems in the experimental domain. The strong hopes of realistic physics, by finding the right Calabi-Yaus, were dampened by the sheer quantities with which such manifolds could be produced. The subsequent evolution of the theory can be seen as a progressive march in which more of the mathematical structure is revealed, but neither the central principle defining string theory, nor firm experimental evidence are produced. The present-day situation has not shifted so very far from this, though there have been extremely important developments coming from the discovery of new dualities and new non-perturbative objects, as we will see. These are leading to a potential definition of the theory, but the problem with an apparent bounty of solutions remains, only in quantities many orders of magnitude larger.

References

1. Alvarez-Guamé, L., & Freedman, D. Z. (1980). Kähler geometry and the renormalization of supersymmetric σ models. *Physical Review D, 22*(4), 846–853.
2. Alvarez-Guamé, L., & Freedman, D. Z. (1981). Geometrical structure and ultraviolet finiteness in the supersymmetric σ model. *Communications in Mathematical Physics, 80*, 443–451.
3. Alvarez-Guamé, L., & Ginsparg, P. (1985). Finiteness of Ricci flat supersymmetric non-linear σ models. *Communications in Mathematical Physics, 102*(2), 311–326.
4. Amati, D., Ciafaloni, M., & Veneziano, G. (1987). Superstring collisions at Planckian energies. *Physical Letters B197*(1,2), 81–88.
5. Amati, D., Ciafaloni, M., & Veneziano, G. (1987). Can spacetime be probed below the string size? *Physical Letters B & 216* (1,2), 41–47.
6. Ashtekar, A. (1986). New variables for classical and quantum gravity. *Physical Review Letters, 57*, 2244–2247.
7. Ashtekar, A. (1988). *New perspectives in canonical gravity*. Naples: Bibliopolis.
8. Aspinwall, P. S. (1999). K3 surfaces and string duality. In S.-T. Yau (Ed.), *Surveys in differential geometry, volume V: Differential geometry inspired by string theory* (pp. 1–95). Boston: International Press.
9. Aspinwall, P. S. (2001) Resolution of orbifold singularities in string theory. In B. Greene, & S.-T. Yau (Eds.), *Mirror Symmetry II* (pp. 355–95). Providence: American Mathematical Society.
10. Aspinwall, P. S., Greene, B. R., Kirklin, K. H., & Miron, P. J. (1987). Searching for three-generation Calabi-Yau manifolds. *Nuclear Physics, B294*, 193–222.
11. Aspinwall, P. S., Greene, B. R., & Morrison, D. R. (1994). Calabi-Yau moduli space, mirror manifolds, and space-time topology change in string theory. *Nuclear Physics, B416*, 414–480.
12. Aspinwall, P. S., Greene, B. R., & Morrison, D. R. (1994). Space-time topology change and stringy geometry. *Journal of Mathematical Physics, 35*, 5321–5337.

13. Bluhm, R., Dolen, L., & Goddard, P. (1987). A new method of incorporating symmetry into superstring theory. *Nuclear Physics, B289*, 364–384.
14. Borcherds, R. E. (1986). Vertex Algebras, Kac-Moody Algebras, and the Monster. *Proceedings of the National Academy of Science, 83*(10), 3068–3071.
15. Borcherds, R. E. (1992). Monstrous Moonshine and Monstrous Lie Superalgebras. *Inventiones Mathematicae, 109*(1), 405–444.
16. Borcherds, R. E. (1992). Sporadic groups and string theory. *First European Congress of Mathematics. Volume I: Invited Lectures* (pp. 411–421). Birkhüser Verlag.
17. Borcherds, R. E. (2002). What is ... the monster? *Notices of the AMS49*(9), 1076–1077.
18. Bousso, R., & Polchinski, J. (2000). Quantization of four form fluxes and dynamical neutralization of the cosmological constant. *Journal of High Energy Physics, 6*, 1–25.
19. Breit, J. D., Ovrut, B. A., & Sergè, G. C. (1985). E_6 symmetry breaking in the superstring theory. *Physics Letters, 158B*(1), 33–39.
20. Candelas, P., Horowitz, G., Strominger, A., & Witten, E. (1985). Vacuum configurations for superstrings. *Nuclear Physics, B258*, 46–74.
21. Candelas, P., Dale, A. M., Lutken, C. A., & Schimmrigk, R. (1988). Complete intersection Calabi-Yau manifolds. *Nuclear Physics B, 298*, 493–525.
22. Candelas, P., & Kalara, S. (1988). Yukawa couplings for a three-generation superstring compactification. *Nuclear Physics, B298*, 357–368.
23. Candelas, P., de la Ossa, X. C., Green, P. S., & Parkes, L. (1991). A pair of Calabi-Yau manifolds as an exactly soluble superconformal field theory. *Nuclear Physics, B359*, 21–74.
24. Candelas, P., & Davies, R. (2010). New Calabi-Yau manifolds with small Hodge numbers. *Progress of Physics, 58*(4–5), 383–466.
25. Castellani, E. (2009). Dualities and Intertheoretic Relations. In M. Suárez, M. Dorato, & M. Rédei (Eds.), *EPSA Philosophical issues in the sciences: Launch of the European philosophy of science association* (pp. 9–19). Heidelberg: Springer.
26. Chodos, A. (1986). Marginalis: String fever. *American Scientist, 74*(3), 253–254.
27. Conway, J. H., & Norton, S. P. (1979). Monstrous moonshine. *Bulletin of the London Mathematical Society, 11*(3), 308–339.
28. Conway, J. H., Curtis, R. T., Norton, S. P., Parker, R. A. & Wilson, R. A. (1985). *Atlas of finite groups*. Oxford: Oxford University Press.
29. Conway, J. H., & Sloane, N. J. A. (1999). Sphere Packings, Lattices, and Groups. New York: Springer.
30. Cremmer, E., & Scherk, J. (1976). Spontaneous compactification of space in an Einstein-Yang-Mills-Higgs model. *Nuclear Physics, B108*, 409–416.
31. Cremmer, E., & Scherk, J. (1976). Spontaneous compactification of extra space dimensions. *Nuclear Physics, B118*, 61–75.
32. Dawid, R. (2013). *String theory and the scientific method*. Cambridge: Cambridge University Press.
33. Deser, S. (1987). Gravity from strings. In L. Brink et al. (Eds.), *Unification of Fundamental Interactions* (pp. 138–142). *Physica Scripta, The Royal Swedish Academy of Sciences*. Stockholm: World Scientific.
34. Distler, J., & Greene, B. (1988). Some exact results on the superpotential from Calabi-Yau compactifications. *Nuclear Physics, 309*(2), 295–316.
35. Dixon, L., Harvey, J. A., Vafa, C., & Witten, E. (1985). Strings on orbifolds. *Nuclear Physics, B261*, 678–686.
36. Dixon, L., Harvey, J. A., Vafa, C., & Witten, E. (1986). Strings on orbifolds (II). *Nuclear Physics, B274*, 285–314.
37. Dixon, L., Friedan, D., Martinec, E., & Shenker, S. (1986). The conformal field theory of orbifolds. *Nuclear Physics, B282*, 13–73.
38. Dixon, L., Kaplunovsky, V. S., & Vafa, C. (1987). On four-dimensional gauge theories from type II superstrings. *Nuclear Physics, B294*, 43–82.
39. Dixon, L., Ginsparg, P., & Harvey, J. (1988). Beauty and the beast: Superconformal symmetry in a monster module. *Communications in Mathematical Physics, 119*, 221–241.

40. Ellis, J. (1986). The superstring: Theory of everything, or of nothing? *Nature*, 323, 595–598.
41. Ferrara, S., Lüst, D., Shapere, A., & Thiesen, S. (1989). Modular invariance in supersymmetric field theories. *Physics Letters B, 225*(4), 363–366.
42. Feynman, R. P. F. (1988). Richard Feynman. In P. C. W. Davies & J. Brown (Eds.), *Superstrings: A theory of everything?* Cambridge: Cambridge University Press.
43. Frampton, P. (1974). *Dual Resonance Models*. Redwood City: The Benjamin/Cummings Publishing Company.
44. Frampton, P. (1986) *Dual resonance models and superstrings*. Singapore: World Scientific.
45. Freund, P. G. O., Oh, P., & Wheeler, J. T. (1984). String-induced space compactification. *Nuclear Physics, B246*, 371–380.
46. Frenkel, I. B., & Kac, V. G. (1980). Basic representations of Affine Lie algebras and dual resonance models. *Inventiones Mathematicae, 62*, 23–66.
47. Frenkel, I. B., Lepowsky, J., & Meurman, A. (1984). A natural representation of the Fischer-Griess monster with the modular function J as character. *Proceedings of the National Academy of Science81*(10), 3256–3260.
48. Frenkel, I. B., Lepowsky, J., & Meurman, A. (1989). *Vertex operator algebras and the monster*. Boston: Academic Press.
49. Freund, P. G. O., & Mansouri, F. (1982). Critical dimensions for strings, bags. *Zeitschrift für Physik C - Particles and Fields, 14*, 279–280.
50. Font, A., Ibáñez, L. E., Lüst, D., & Quevedo, F. (1990). Strong-weak coupling duality and non-perturbative effects in string theory. *Physics Letters B, 249*(1), 35–43.
51. Font, A., Ibáñez, L. E., Lüst, D., & Quevedo, F. (1990). Supersymmetry breaking from duality invariant gaugino condensation. *Physics Letters B, 245*(3,4), 401–408.
52. Galison, P. (2004). Mirror symmetry: Persons, values, and objects. In M. Norton Wise (Ed.), *Growing explanations: Historical perspectives on recent science* (pp. 23–63). Durham: Duke University Press.
53. Gannon, T. (2006). Modular Forms and Physics. *Moonshine beyond the monster: The bridge connecting algebra*. Cambridge: Cambridge University Press.
54. Garfield, E. (1986). The most-cited 1985 physical-sciences articles: Some knots in superstrings untied and quasicrystals not so quasi anymore. *Essays of an Information Scientist, 10*, 328–341.
55. Gell-Mann, M. (1986). Supergravity and superstrings. In M. M. Block (Ed.), *First Aspen Winter Physics Conference* (pp. 325–335). New York: The New York Academy of Sciences.
56. Gell-Mann, M. (1988). Is the whole universe made out of superstrings? In K. Winter (Ed.), *Festi-Val - Festschrift for Val Telegdi* (pp. 119–40). New York: Elsevier.
57. Georgi, H., & Glashow, S. L. (1974). Unity of all elementary particle forces. *Physical Review Letters, 32*, 438–441.
58. Gepner, D. (1987). Exactly solvable string compactifications on manifolds of $SU(N)$ Holonomy. *Physics Letters B, 199*(3), 380–388.
59. Gepner, D. (1988). Yukawa couplings for Calabi-Yau string compactification. *Nuclear Physics, B311*, 191–204.
60. Ginsparg, P. (1987). On toroidal compactification of heterotic superstrings. *Physical Review D, 35*(2), 648–654.
61. Giveon, A., Porrati, M., & Rabinovici, E. (1994). Target space duality in string theory. *Physics Reports, 244*, 77–202.
62. Goddard, P. (1989). Gauge symmetry in string theory. *Philosophical Transactions of the Royal Society of London A, 329*(1605), 329–342.
63. Goddard, P. (2012) From dual models to relativistic strings. In A. Capelli et al. (Eds.), *The birth of string theory* (pp. 236–261). Cambridge: Cambridge University Press.
64. Goddard, P. (2013). *Algebras, groups, and strings*. ESI: 12–15.
65. Goroff, M., & Sagnotti, A. (1986). The ultraviolet behavior of Einstein gravity. *Nuclear Physics, B266*, 709–736.
66. Green, M. B., Schwarz, J. H., & Brink, L. (1982). $N = 4$ Yang-Mills and $N = 8$ supergravity as limits of string theories. *Nuclear Physics, B198*, 474–492.

67. Greene, B. R., & Kirklin, K. H. (1987). On the equivalence of the two most favoured Calabi-Yau compactifications. *Communications in Mathematical Physics, 113*, 105–114.
68. Greene, B. R., & Plesser, M. R. (1990). Duality in Calabi-Yau moduli space. *Nuclear Physics, B338*(1), 15–37.
69. Greene, B. R. (1997). Constructing Mirror Manifolds. In B. Greene, & S. -T. Yau (Eds.), *Mirror Symmetry II* (pp. 29–70). Providence: American Mathematical Society.
70. Grisaru, M., van de Ven, A., & Zanon, D. (1986). Four-loop β-function for the $N = 1$ and $N = 2$ supersymmetric non-linear sigma model in two dimensions. *Physics Letters B, 173*(4), 423–428.
71. Gross, D. J., & Witten, E. (1986). Superstring modifications of Einstein's equations. *Nuclear Physics, B277*(1), 1–10.
72. Gross, D. J., & Periwal, V. (1988). String perturbation theory diverges. *Physical Review Letters, 60*(1), 2105–2108.
73. Gross, D. J., Harvey, J. A., Martinec, E., & Rohm, R. (1985). Heterotic string theory (I) The free heterotic string. *Nuclear Physics, B256*, 253–284.
74. Gross, D. J., Harvey, J. A., Martinec, E., & Rohm, R. (1985). Heterotic string. *Physical Review Letters, 54*(6), 502–505.
75. Gross, D. J. (1989). Strings at SuperPlanckian energies. *Philosophical Transactions of the Royal Society of London A, 329*(1605), 401–413.
76. Horowitz, G. T. (1988) String theory without space-time. In A. Ashtekar, & Stachel, J. (Eds.), *Conceptual problems of quantum qravity* (pp. 299–325). Birkhaüser.
77. Hull, C. M., & Witten, E. (1985). Supersymmetric sigma models and the heterotic string. *Physics Letters B, 160*(6), 398–402.
78. Isham, C. J., R. Penrose, & Sciama, D. W. (Eds.). (1975). *Quantum gravity: An oxford symposium*. Oxford: Oxford University Press.
79. Isham, C. J., Penrose, R., & Sciama, D. W. (Eds.). (1981). *Quantum gravity: A second oxford symposium*. Oxford: Oxford University Press.
80. Jacobson, T., & Smolin, L. (1988). Nonperturbative quantum geometries. *Nuclear Physics, B299*(2), 295–345.
81. Kachru, S., Kallosh, R., Linde, A., & Trivedi, S. P. (2003). de sitter vacua in string theory. *Physical Review D, 68*, 046005-1–10.
82. Kaiser, D. (Ed.). (2005). *Pedagogy and the practice of science*. Cambridge: MIT Press.
83. Kawai, H., Lewellen, D. C., & Henry Tye, S.-H. (1987). Four-dimensional type II strings and their extensions: Type III strings. *Physics Letters B, 191*(2), 63–69.
84. Kerner, R. (1968). Generalization of the Kaluza-Klein theory for an arbitrary non-Abelian Gauge group. *Annales de l'institut Henri Poincaré, 9*(2), 143–152.
85. Kikkawa, K., & Yamasaki, M. (1984). Casimir effects in superstring theory. *Physics Letters B, 149*, 357–360.
86. Lerche, W. D., Lüst, & Schellekens, A. N. (1987). Chiral four-dimensional heterotic strings from self-dual lattices. *Nuclear Physics, B287*, 477–507.
87. Lerche, W., Schellekens, A. N., & Warner, N. P. (1989). Lattices and strings. *Physics Reports, 177*(1 & 2), 1–140.
88. Lerche, W., Vafa, C., & Warner, N. P. (1989). Chiral rings in $N = 2$ superconformal theories. *Nuclear Physics, B324*, 427–474.
89. Lüst, D. (1986). Compactification of ten-dimensional superstring theories over Ricci-flat coset spaces. *Nuclear Physics B, 276*(1), 220–240.
90. Mandelstam, S. (1991). The n loop string amplitude: Explicit formulas, finiteness and absence of ambiguities. *Physics Letters B, 277*, 82–88.
91. Matsubara, K. (2013). Underdetermination and string theory dualities. Synthese. *Realism, 190*(3), 471–489.
92. Matsubara, K. (2013). *Stringed along or caught in a loop?: Philosophical reflections on modern quantum gravity research*. PhD Thesis. Uppsala University.
93. Mayo, D. G. (1996). *Error and the growth of experimental Knowledge*. Chicago: University of Chicago Press.

94. Moffat, J., et al. (1986). Are superstring theories finite? In J. M. Cameron (Ed.), *1st Lake Louise winter institute: New frontiers in particle physics* (pp. 610–615). Singapore: World Scientific.

95. Myrvold, W. C. (2003). A Bayesian account of the virtue of unification. *Philosophy of Science, 70*, 399–423.

96. Narain, K. S. (1985). New heterotic string theories in uncompactified dimensions <10. *Physics Letters B, 169*(1), 41–46.

97. Narain, K. S., Sarmadi, M. H., & Witten, E. (1987). A note on toroidal compactification of heterotic string theory. *Nuclear Physics, B279*, 369–376.

98. Narain, K. S., Sarmadi, M. H., & Vafa, C. (1987). Asymmetric orbifolds. *Nuclear Physics, B288*, 551–577.

99. Nemeschansky, D., & Sen, A. (1986). Conformal invariance of supersymmetric σ-models on Calabi-Yau manifold. *Physics Letters B, 178*(4), 365–369.

100. Pope, C. N., Sohnius, M. F., & Stelle, K. S. (1987). Counterterm counterexamples. *Nuclear Physics, B283*, 192–204.

101. Ramond, P. (1987). Field theory of strings. In L. Brink et al. (Eds.), *Unification of fundamental interactions* (pp. 104–108). *Physica Scripta, The Royal Swedish Academy of Sciences*. Singapore: World Scientific.

102. Rickles, D. (2011). A philosopher looks at dtring dualities. *Studies in the History and Philosophy of Modern Physics, 42*(1), 54–67.

103. Rickles, D. (2013). AdS/CFT duality and the emergence of spacetime. *Studies in the History and Philosophy of Modern Physics, 44*(3), 312–320.

104. Rickles, D. (2013). Mirror symmetry and other miracles in superstring theory. *Foundations of Physics, 43*, 54–80.

105. Ross, G. G. (1989). The (Low-Energy) physics of the superstring. *Philosophical Transactions of the Royal Society of London A, 329*(1605), 373–393.

106. Rovelli, C., & Smolin, L. (1987). A new approach to quantum gravity based on loop variables. In B. R. Lyer (Ed.), *Highlights in gravitation and cosmology*. Cambridge: Cambridge University Press.

107. Sakai, N., & Senda, I. (1986). Vacuum energies of string compactified on Torus. *Progress in Theoretical Physics, 75*, 692–705.

108. Schellekens, A. N. (1986). *Four dimensional strings*. CERN Preprint: CERN-TH.4807/87.

109. Schellekens, A. N. (Ed.). (1989). *Superstring construction*. Elsevier: North Holland.

110. Schwarz, J. H. (1987). Superstrings–An overview. In L. Durand (Ed.), *Second aspen winter school on physics* (pp. 269–276). New York: New York Academy of Sciences.

111. Schwarz, J. H. (1989). The search for realistic superstring vacuum. *Philosophical Transactions of the Royal Society of London A, 329*(1605), 359–371.

112. Schwarz, J. H., & Sen, A. (1993). Duality symmetries of 4D heterotic strings. *Nuclear Physics, 312*, 105–114.

113. Schwarzschild, B. (1985). Anomaly cancellation launches superstring Bandwagon. *Physics Today, 38*(17), 17–20.

114. Segrè, G. C. (1987). Superstrings and four-dimensional physics. In H. Latal & H. Mitter (Eds.), *Concepts and trends in particle physics* (pp. 123–222). Berlin: Springer.

115. Sen, A. (1985). Naturally light Higgs Doublet in supersymmetric E_6 grand unified theory. *Physical Review Letters, 55*(1), 33–35.

116. Schimmrigk, R. (1987). A new construction of a three-generation Calabi-Yau manifold. *Physics Letters B, 193*(3), 175–180.

117. Sen, A. (1982). Gravity as a spin system. *Physics Letters B, 119*, 89–91.

118. Shapere, A., & Wilczek, F. (1989). Selfdual models with theta terms. *Nuclear Physics, B320*, 669–695.

119. Smolin, L. (2006). The case for background independence. In D. Rickles, S. French, & J. Saatsi (Eds.), *The structural foundations of quantum gravity* (pp. 196–239). Oxford: Oxford University Press.

120. Staley, K. (2004). *The evidence for the top quark: Objectivity and bias in collaborative experimentation*. Cambridge: Cambridge University Press.
121. Strominger, A., & Witten, E. (1985). New manifolds for superstring compactification. *Communications in Mathematical Physics, 101*(3), 341–361.
122. Strominger, A. (1986). Superstrings with Torsion. *Nuclear Physics, B274*, 253–284.
123. Strominger, A. (1987). Lectures on closed string field theory. In L. Alvarez-Gaumé et al. (Eds.), *Superstrings '87: Proceedings of the Trieste Spring School, 1–11 April 1987* (pp. 311–340). Singapore: World Scientific.
124. Strominger, A., Yau, S.-T., & Zaslow, E. (1996). Mirror symmetry is T-duality. *Nuclear Physics, B479*(1–2), 243–259.
125. Todorov, A. N. (1980). Applications of the Kähler-Einstein-Calabi-Yau metric to moduli of $K3$ surfaces. *Inventiones Mathematicae, 61*(3), 251–265.
126. Tuite, M. P. (1992). Monstrous moonshine and orbifolds. *Communications in Mathematical Physics, 146*(2), 277–309.
127. Vistarini, T. (2013). *Emergent spacetime in string theory*. PhD Thesis. University of Illinois at Chicago.
128. Witten, E. (1985). Fermion quantum numbers in Kaluza-Klein theory. In R. Jackiw et al. (Eds.), *Shelter Island II* (pp. 227–277). Cambridge: MIT Press.
129. Witten, E. (1985). Symmetry breaking patterns in superstring models. *Nuclear Physics, B258*, 75–100.
130. Witten, E. (1986). Interacting field theory of open superstrings. *Nuclear Physics, B276*(2), 291–324.
131. Witten, E. (1986). Non-commutative geometry and string field theory. *Nuclear Physics, B278*(2), 253–294.
132. Witten, E. (1989). Higher symmetry in string theory. *Philosophical Transactions of the Royal Society of London A, 329*(1605), 349–357.
133. Yau, S.-T. (1978). On the Ricci curvature of a compact Kähler manifold and the Monge-Ampére equation, I. *Communications on Pure and Applied Mathemantics, 31*(3), 339–411.
134. Yau, S.-T. (1985). Compact three dimensional Kähler manifolds with zero Ricci curvature. In W. A. Bardeen, & A. White (Eds.), *Proceedings of the Symposium on Anomalies, Geometry and Topology: Argonne* (pp. 395–406). Singapore: World Scientific.
135. Yau, S. -T. (1998). *Mirror Symmetry I*. Providence: American Mathematical Society.
136. Yau, S. -T., & Nadis, S. (2010). *The shape of inner space: String theory and the geometry of the universe's hidden dimensions*. New York: Basic Books.

Chapter 10
A 'Second Superstring Revolution' and the Future of String Theory

> *String theory is not a theory of strings.*
>
> Robbert Dijkgraff

The close of the 1980s and the beginning of the 1990s didn't have the same degree of excitement as the mid-1980s. Though superstrings were pursued with vigour following the various anomaly cancellation results and the construction of the heterotic superstring, there were still many nagging doubts as to the basic structure of the theory—the existence of an apparent 'super-plurality' of theories compounded these. As Joseph Polchinski puts it: "[s]tring theory went through this tremendous wave of activity in the 1984 to 1987–1988 period. From 1988 to 1995, there was a perception that it had slowed down. Now in retrospect, huge amounts of stuff were done in those days: mirror symmetry, D-branes, Neveu-Schwarz branes, supergravity. Huge amounts of stuff being done, but nobody knew that it all fit together".[1] Many of these doubts were eventually eased, to a large extent, by a cluster of events in which the notions of D-branes and duality are centre stage.

This final chapter covers these recent developments, and brings the story near to the present day. Naturally, since the dust has yet to settle on literally tens of thousands of papers, I will have much less to say and simply sketch some key discoveries and events rather than attempting to describe their precise historical development. In particular: the recognition of the importance of D-branes in string theory, the existence of dualities, the eleven-dimensional low-energy limit giving $E_8 \otimes E_8$ superstrings, the role of black holes in string theory and the counting of microstates (giving the correct Bekenstein-Hawking entropy), and the interpretation of the 'string landscape' (with the controversial aspects that go along with it). Characterising these new developments is an exploration of non-perturbative $g_s \to \infty$ aspects of the theory in a bid to gain a better understanding of what string theory actually *is*. This had the effect of bringing to light various features that were simply not visible in perturbative,

---

[1] The quotations in the following sections are taken from an interview of Prof. Joseph Polchinski conducted by the author on March 18th, 2009. The complete transcript is available online at: http://www.aip.org/history/ohilist/30538.html.

D. Rickles, *A Brief History of String Theory*, The Frontiers Collection,
DOI: 10.1007/978-3-642-45128-7_10, © Springer-Verlag Berlin Heidelberg 2014

weak coupling behaviour. This bundle of developments is often called the 'second superstring revolution'.[2]

10.1 Dualities, D-Branes, and M-Theory

The introduction of D(irichlet) branes is often traced to Polchinski's 1995 paper "Dirichlet Branes and Ramond-Ramond Charges". However, there was a slow steady progression leading to the appreciation of the concept's importance, during which they were essentially introduced many times over, since the late 1980s. They were in fact fully introduced, more or less as they are understood today (though without the additional background information that grounds their present importance), and also named, in a 1989 paper written by Polchinski, together with his students, Jin Dai and Robert Leigh: "New Connections Between String Theories".[3] This paper is also one of the first to really promote the importance of dualities in string theory as a way of understanding what it really is and what objects it really contains. However, the paper was relatively obscure until after 1995.[4] Michael Green [28] also had the concept at around the same time. Once again, we find that conditions needed to have 'ripened' in the appropriate way, so that the value, in this case of the D-brane idea, was properly appreciated and its potential utility within a range of problems (especially involving dualities and black hole physics) better understood.

---

[2] The terminology of 'second superstring revolution' has its origins in a talk delivered at the Sakharov Conference in Moscow (in May, 1996) by John Schwarz (see http://arxiv.org/abs/hep-th/9607067). In fact, I think the term 'revolution' is rather more appropriate here than in the 1984 case since there are *genuinely* radically new concepts that emerge from this work. After 1994–1995 the way string theory was understood was dramatically and permanently altered. This goes beyond a simple 'high impact event'. But, still, I'm not sure that the second revolution qualifies as a revolution either, at least not in the sense of Kuhn's elucidation of the concept. There was a structure being investigated and tools and concepts were invoked to better understand that same structure. It wasn't a case of *overturning* some pre-established framework—indeed, most of the essential concepts were discovered by the mid-1980s onwards, but were simply not integrated.

[3] Here they demonstrate that a theory with both open and closed strings in a spacetime with compactified dimensions, is equivalent (dual) to a theory of open strings in which their endpoints have been fixed to single hyperplane: a D-brane. Polchinski also discussed some of these ideas in his talk at *Strings 1989* (held in March of that year). Again the key concepts and terminology of D-branes are clearly present. For example, he writes (invoking T-duality in which $\langle m, n, r \rangle \rightarrow \langle n, m, \frac{\alpha'}{r} \rangle$): "Open strings can't wind, so there are no states to get light as $r \rightarrow 0$. From the point of view of the open strings, the compactified dimension does not reappear. Indeed, one finds that the vanishing of the normal derivative of X [the compactified coordinate] implies the vanishing of the tangential derivative of X': the string endpoints are fixed on a hyperplane. This hyperplane is actually a dynamical object, the Dirichlet-brane, with a calculable tension $T' = T/\pi g^2$, where g is the open string coupling. Far away from the D-brane in the dual theory one finds only closed strings" [51, p. 436].

[4] There were just three citations in 1990, followed by 5 in 1991, 2 in 1992, 0 in 1993, and 3 in 1994. Then 12 citations in 1995, followed by 89 in 1996, once it was better adapted to the research landscape.

Before we get to D-branes' successful uptake, there is another important and related discovery, by Witten, also from 1995 [77]. This is the discovery that eleven-dimensional supergravity theory is a low energy limit of the ten-dimensional Type IIA superstring theory (or a strong coupling limit of Type IIA supergravity in ten dimensions). The new ideas came from a study of the non-perturbative behaviour of superstring theories, where the coupling constant is very large. A vital part of this project was the discovery (again, a somewhat protracted evolution, and not yet complete) of *S-duality* operating within string theories, relating strongly coupled and weakly coupled heterotic string theories (or sectors of the same theory) in four dimensions via the (non-perturbative, modular) $SL(2, \mathbb{C})$ mapping.

The notion of S-duality was introduced into the string theory literature by Anamaria Font, Dieter Lüst, Luis Ibáñez, and Fernando Quevedo in 1990 [22].[5] They left it as a conjecture that there was strong-weak coupling S-duality in the compactified heterotic string theory, generalising David Olive and Claus Montonen's own conjecturing of electric-magnetic duality [47]. David Olive, together with his PhD student Claus Montonen, had conjectured in 1977 that, when quantized, the magnetic monopole soliton solutions constructed by 't Hooft and Polyakov,[6] form a gauge triplet with the photon. This correponds to a Lagrangian similar to the original Georgi-Glashow one (as they say, the simplest Lagrangian containing $U(1)$ (Dirac) magnetic monopoles that arise as solitons), but with magnetic replacing electric charge. They referred to this new symmetry as "dual invariance," which simply means that the physical predictions will be unchanged regardless of whichever action one uses to extract those predictions.[7] The duality has curious implications:

> In the original Lagrangian, the heavy gauge particles carry the $U(1)$ electric charge, which is a Noether charge, while the monopole solitons carry magnetic charge which is a topological charge. In the equivalent "dual" field theory the fundamental monopole fields, we conjecture, play the rôle of the heavy gauge particles, with the magnetic charge being now the Noether charge (and so related to the new $SO(3)$ gauge coupling constant) [47, p. 117].

The dual invariance involves an S-duality mapping $e \to 1/e$ (where e is the square root of the fine structure constant), also interchanging the 'elementary' excitations

---

[5] See also Schwarz and Sen's 1993 paper [60] and Ashoke Sen's paper from 1994 [64]. Chris Hull and Paul Townsend [38] had labeled the Type II superstring version of S-duality "U-duality" (where it is seen to be combined with T-duality). Several other developments are contained in Font et al. [22]. For example, they show that S-duality follows from a duality between the elementary heterotic strings and the compactified (wrapped) NS 5-brane—the latter 'heterotic 5-brane' had already been conjectured by Michael Duff ([19], section 6.1) (and also Andrew Strominger, in a UCSB preprint). They also discussed the possibility that S-duality can be given a *geometrical* interpretation, involving the compactification of 11-dimensional supergravity—later confirmed in the work forming the beginnings of M-theory. Finally, they also discussed the possibility that the heterotic string can obtained from a 11-dimensional membrane, compactified to 10 dimensions— again, later confirmed by the work in M-theory, as we will see below).

[6] As modified by Manoj Prasad and Charles Sommerfield in 1975 [56], and Eugène Bogomol'nyi (building on Prasad and Sommerfield's work) in 1976 [6] (later called BPS states).

[7] The impact of this paper has been highly significant in recent years (with a sharp rise in 1995, in fact), as can be seen from Fig. 10.1.

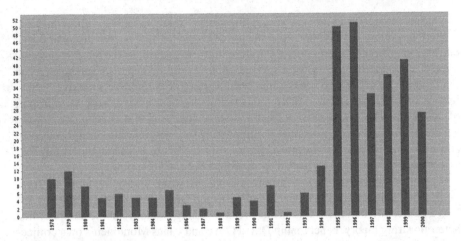

Fig. 10.1 Graph showing the number of publications referring to David Olive and Claus Montonen's paper "Magnetic Monopoles as Gauge Particles?" The spike in 1995, following the uptake of S-duality in superstring theory, is clearly visible. *Image source* Thompson-Reuters, *Web of Science*

(visible in perturbation theory) and the non-perturbative 'solitonic' (composite) excitations (that is, the electric and magnetic charges as in the above quotation). The Olive-Montonen *conjecture* was that there should exist a dual electromagnetic quantum field theory in which the roles of the elementary excitations and the composite solitons are exchanged. In order to *test* a 'strong/weak' S-duality conjecture like this, one clearly needs to probe throughout all values of the coupling, from weak to strong. Fortunately, the fact that the electrically charged particles and magnetically charged monopole solutions are (supersymmetric) BPS states, mentioned above, means that they benefit from the stability of such states under renormalization of the coupling constant.[8]

Strictly speaking, this work lies outside of string theory. Yet, part of Witten's concern, in [77], was precisely to build a watertight case for the existence of S-duality in string theory—and also, crucially, to expand S-duality beyond the four-dimensional case. There was a sense, apparent from the opening lines of Witten's paper,[9] that these ideas were pointing towards a deeper understanding of string theory. Ten years earlier, in his opening talk at the conference on Unified String Theories, in Santa Barbara, David Gross [29] had bemoaned the lack of a non-perturbative treatment of string theory: it was only known at weak coupling and, thanks to Gross' own calcu-

---

[8] Also, they exhibit a direct dependence between their masses (and tension) and the coupling strength, matching the central charge.

[9] Namely, "Understanding in what terms string theories should really be formulated is one of the basic needs and goals in the subject" [77, p. 85]. It is in this same paper that the terminology of a "web of connections between the five string theories and eleven-dimensional supergravity" (ibid., p. 87) first makes an appearance, though the concept had been suggested several times before, not least in [12].

lations, it was known that the perturbation series was not Borel summable. S-duality was precisely the tool that was needed to open up the non-perturbative regimes since it allows one to transform between theories at g and at $1/g$, as with the electric-magnetic case above. In the string theoretic case, because all string theories contain gravity, the BPS solitons are a kind of black hole solution (of a supersymmetric theory) originally known as an "extreme" (now called *extremal*) Reissner-Nordström (charged) black hole. It turned out that D-branes provided a key component to make the concepts work in the string theoretic context.[10]

David Fairlie and Edward Corrigan had in fact discovered something very close to D-branes way back in 1975:

> One of the goals of the earlier period was to construct an off-shell theory, so that the string states could couple to currents. Edward Corrigan and I had a solution to this problem motivated by the analogue approach, by the introduction of Dirichlet boundary conditions; we used both the analogue and the operator methods to construct amplitudes with the correct properties. A general bosonic state can be expressed as $x^\mu(\sigma, \tau) = q + 2ip\tau + \sum a_n^\mu \exp(in\tau)\cos(n\sigma)$. ... One of the features of our paper was that the string would stop at a finite point of spacetime and latch on to a current, or a zero-brane in present day jargon. In our idea, the stopped strings would then interact with currents [21, p. 289].

What was missing from Fairlie and Corrigan's approach was, as Fairlie writes, the idea "to impose Dirichlet boundary conditions[11] in only a subset d of the dimensions" [21, p. 290]. This paper was, of course, written as work on dual strings was entering its quietest period.

Michael Green [28] considered (toroidal) compactification of arbitrary numbers of target-space dimensions of a theory of orientable open strings. He begins by defining scattering amplitudes for interacting (oriented) closed and open strings using the sum-over-worldsheets idea, with worldsheet boundaries which have embedding

---

[10] The extremality of the black holes in question refers to the fact that their masses are as small as is allowed for specified electric charges. They are in this sense already similar to D-branes: in this context they are D0-branes, like point particles. However, it is also possible for there to be 'black brane' (BPS) solutions, which have higher dimension. These would later prove crucial in stabilizing the size and shape moduli of the theory and in the string theoretic computation of black hole entropy (which suggested a resolution of the information paradox), which in turn prepared the ground for the Maldacena (AdS/CFT) conjecture.

[11] Dirichlet boundary conditions simply refer to a condition to cancel boundary terms associated with the ends of open strings, telling us how the end points behave: $X^\mu|_{\sigma=0,\pi} = 0$ (where we adopt the usual convention of parameterizing the string by having the spatial coordinate σ take values in the interval $[0, \pi]$, and where $X^\mu(\sigma, \tau)$ are the fields describing the position of the string point (σ, τ) in the target spacetime). Depending on which values of μ one includes, one will have boundaries of different dimensionalities. Of course, for closed strings we find $X^\mu(0, \tau) = X^\mu(\pi, \tau)$, with end points identified. (Dirichlet conditions are contrasted with Neumann conditions, which demand that the normal derivative of $X^\mu(\sigma, \tau)$ vanishes: $\partial_\sigma X^\mu|_{\sigma=0,\pi} = 0$. Such a condition allows the end points to move freely, and in fact do so at the speed of light.) Clearly, however, the Dirichlet conditions *fix* the string end points to a particular location. The $X^\mu = 0$ constraint surface corresponds to a D-brane. As we will see below, Polchinski's breakthrough involved viewing these surfaces as physical, dynamical entities that correspond to expected nonperturbative effects in string theory (weighted by terms $e^{-O(1/g_s)}$).

coordinates into target spacetime that satisfy Neumann boundary conditions (roughly describing trajectories of open-string end-points). Compactifying onto a torus of scale R he argues the theory is equivalent to a theory with "Dirichlet boundaries" (boundaries at fixed positions) on the T-dual torus, with scale α'/R. He finds angular variables associated with boundaries (and conjugate to the boundary's winding numbers in the dual Neumann theory) which he interprets as the positions of the end-points of the strings. This duality is, he points out, an open-string version of T-duality: it implies that one can adopt the conventional Neumann picture of free open-string end-points or adopt the alternative picture of Dirichlet boundaries in the dual torus.

Polchinski [52] gave a modified formulation in which the boundaries (D-branes) carry Ramond-Ramond [RR][12] charges and are weighted by $e^{-O(1/g_s)}$ terms—as mentioned, the D-branes refer to 'defects' (with their own dynamics) characterised by the open strings attached to them. It turned out that these corresponded exactly to "particles" that Witten [77, p. 97] and others had predicted to exist in the Type II theory as a result of his investigations into the web of dualities (see below). They were, then, viewed as central to the enterprise of extracting information about the nonperturbative sectors of string theory, tightening up the web of dualities linking the various perturbative string theories, and figuring out what are the theory's fundamental degrees of freedom. This involved a kind of interlocking effect in which what had been viewed as separate skirmishes on apparently disconnected problems were seen to be joined together in a unified way.[13]

Polchinski describes his own, fairly long and winding journey to D-branes (and their acceptance as important pieces of the string theory puzzle) as follows[14]:

> The first piece was this paper ... with Yunhai Cai [50]. So there's the Green-Schwarz result that the anomalies in string theory cancel only in SO(32), and we wanted to understand in detail how that happened because we had already, which turned out in the end to be fallacious that we should be able to cancel them for any group. ... We in the end understood that the anomaly arose because a certain closed string field, a Ramond-Ramond field ... [a field in a

---

[12] There are multiple 'sectors' of states in superstring theories, defined by the boundary conditions. For the closed superstring one has: NSNS, NSR, RNS, and RR (where 'NS' = 'Neveu-Schwarz' and 'R' = 'Ramond'). The NSNS and RR sectors describe bosons, while the NSR, RNS sectors describe fermions. As mentioned, it turns out that D-branes are sources of RR-charge.

[13] Polchinski expresses it as follows: "duality at the time had seemed to be a very sporadic and random thing [and Witten] explained how every single string theory had a strongly coupled dual, and how you would figure out what it is ... [s]o suddenly it became a framework and not just some oddity" (interview with the author).

[14] Note that Polchinski had written an earlier paper on "supermembranes" together with his students James Hughes and Jun Liu while at Texas (with Michael Duff's presence at College Station, supermembranes were something of a specialty there). This develops the idea of four-dimensional membranes in a six-dimensional supersymmetric gauge theory. There are no D-brane elements as such, but they close with an interesting speculation about our four-dimensional space-time being a membrane solution "lying in some higher-dimensional field theory" [37, p. 373]. The problem they raise with taking this seriously is that they are unable to obtain, from the underlying theory, the necessary spin-1, and spin-2 fields living on the membranes so that the gravitational force is not right in the membrane world. This has a strong whiff of Lisa Randall and Raman Sundrum's "alternative to compactification" model, where they focus on a 3-brane in five dimensions [57].

certain sector of the string] ... had an equation of motion that couldn't be satisfied. Now in the modern language, we had discovered that Dirichlet nine-branes carry Ramond-Ramond charge, so this is an important fact in the modern language, and it is fundamental to every talk you hear. But in those days, the idea of branes was not around; and secondly, *the importance of Ramond-Ramond charge was not around*, so we had resolved why would couldn't catch the anomaly, but anyway, it just sat there. (Interview with the author, emphasis mine.)

As David Gross notes, "[n]ew ideas in physics sometimes take years to percolate into the collective consciousness" [30, p. 9106]. This is certainly a case in point. Polchinski adds that they hadn't pushed the idea as much as they might have because "we also believed that the heterotic string was the theory of the world. This was just an exercise that we were doing" (ibid.). The second step involved duality symmetries in a more central way:

The second predecessor work was this work with Dai and Leigh. The title is "New Connections Between String Theories", early 1989. I wanted to call it "Fun With Duality", but Rob Leigh was a serious guy and wouldn't let me do that.[15] So again, the T-duality was around, and by that time people were talking about it quite openly as evidence that strings had a minimum length size. ... And the whole focus on heterotic string, everybody in the world was working on heterotic string because that was the one that seemed to be connected most closely to nature, and nobody ever asked what happens if you apply it to any of the other string theories: Type I, IIA, IIB. And it turned out these had interesting answers, because if you apply it to open string theories[16], then there's the story that the T-duality involves the winding modes of the closed string, but the open string doesn't have them, and in the end the only way you get a consistent picture is that the T-dual of the open string theory is a theory with a D-brane in it, and so in particular D9-branes are dual to D8-branes, D7-branes, D6-branes, and so on through a series of T-dualities. ... In that paper we named D-branes and also orientifolds, which is another word you hear a lot these days. No one had ever asked what's the T-dual of an unoriented theory, and again it's non-trivial—instead of being a smooth space now it has an object in it, but it's sort of one of these orientifold planes. (ibid.)

These dualities were used to *reduce* the number of theories by establishing equivalences between them. They found that IIA, IIB, and Type I string theories were all dual—in terms of my earlier classification of pluralities, the Type 1 plurality had been reduced from 5 to 2 elements, and with it five families of ground states reduced to two families.[17]

A simple thought experiment led to the discovery of these dualities and with them the D-brane concept:

---

[15] Polchinski did, however, manage to use this title in his *Strings '89* talk, for subsection II's heading [51, p. 435]!

[16] In fact, Kikkawa and Yamasaki [41, pp. 359–360] *did* briefly consider T-duality for open strings, but they found that the tension energy contribution was missing so that rather than achieving a minimum in a symmetric potential (i.e. for the effective potential), as discussed above (p. 178), $a_i = \sqrt{\alpha'}/R_i$ goes to zero. They note that the model consisting solely of open strings is, in any case, "unnatural" given the splitting and joining mechanism.

[17] Given this reduction Polchinski had wanted to write a follow up paper entitled "There is Only One String Theory" (interview with the author). Clearly this (and others like them around at the same time) were significant steps on the way to Witten's more systematic speculations about a single, unifying *M*-theory underlying all string theories.

> We know if we put a closed string in a small box we get T-duality in a big box. What if we just repeat this to the other string theories? And it partly began as a way to keep my students occupied, but it became a really interesting question when the answer wasn't obvious. So we didn't pre-suspect, but we discovered that these theories were all dual to each other. (ibid.)

The thought experiment in question involves a classic device in physics, namely putting a system in a box. Putting a quantum system in a box implies that you must have integer numbers of wavelengths in the box. Shrinking the dimensions of the box reduces the wavelength (and simultaneously increases the energy, in inverse proportion). Erwin Schrödinger had been aware of this implication at least as early as 1939 [58] and had used it to argue that the universe must be closed like a large box in order to provide an explanation for the atomicity of matter and light. Schrödinger also considered what happened as one varied the radius, as in the case of an expanding universe, but Polchinski et al. considered what happened as one shrank the box to a point. In this case a restricted set of states survive at the limit, as the box vanishes. This much is true for point-particles, but we are dealing with strings, and they interact quite differently with compact dimensions:

> For a closed string: now there's the center of mass motion of the string. There's a wave function for the center of mass of this string, which does the same things we just said. If there is any center of mass momentum, the energy gets very large, so we only have zero center of mass momentum. But a closed string can do something a particle can't; it can wind, and it can wind many times before connecting back to itself. And as you make the box smaller, these states don't have much energy because the string is not very long. What you find is that if you calculate the energies of the states in a very, very small box, the energy of the winding states in a very small box are exactly equal to the energy of the momentum states in a very large box. So this was the point of Sakai and Senda, and then again Frank Wilczek, Strominger, and the others explained it wasn't just the spectrum but the interaction as well. You *cannot* shrink the box. It's interesting, because it's a thought experiment where you have the mathematics—you can do all the calculations, but then at the end of the day you have to look at it and say, "Hey, the physics is this." And the physics is that you try to make the box smaller, but past a certain point what happens is a new spacetime emerges and the box gets bigger. That's T duality. But if you do this with open strings, they can't wind, and so what happens is when the box gets big, you have both open and closed. The box gets big, there's a D-brane in there. And if you have unoriented strings, strings that can wind but they don't have a direction—this actually is the part that puzzled us for the longest time—then you get a box with a wall an O-plane. So in some sense, these pictures were completely implicit in the original discussion. They were implicit in the technology of string theory, but no one had ever asked what is the actual physical picture that goes with the mathematics. (ibid.)

Though all of the central features of D-branes, along with their potential role in dualities, were in the published literature, the focus of the majority of string theorists at the time was to see how far one could get with Calabi-Yau compactifications of heterotic strings. We saw in the previous chapter that when attempts were made to generate realistic physics from non-heterotic strings problems quickly emerged. The question that caused a shift (itself a consequence of converging duality results pointing to a unique underlying theory) was: what are the objects that carry Ramond-Ramond charge that are demanded by S-duality? This created an explanatory gap that could be exactly filled by D-branes.

The explanatory gap, and the vision of a unified string theory that demanded it be filled, was presented by Witten at the *Strings 1995* conference, at the University of Southern California: http://physics.usc.edu/Strings95/.[18] The original talk was entitled "Some Comments On String Dynamics" and focused on determining the strong coupling behavior of various string theories in various dimensions. This was refashioned soon after into the paper "String Theories in Various Dimensions". In this he had written:

> Apart from anything else that follows, the existence of particles with masses of order $1/\lambda$, as opposed to the more usual $1/\lambda^2$ for solitons, is important in itself. It almost certainly means that the string perturbation expansion—which is an expansion in powers of λ^2—will have non-perturbative corrections of order $\exp(-1/\lambda)$, in contrast to the more usual $\exp(-1/\lambda^2)$. ... The fact that the masses of RR charges diverge as $\lambda \to 0$—though only as $1/\lambda$—is important for self-consistency. It means that these states disappear from the spectrum as $\lambda \to 0$, which is why one does not see them as elementary string states [77, p. 91].

Witten's paper itself caused a flurry of activity, as can be discerned from Fig. 10.2. At the conference, both Green and Polchinski knew about the objects Witten was describing: "Mike Green and I ... looked at each other and said, 'He must be talking about D-branes' " (interview with the author). Neither Green nor Polchinski were moved to immediate action, however. For Polchinski, the feast of ideas was a little too rich, and required some digestion, in the form of set of 'homework problems' to work through (with D-branes and open strings fairly low down on his list). The other reason was that the results were already out there, in his papers with Cai (showing that 9-branes carry Ramond-Ramond charge) and with Dai and Leigh (showing that 9-branes are T-dual to all of the other branes). This stance changed in August of 1995:

> I started working through Ed's dualities for open strings, and actually I thought I found a contradiction. I thought I found that one of them was impossible, so I emailed Ed and we worked on it together. That was actually the first time I ever collaborated with him. But in the course of that collaboration, and also working through my homework problems, it suddenly was obvious. It's one of these like the renormalization where all of the pieces were there, and suddenly you *know* that they're all there. So I emailed Ed and said, "Oh, by the way, these D-branes carry a Ramond-Ramond charge, and they have these other properties". I thought

---

[18] At the same conference Michael Green's talk on "Boundary Effects in String Theory" was devoted to D-branes, D-instantons, and stringy non-perturbative effects. Referring to Polchinski's work, he states quite explicitly that "There are ... soliton-like 'D-brane' configurations whose rôle in the context of superstrings has not yet been illuminated ... [that] might provide solitonic states that are needed if the suggested non-perturbative equivalence of the type 1 and heterotic theories is correct" (see the conference talk: http://physics.usc.edu/Strings95/Proceedings/pdf/9510016.pdf, p. 10). Chris Hull's talk was on "Duality, Enhanced Symmetry, and 'Massless Black Holes'. He writes of the "unexpected equivalences between string theories that look very different in perturbation theory [resulting] from different perturbation expansions of the same theory" pointing to cases in which "the strong coupling limit of a given theory with respect to a particular coupling constant is described by the weak coupling expansion of a dual theory, which is sometimes another string theory and sometimes a field theory" (http://physics.usc.edu/Strings95/Proceedings/pdf/hull.pdf, p. 1). I take this to show, similarly to the events surrounding the anomaly cancellation results (though much more so), that the field as a whole was poised at a critical point making it particularly receptive to the kind of unifying framework Witten proposed.

Fig. 10.2 Graph showing the impact of Witten's paper "String Theories in Various Dimensions" in terms of the number of referring publications. *Image source* Thompson-Reuters, *Web of Science*

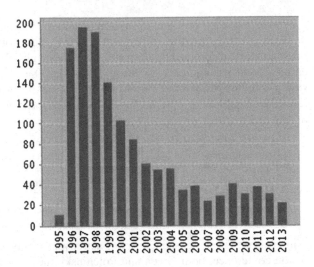

it was neat. But I was not prepared for the response. He appreciated much more than I did how important this was. (Interview with author.)

Judging from Witten's original *Strings 1995* talk, and the other talks that were given there, I think it's fair to say that the full implications took some time to ferment in both Polchinski's and Witten's minds. Once it had, however, and their two papers were published, the impact was extremely dramatic:

> Within weeks of my paper, Vafa and Douglas and Sen had all pointed out important implications. I don't know of any episode like it in my experience where there had been such a change in a field. It's weird, because although I felt like I pulled the cork out of the dam, I didn't have any sense—it just blew me away. Why are they so important? Well, of course we suspected for a long time, and it was clear in my—I mean in my Les Houches lectures, I explained why string theory is not a theory of strings, and this was before any of this happened. It's clear that whatever the fundamental formulation of string theory is, D-branes are closer to it than strings. If you . . . ask today for what is our most complete formulation of string theory, either matrix theory, the Banks et al. one, or AdS/CFT duality, in both of those it's the degrees of freedom on branes that are the fundamental degrees of freedom. So it's pretty remarkable that there's all this stuff underlying string theory. (Interview with author.)

The impact of Polchinski's paper closely matches Witten's, and would have belonged to a pattern of co-citation (see Fig. 10.3).

The importance of the D-brane (re)discovery can be seen as involving an earlier argument of Steven Shenker [66], couched in Matrix theory, in which he shows that the $e^{1/g}$ versus e^{1/g^2} behavior should be generic in string theory because string perturbation theory generically behaves like $(2g)!$ at genus g. Hence, since the perturbation theory diverges faster, this suggests that non-perturbative effects are likely to be much larger in string theories than in low-energy field theory. There were natural links to black hole physics stemming from this argument, as we will see in the next section, but first let us consider Witten's M-theory proposal in more detail.

Fig. 10.3 Number of publications referring to Polchinski's paper [53] following its publication in 1995. *Image source* Thompson-Reuters, *Web of Science*

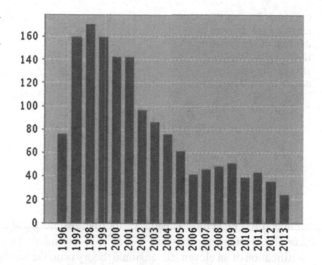

As mentioned, Witten had introduced the notion of '*M*-theory' during a talk at a conference at the University of Southern California in 1995. *M*-theory is a conjectured theory postulated to explain the web of dual theories and provide further insights into the non-perturbative aspects of string theory. It would unify the disparate string theories, and bring order (and hopefully) uniqueness to string theory.[19]

One of the remarkable aspects of this web of dual theories (see Fig. 10.4), is that it involves theories of different dimensionalities, both ten- and eleven-dimensional. One can derive the various theories by compactifying *M*-theory on specific manifolds, and by exploiting the existence of dualities interconnecting them. For example, Paul Townsend demonstrated, early in 1995, that *M*-theory compactified onto the circle S^1 (or considering the behaviour of the eleven-dimensional theory on $\mathbb{R}^{10} \times S^1$), yields Type IIA superstrings. In his own words, since the *M*-theory concept had not yet been presented: "the type IIA ten-dimensional superstring theory is actually a

---

[19] The first appearance of the term in print appears to be [62]. In a later popular article of Witten's we find the oft-quoted explanation of the letter '*M*': "M stands for magic, mystery, or matrix, according to taste" [80, p. 1129]. There's an ambiguity over the proper domain of *M*-theory, with e.g. Greene, Morrison, and Polchinski [27, p. 11039] assuming that *M*-theory simply refers to one of several limit points of a large parameter space that also includes the five superstring theories (i.e. the space of string vacua), while Witten appears to suggest that *M*-theory denotes the framework underlying *all* six of these limit points. On the former approach, both *M*-theory and the five superstring theories offer ways of describing whatever structure admits the large parameter space; on the latter approach the five superstring theories and eleven-dimensional supergravity offer ways of describing *M*-theory (the underlying, unknown structure) so that each limiting theory provides a physical scenario (fixed by values of the parameters) that is nomologically possible in *M*-theory, for certain settings. For example, the eleven-dimensional supergravity theory is seen to describe the long wavelength limit (see [78, p. 383]). The standard view that has emerged is that eleven-dimensional supergravity constitutes another low-energy limit of *M*-theory. Of course, whatever stance we adopt, it is clear that the underlying theory cannot be ten-dimensional, but must be an eleven-dimensional quantum theory (or perhaps something that completely transcends these old categories).

Fig. 10.4 Edward Witten's much copied diagram of the web of string theories linked by dualities, understood as limiting cases of a deeper theory. *Image source* [80, p. 1128]

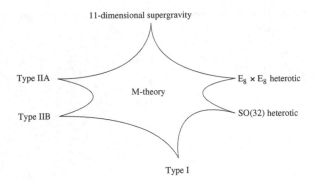

compactified eleven-dimensional supermembrane theory" [76, p. 184]. Later that year, Petr Hořava and Edward Witten [33] extended this result, showing that compactification of an eleven-dimensional theory onto the orbifold $S^1/\mathbb{Z}_2$ (or considering the behaviour of the eleven-dimensional theory on $\mathbb{R}^{10} \times S^1/\mathbb{Z}_2$), yields heterotic $E_8 \otimes E_8$ superstrings.[20]

Hořava and Witten go on to consider further interconnections in M-theory's web, due to the duality relations holding between the various theories. By using dualities between the heterotic $E_8 \otimes E_8$ and Type IIA theories (related to the eleven-dimensional theory as above), they are also able to include heterotic and Type I superstrings with $SO(32)$ gauge group in M-theory's reach. The method involves a further compactification of M-theory, this time of the tenth dimension, onto a circle, so that one is considering the behaviour of M-theory on $\mathbb{R}^{10} \times S^1 \times S^1/\mathbb{Z}_2$. They explain the existence of the duality holding between these theories using the classical symmetries of M-theory thus compactified, with T-duality transformations taking one to the theories of $SO(32)$ strings.[21]

Type IIB theories were reached in a similar way by John Schwarz, using T-duality transformations [61], showing that IIB strings compactified on a circle correspond to the eleven-dimensional supergravity theory compactified on a torus. As Schwarz

---

[20] The eleven-dimensional theory reduces to ten-dimensional strings since it contains 2-membranes which when compactified onto small circles appear as strings (see [18] for the earliest discussion of this idea, back in 1987). Of course, another way of putting this reverses the direction from reduction to emergence, so that at strong coupling (beyond perturbation theory) the ten-dimensional string theory 'gains' an additional dimension, with S^1's radius increasing (as the two-thirds power) with the coupling strength—the ten-dimensional appearance of IIA strings is thus an artefact of perturbation theory. Note that this chapter [33] includes Witten's own introduction of the M-theory concept into print, with the words: "The most ambitious interpretation of these facts is to suppose that there really is a yet-unknown eleven-dimensional quantum theory that underlies many aspects of string theory... As it has been proposed that the eleven-dimensional theory is a supermembrane theory but there are some reasons to doubt that interpretation, we will non-committally call it the M-theory, leaving to the future the relation of M to membranes" [33, p. 507]. Hence, the M, and some of the central concepts surrounding M-theory, owe much to Townsend's earlier efforts.

[21] Witten had already examined this heterotic–Type I string duality in [77], and then later with Joseph Polchinski [54], both in 1995.

notes, "[t]he remarkable thing about this kind of reasoning is that it works even though we don't understand how to formulate the M theory as a quantum theory" [62, p. 97]. Again, as we have seen previously, the curious behaviour of aspects of spacetime in string theory appears to point beyond a simple picture in which spacetime is fundamental. For example, at a lecture presented at the 29th International Ahrenshoop Symposium (in Buckow, Germany in 1995), Schwarz is led to state:

> The remarkable role of duality symmetries and their geometrically non-intuitive implications suggest to me that the theory might look very algebraic in structure without evident geometric properties so that no space-time manifold is evident in its formulation. In this case, the existence of space-time would have to emerge as a property of a class of solutions. Other solutions might not have any such interpretation [63, p. 3].

There remained a problem with the notion of *M*-theory: it was a rather abstract promissory note that required a precise construction or definition, which should result in a non-perturbative formulation of the theory. Several such attempts were made. We have seen how the eleven-dimensional theory was offered up as a potential source for a definition of the theory, in which the string theories are defined via various reductions.[22] A closely-related approach is the Matrix model[23] of Banks,

---

[22] For example, Dijkgraaf, Verlinde, and Verlinde, write that "[b]y definition, M-theory is the eleven-dimensional theory that via compactification on a circle S^1 is equivalent to ten-dimensional type IIA string theory" [16, p. 43]. Of course, this really shows us the various limits of *M*-theory again, rather than pinning down the theory itself.

[23] It is interesting to note that in August of 1991, Paul Ginsparg established the first database ('hep-th' for 'high energy physics—theory'—a name Ginsparg attributes to Steven Shenker: [26, p. 4]) of what is now called "arXiv" (initially it was xxx.lanl.gov, but was renamed in 1998) to function as a repository for papers discussing the matrix model and "intended for usage by a small subcommunity of less than 200 physicists" (http://people.ccmr.cornell.edu/~ginsparg/blurb/pg96unesco.html.). There were just 160 initial users "assembled from pre-existing e-mail distribution lists in the subject of two-dimensional gravity and conformal field theory" [25, p. 159]—this mailing list might have been drawn from Joanne Cohn's list (she had attempted a 'manual' emailing approach to electronic distribution in 1991). The first chapter, http://xxx.lanl.gov/abs/hep-th/9108001, deposited on August 14th, was James Horne and Gary Horowitz's "Exact Black String Solutions in Three Dimensions". Another string theorist, Wolfgang Lerche, put Ginsparg onto Tim Berners-Lee's (then at CERN) new computer program: WorldWideWeb.app. (For more details on the history of arXiv, I refer the reader to Ginsparg's paper celebrating the 20th anniversary of the archive: [26].) As N. David Mermin would wryly remark, the archive might constitute string theorists' "greatest contribution to science" [46, p. 9]. In fact, string theory (and related areas) boast a surprising number of 'computer firsts': Green, Schwarz, and Witten's textbook was the first to be delivered camera-ready in TeX—as he recalls, given the slowness of the computers in those days, "[e]very time I TeX'd a chapter, it would take about five minutes" (http://oralhistories.library.caltech.edu/116/1/Schwarz_OHO.pdf). Ginsparg also points out that he and Lance Dixon were the first to add their email addresses to a preprint. This suggests a curious possibility: is it possible that a scientific field might be pushed, to a fairly large extent, by the availability and early exploitation of easier and wider readership and easier (and wider) methods of communicating? Jokingly, John Schwarz claimed that string theorists' "main use of computers is likely to be to produce prettier preprints [but] more disturbingly is that we can also produce them faster" adding that "[i]f present trends continue we could reach a situation in which certain theorists turn out preprints as fast as the rest of us can read them" [59, p. 201]. If one is to think seriously about scientific revolutions in the period that coincides with the development of such tools as email, the internet, archive systems, and so

Fischler, Shenker, and Susskind[24] [3], which essentially reverses the direction, so that one considers how the eleven-dimensional theory *emerges* from a strongly-coupled limit of IIA theory. This leads to a definition of M-theory as the eleven-dimensional theory on a flat ("decompactified"), infinite background spacetime. The *matrices* in the name refer to the $N \times N$ matrices $X^i (i = 1, \cdots, 9)$ providing the coordinates of N interacting Dp-branes in the target space.[25] The eleven-dimensional theory then emerges in the $N \to \infty$ limit.[26] This expedient allows one to probe the nonperturbative spectrum of a string theory.

The central problem with these was the same as with the definitions of string field theory: they were specified against some fixed background. What was (and still *is*) required, however, is a background independent formulation in which both the properties of the objects *and* the spacetime in which they propagate are determined by the theory. However, the idea of 'emergent dimensions' (with correspondences between radii of one theory and couplings of another theory) is a direct descendent of the same feature one finds in the so-called 'Maldacena conjecture' linking a gauge theory in four-dimensional Minkowski space with a string theory in $AdS_5 \times S^5$—it is, in turn a descendent of a *potpourri* of ideas, including 'the holographic principle' (a fact acknowledged in the Matrix model paper of Banks et al., [3, p. 5112]).

Schwarz had predicted (amongst many other predictions for the future of string theory) that "*It will be understood why six dimensions are compactified and three are not*" [59, p. 200]. While not exactly resolved by the developments in non-perturbative string theory, the meaning of the terms in the question itself have been transformed.

(Footnote 23 continued)

on, one must consider the possible influences (and perhaps biases) they introduce. In fact, Roger Penrose [49] has argued that there might be a kind of path-dependence effect (along the lines of that found in the competition between VHS and Betamax video standards), whereby the spread of email and internet access, and with it the easy establishment of connections, allows for the spread of ideas so that a dominant trend can spread and become more entrenched because of such networks *even when the theories do not have standard experimental evidence supporting them* (cf. [75, p. 157]). Having said this, string theory might also have been amongst the last to use a 'human computer' to check results (and save then precious computing time): Michael Green pointed out (in a talk at a workshop in honour of John Schwarz) that CERN's Wim Klein (a calculating prodigy that CERN had discovered doing calculations in Circus shows) would check difficult calculations for them and others—for more on Klein, see: http://home.web.cern.ch/cern-people/updates/2012/12/remembering-wim-klein.

[24] The Hamiltonian construction of their model depends on Susskind's old tool of the 'infinite momentum frame,' that he had used in his earliest studies of the dual resonance model that had introduced the string and worldsheet concepts.

[25] Note, that these coordinates (being described by matrices) are non-commutative, which has been interpreted as implying a specific kind of 'quantum geometry' (see, e.g., [79]). However, as the authors of [3] admit, the microscopic degrees of freedom are not known, therefore it is hard to make this claim precise—interestingly, in 1988, Joseph Atick and Edward Witten had speculated about a "new version of Heisenberg's principle [involving] some non-commutativity where it does not usually arise" noting that it "may be the key to the thinning of the degrees of freedom that is needed to describe string theory correctly" [2, p. 314].

[26] There are clear elements of this approach that hark back to 't Hooft's $1/N$ expansion from 1974.

Strominger, in his lecture[27] on "Black Holes and String Theory" argues that "the notion of space ... and time ... and dimension are not absolute." He compares the situation to the phases of H_2O, and their temperature-dependence: in various regimes, water switches between solid, liquid, and steam. Just as we don't have any problems making sense of this, so in the case of the dimension of spacetime, there is a dependence on the energy of the system: it becomes another dynamical parameter, in much the same way that the metric of spacetime is made dynamical in classical general relativity. This is, of course, one of the most radical implications of the more recent work on string theory, and is still being unpacked, though it clearly points towards problems with upholding 'locality' at a fundamental level.[28]

10.2 Black Holes, Information, and the AdS/CFT Duality

The study of strings in a more general class of non-compactified backgrounds began soon after the construction of the heterotic string.[29] Thinking about how string theory bears on the physics of black holes is, of course, a perfectly natural course of action given string theory's claims to provide a theory of quantum gravity reproducing the classical equations of general relativity in the low energy limit. And indeed, in the low energy limit of string theories one can find solitons corresponding to black hole solutions. The D-brane technology allowed certain kinds of black holes to be constructed as configurations of coincident D-branes. However, the initial phase of string theoretic black hole research involved the study of strings on black hole *backgrounds*, rather than their construction and microscopic degrees of freedom.

In 1987, de Vega and Sánchez [14] studied the problem of a bosonic string in a D-dimensional Schwarzschild background, thus allowing for the study of strings on black hole spacetimes. Curtis Callan, Robert Myers, and Malcolm Perry suggested in 1988 that string theory might be useful for resolving some of the paradoxes that arise when considering black hole evaporation [8]. The idea is that the solutions of classical general relativity and string theory, though approximately identical at low energies (small curvatures), will differ at higher energies (strong curvatures). Given the improved ultraviolet behaviour of string theories, the hope was that it would forbid the formation of the singularities generic in Einstein's equations. However,

---

[27] Specifically, in answer to an audience question on whether string theory is eleven-dimensional: http://athome.harvard.edu/programs/sst/video/sst1_7.html.

[28] Interestingly, strikingly similar results—suggesting that locality is not fundamental, but must instead emerge from the physical degrees of freedom—can be found in a variety of approaches to quantum gravity, indicating that locality is very likely to be relativised in the physics of the future.

[29] In fact, so far as I can tell, Claud Lovelace [44] appears to have been the first to consider the behaviour of strings on curved space *before* the construction of heterotic strings, and even before the consistency proofs of Green and Schwarz. Lovelace was interested in the case of compactification on a hypersphere, rather than a hypertorus, as a way of generating a non-Abelian gauge theory. He argued, however, that compact Ricci-flat manifolds are restricted to Abelian symmetries, which restricts the compactification to those on a hypertorus once again.

Callan, Myers, and Perry focused on the reduced temperatures of black holes in the context of string theory (in comparison with solutions of Einstein's equations for black holes of the same mass), which they show to hold for multiple cases including heterotic strings in four-dimensions.[30]

Stephen Hawking and Jacob Bekenstein had demonstrated in the mid-1970s that black holes behave like thermodynamic objects, with a temperatures (emitting 'Hawking radiation' at $T_H = \hbar c^3 / 8\pi kGM$, with M the mass of the black hole) and entropies (of $S_B = A/4G_N$, where A is the area of the black hole's horizon).[31] The black hole will radiate its energy away, losing mass, eventually evaporating, leaving some kind of Planck scale remnant, or nothing at all. There is a paradox surrounding the quantum mechanical description of what happens to information that goes in to black holes given this Hawking evaporation. It appears that one could throw *pure* quantum states into a black hole and get mixed (thermal) states out, apparently in violation of unitarity, and resulting in a loss of information—this is often labeled the 'black hole information paradox.' Hawking believed that his analysis demonstrated that quantum mechanics is *violated* by evaporating black holes implying that the time-evolution of such a process had to be grounded in something else. In the context of quantum mechanics, entropy has a very specific combinatorial characterisation given in terms of the number of different quantum states a system might occupy. It seems natural to think, therefore, that if black holes are to be assigned an entropy then there ought to be an associated set of microstates. It became a challenge for any approach to quantum gravity to try and derive these microstates, and have them match the famous figure of Bekenstein. A second challenge was to see if Hawking's radical conclusion was correct. This set of ideas became a kind of thought laboratory for testing string theory, and other approaches to quantum gravity—it would also serve as a kind of testing zone for the newly incorporated D-branes.

One of the major breakthroughs, made possible by the discovery of D-branes, was the first calculation of the Bekenstein entropy for black holes, by counting their quantum states.[32] Though there were earlier attempts to compute black hole

---

[30] Their analysis is based on Huang and Weinberg's demonstration that the Veneziano model possesses a highest possible temperature, namely the Hagedorn temperature (i.e. that beyond which there is an exponential rise in the density of particle states: adding heat creates particles that increase entropy, rather than increasing temperature) [36]. This chapter of Huang and Weinberg's is an interesting early application (just 2 years after the Veneziano formula had been written down) of the dual model to cosmological and gravitational contexts: I believe it constitutes the first such paper. What Callan, Myers, and Perry showed was that black holes also have a maximum temperature around the Hagedorn temperature.

[31] Bekenstein's reasoning was highly intuitive: given the irreversible growth of a black hole's surface area, and given entropy's similar irreversibly growing nature, the possibility is open to write the black hole entropy as a (monotonically increasing) function of this area. Bekenstein credits John Wheeler with suggesting the choice of ascribing a unit of entropy k to something of the order of the square of the Planck length (see [5, p. 44]). This is, of course, related to the holographic principle which describes the non-extensivity of physics within some boundary (or 'in the bulk'): the degrees of freedom on the boundary suffice to determine the bulk physics, which contains surplus, unphysical degrees of freedom.

[32] We should also mention here that D-branes (though they called them *p-branes* in this case) were used in the context of black hole physics *before* their dramatic rise to fame in 1995. In a 1991

entropy [72], Strominger and Vafa [71] were the first to make the link, in 1996, for a highly idealised situation involving five-dimensional extremal black holes.[33] The microstates are then enumerated by counting the degeneracy of the BPS soliton bound states. The combinatorial aspects come about through Polchinski's identification of D-branes as sources of BPS states carrying Ramond-Ramond charge.[34] Hence, the problem reduces to counting bound states of D-branes. As an example, one can take the five-dimensional (extremal) Reissner-Nordström black hole. This is a solution of the equations of the classical supergravity limit of IIB string theory, with five directions compactified onto a five-torus T^5, with the black hole's 'charges,' M, N, P determined by the ten-dimensional theory. In terms of D-branes, however, one directly compactifies the IIB theory onto T^5 around which M D5-branes are wrapped, along with N D1-branes (i.e. strings) wound around the circle, $S^1 \subset T^5$, which determine a quantized momentum P (á la Kaluza-Klein compactification). One can *count* the states of the D-brane construction of the black hole solution by quantizing the open strings (with the momentum P) linking D1-branes to the D5-branes, which gives the simple expression: $S_D = 2\pi\sqrt{NMP} = S_B$.

Strominger and Vafa pointed out the potential relevance of their work to the black hole information paradox [71, p. 103]. They suggest that D-brane technology might be used to directly compute the low-energy scattering of quanta by an (extremal) black hole, to check for unitarity or its violation.[35] They note that S-type dualities could be utilised to make this a possibility, turning a strongly coupled problem to a weakly coupled one. Studying the Hawking radiation in terms of open string excitations, one finds that unitarity is indeed preserved.[36] This simple suggestion highlights just how interconnected the physics of strings (D-branes), black holes, and dualities was, and still is.

Many people followed Strominger and Vafa's approach, including Curtis Callan and Juan Maldacena [9] who derived the entropy, radiation rate, and Hawking temper-

---

(Footnote 32 continued)

investigation of black hole solutions in ten-dimensional string theory, Horowitz and Strominger [34] show that there are extended black hole solutions (extended objects surrounded by an event horizon) that correspond to magnetically charged string soliton solutions (including 5-brane solutions). There was at this time, in fact, a fairly thriving industry studying black hole solutions via branes.

[33] Note, however, that the extremal black holes involve a zero-temperature approximation, and so are not thermal objects: of course, no Hawking radiation is possible in this zero-temperature limit. One can consider 'near-extremal' cases by perturbing around the extremal solution. In such cases, one can generate a small amount of Hawking radiation.

[34] I already referred to a kind of correspondence between D-branes and black holes: they share charge, mass, and tension. The D-brane based computation of black hole entropy confirmed this link.

[35] In D-brane terms, one can visualise black hole evaporation by picturing a surface (the D-brane) onto which a separate pair of open strings is stuck, which collide and join to form a closed string, which is then emitted off the surface as gravitational radiation as it becomes 'unstuck'.

[36] Amusingly, Schwarz 'predicted' in 1986 that there would be no loss of coherence in the string theoretic context, despite not having the tools available to do the analysis [59, p. 199].

ature from a similar analysis, generalised to non-extremal five-dimensional Reissner-Nordström black holes.[37]

These notions, and the idea of utilising a large-N limit of coincident D-branes with dual descriptions, led to what might (given past naming conventions) be labeled a *third* superstring revolution: the gauge/gravity duality encapsulated in the AdS/CFT correspondence (where 'AdS' = 'anti-deSitter')—though, strictly speaking, it is more of an aftereffect of the cluster involving dualities, black holes, D-branes, and M-theory. The AdS-CFT correspondence (otherwise known as the Maldacena conjecture) is a radical duality based on the black-hole—D-brane (or open string/closed string) correspondence, and on an examination of their different limits. It involves the claim that a quantum theory with gravity is equivalent (in the sense of duality from previous chapters) to a quantum gauge theory without gravity: a string theory on anti-de Sitter space AdS_5[38] possesses equivalent physically observable properties to a conformal field theory defined on the (conformal) boundary $\partial \mathscr{S}_{\text{AdS}_5}$. The degrees of freedom of one theory are transformed (by the duality) into the degrees of the other theory. As Polchinski puts it:

> This entropy counting is neat, but the gauge/gravity duality is amazing, because it really says that gravity and string theory are not anything new; they've always been present in the framework of quantum field theory or gauge theory, if we simply knew how to read the code, and Maldacena told us how to read the code. This has many implications. One is it does resolve the information problem at least implicitly, because it shows that you can formulate the quantum mechanics of the black hole in terms of the gauge theory which is purely quantum mechanical—it satisfies the ordinary laws of quantum mechanics. It shows that Hawking was wrong about the breakdown of the laws of quantum mechanics. What does break down in some sense is locality. The fundamental degrees of freedom in the gauge theory are not local in space time.[39]

---

[37] Strominger, together with Maldacena and Witten [69], extended the analysis to the case involving compactification of $M \times S^1$ (where M is Calabi-Yau 3-fold). The microscopic degrees of freedom of black holes are then represented by fivebranes wrapping around $P \times S^1$ (with P a four-cycle in M). This brought the analysis of black holes back into the fold of M-theory. A series of progressive refinements and generalizations were made to the study of quantum black holes, but to discuss them would introduce an explosion of new literature.

[38] This is actually part of a product space with $\mathscr{S}_5$, an Einstein manifold (i.e. a solution of the Einstein field equations) of positive cosmological constant: it needs to be S^5 to get the symmetries of the gauge theory out correctly. Anti-de Sitter space is essentially like hyperbolic space with an additional time coordinate. It has a boundary and so one has to say what the boundary conditions are in any theory defined on this space. Of course, the scheme is not realistic: anti-de Sitter space has a negative cosmological constant, and in our universe it is apparently positive. There has been work on more realistic theories involving a dS/CFT correspondence (e.g. [70]), but this is still very much work in progress.

[39] There is a sense in which this feature is the non-perturbative counterpart of the kinds of conceptual problems that emerged in the 1980s through the consideration of spacetime/string interactions. The earlier predictions about profound changes that might be in store for the understanding of spacetime in string theory seem to have been realised to a large extent by the non-locality implications of the AdS/CFT conjecture.

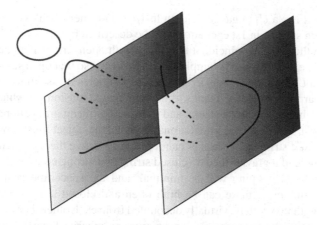

Fig. 10.5 D-branes: with open strings on the same surface and stretched between distinct surfaces. Open strings correspond to gauge particles, closed strings to gravitons

This descriptive freedom in the languages (based around different degrees of freedom) one can use to describe the physical situation[40] of multiple Dp-branes (gauge fields on flat worldvolumes versus gravitating objects embedded in string theory backgrounds) forms the core of the gauge-string duality in the AdS/CFT correspondence—one often speaks of a 'dictionary' for translating between languages. The AdS/CFT duality is, however, still restricted to a supersymmetric cousin of QCD.[41]

The duality, first presented in [45], involves the fact that at weak coupling D-branes don't warp spacetime geometry: they have a tension that is inversely proportional to the string coupling constant describing the strength of interactions. Therefore, at weak coupling (i.e. in the perturbative expansion for which $g_s \ll 1$) they will be unobservable.[42] At strong coupling D-branes *can* warp geometry, generating horizons just as black holes do: they have a tension that contributes to the stress-energy tensor which, if strong enough, will warp spacetime geometry near the D-brane. Strings can be bounded by pairs of D-branes (see Fig. 10.5) and when this happens the strings become massless (and are able to mimic gluons). The open string excitation spectrum contains a massless spin-1 particle, so that a Dp-brane with open

---

[40] Also clearly harking back to the earlier discussion of perturbative dualities (e.g. on 194).

[41] The conformal symmetry means, of course, that this supersymmetric theory is non-confining, since it is scale invariant: once one sets the coupling strength it remains at that strength independently of energy scale, unlike the QCD case. The supersymmetry is needed to stabilise the theory at high coupling. Hence, this cousin of QCD is a fairly distant one in that it does not possess asymptotic freedom and, as such, can provide only qualitative estimations of the non-perturbative behaviour of QCD proper.

[42] Though they will still have the description as a surface (in the full ten-dimensional spacetime) on which open strings are confined. Closed strings are free to move away from this surface (in the bulk). If open strings join to form a closed string then they too can move off the D-brane—physically, this corresponds to gravitational radiation being emitted from a photon.

strings attached has a $U(1)$ gauge field on its $(p + 1)$-dimensional world-volume—
hence, the open strings, in 1st excitation, will be described by a $(p + 1)$D $U(1)$ gauge
theory. When there are N coincident D-branes (with open strings held between any
pair of them) the gauge group is 'amplified' up to $U(N)$ (so that open strings are now
described by a $U(N)$ gauge theory).[43] A stack of D-branes is essentially like a black
hole again, warping geometry, thanks to the aggregated tension. In which case one
ceases to talk about a 'D-brane stack,' and speaks of a 'mean gravitational field' (or
a 'black brane'). But given the gravitational aspects, the object must now be part of
a theory of *closed* strings.[44] D-branes can represent a gauge theory (open strings) at
weak coupling, and a gravity theory (closed strings) at strong coupling.

The geometrical warping will be minimal, and the spacetime near flat, when
$Ng_s \ll 1$. In this case, there can be both open and closed strings, but with low
coupling strength they will be virtually decoupled from each other. The closed strings
that decouple from the open strings give a picture of linearised, perturbative gravity.
The open strings stuck to the D-brane, as we have seen (in the case of their low
energy modes), are described by a gauge theory restricted to the D-brane (or D-brane
stack). If we increase the coupling strength so that $Ng_s \gg 1$ then the gravitational
effect of the D-branes on the spacetime metric becomes non-negligible, leading
to a curved geometry and, in fact, a black hole geometry (or a black brane). By
analogy with a standard Reissner-Nordström black hole, this geometry is $AdS_5 \times S^5$
(cf. [35, p. 174]):

$$ds^2 = \frac{r^2}{R^2}\eta_{\mu\nu}dx^\mu dx^\nu + \frac{R^2}{r^2}dr^2 + R^2 d\Omega_5^2 \qquad (10.1)$$

Of course, strings sitting near the event horizon will be red shifted from the point
of view of distant observers, and so will appear to have low energies. In the limit
of low energies (ignoring massive states) the strings near the event horizon will
decouple from the strings on the (flat) conformal boundary. Putting these two scenarios together, it follows that at weak coupling the physics is described by a gauge
theory on flat space and at large coupling is described by a closed string theory on
$AdS_5 \times S^5$. Maldacena conjectured that there was a duality linking these two descriptions together, by varying the 't Hooft parameter $\lambda_{'tHooft} \equiv g^2_{Yang-Mills}N_{colours}$, so that
it was really one theory being viewed from different regions of parameter space.[45]
The gauge theory—which is largely understood, e.g. in terms of observables and their

---

[43] The gauge theoretic aspects arise from the fact that, as we have seen, the degrees of freedom on
the brane are matrix valued, where the indices of the matrix M_{ij} refer to the endpoints of the open
string (if the endpoints lie on different branes then $i \neq j$).

[44] Polyakov refers to the slick manoeuvre of switching from a D-brane to a gravitational description
as "a little like replacing the famous cat by its smile" [55, pp. 548–549].

[45] The holographic nature of the duality is evident from the fact that one is dealing with boundary
data in the string theory. It is the boundary data that delivers the gauge field theory. The gauge
theory lives on the $r \to \infty$ conformal boundary of AdS_5, with the string theory defined throughout
the $r < \infty$ interior, i.e the bulk. This radial dimension (a 5th spatial dimension in the case of the
string theory) is converted into an energy (or renormalization group) scale in the field theory on the

Hamiltonian evolution, and so on—includes in its (boundary) degrees of freedom all the information of the dual gravity theory (in the bulk).

Note that the apparent puzzle concerning the difference in dimensions of the two theories dissolves once one realises that they do not function as spacetime dimensions in both theories. The five dimensions of the string theory that appear to be missing in the gauge theory (from S^5) are retained as 'internal' degrees of freedom of the gauge particles: the full ten-dimensional spacetime coordinates of the string theory appear in the Yang-Mills theory as ten bosonic fields split between six scalar fields to describe D3-brane motion (one of these being the radial direction, and the other five being angles in transverse spatial directions that come from the matrix description of the branes) and four vector fields describing the low energy modes of the open strings stuck to the flat spacetime volume traced out by the D3 brane. The five angles map onto the 5-sphere component of the full product space while the Minkowski spacetime coordinates of the D3 brane (i.e. the worldvolume) and the radial direction map on to AdS_5. The symmetries are preserved between the theories by a mapping from the conformal symmetry of the Yang-Mills theory to isometries of the metric. Note that this also resolves a problem with the attempt to reconstruct the interior data from the boundary data (or vice versa) since *local* events on the boundary (i.e. observables that are close) can be far apart in the interior, but the energy-distance anti-correlation can (at least partially) account for this behaviour. This is how the four-dimensional gauge theory can encode the ten-dimensional gravity theory.

Given that one can describe black holes in string theory, using branes, one can ask about the system on the other side of the duality. Consulting the 'duality dictionary,' one finds a plasma of hot gluons: a thermal system. The Hawking radiation can likewise be translated into standard evaporation in the gluon system. This leads to an intuitive resolution of the Hawking information loss puzzle: the information is preserved, though it leaks out as in the evaporation of the gluon system. This is directly applicable to the black hole information puzzle: given the gauge theory/gravitational theory duality, if the former has no information loss (since the mechanism of such loss is absent in such a theory), then there cannot be information loss in the latter case.[46] Finally, and perhaps more importantly, the correspondence provides, in a similar way to the resolution of the information paradox, what might be the first non-perturbative definition of string theory. Gauge theories are well understood, and if there really is a strong-weak equivalence, then it can be used to tell us what the dual theory is. This is, more or less, where research on string theory stands today: trying to make better sense of the duality (and *prove* it in more certain terms) in order to better understand string theory.

(Footnote 45 continued)

boundary such that events at distances far from the boundary correspond to IR processes and those near the boundary correspond to UV processes.

[46] There is a slight similarity to Gepner's resolution of the problem of the link between Calabi-Yau manifolds and solutions of the equations of string theory using conformal field theories (see p. 192). In that case, given the CFT-CY correspondence, if a CFT can be shown to be a solution then so must its corresponding Calabi-Yau manifold/s.

10.3 From Landscape Gardening to Anthrobatics

String theory gained a strong grip on the public's imagination when, in 1988, a radio series on string theory, *Desperately Seeking Superstrings*, was broadcast, by the BBC. The show was conceived by Paul Davies who built it up around several interviews of famous physicists, some pro-string and some anti-string.[47] There continues to be a controversy about whether string theory deserves the preferential treatment it appears to receive: string theory is claimed to receive more funding and more fresh graduate students than it has earned the right to. As we have seen, the public controversy over superstring theory really began in 1985, with a spate of letters and notes bemoaning what was seen as a bad precedent for physics, jeopardising the historically-close bond between theory and experiment. A year later, Ginsparg and Glashow wrote:

> Contemplation of superstrings may evolve into an activity as remote from conventional particle physics as particle physics is from chemistry, to be conducted at schools of divinity by future equivalents of medieval theologians. For the first time since the Dark Ages, we can see how our noble search may end, with faith replacing science once again. Superstring sentiments eerily recall "arguments from design" for the existence of a supreme being [24, p. 7].

Of course, this was based on the fact that just as the heterotic string theory of Gross, Harvey, Martinec, and Rohm promised closer contact with the real world of low energy physics, the number of possible compactifications quickly dashed such hopes, spoiling predictive capabilities in the process.

The notion of a 'landscape' of string theories took some time to emerge, and came in several forms over a period of two decades.[48] We saw that, initially, in 1985, it was argued by K. S. Narain that there existed *infinitely* many (tachyon-free) heterotic string theories in $D < 10$. Narain associates the distinct string theories with the points of a coset manifold $M_d = SO(26 - d, \ 10 - d)/SO(26 - d) \otimes SO(10 - d)$. He also raises the question of which point, if any, nature selects, and why [48, p. 11]. These were not really viable models for our world. But they pointed very clearly to the fact that the consistency principles of string theory were not restrictive enough: there was still lots of freedom in the construction of consistent theories. Likewise, Andy Strominger, in 1986, wrote that:

> With the inclusion of non-zero torsion, the class of supersymmetric superstring compactifications has been enormously enlarged. It is barely conceivable that all zero-torsion solutions could be classified, and that the phenomenologically acceptable ones (at string tree level) might then be a very small number, possibly zero. It does not seem likely that non-zero torsion solutions, or even just the subset of phenomenologically acceptable ones, can be classified in the foreseeable future. As the constraints on non-zero torsion solutions are relatively weak, it does seem likely that a number of phenomenologically acceptable (at string tree level!) ones can be found... While this is quite reassuring, in some sense life has been made too easy. All predictive power seems to have been lost. All of this points to the overwhelming

---

[47] This was released as a book in 1992: *Superstrings: A Theory of Everything?* [13].

[48] The 'landscape' terminology was introduced into fundamental physics by Leonard Susskind [73].

need to find a dynamical principle for determining the ground state, which now appears more imperative than ever [68, p. 28].

Lerche, Lüst, and Schellekens are not so negative about such a large number of solutions: "Despite the presumably gigantic number of models that may exist, the possibilities are thus still severely limited in comparison with field theory in four dimensions" [42, p. 504]. That is, having a finite space of possibilities, however large, at least signals some promise of control and understanding.

In June 1986, Gell-Mann [23, p. 206] raised the spectre of anthropic reasoning (hanging over these expressions of horror at the vastness of the space of vacua), with respect to the question of "how Nature chooses among the physically inequivalent superstring theories, if we assume that one of them is right". He considers three options, one of which is the idea that "Nature has arbitrarily chosen the one that agrees with our observations" noting that this "seems unpleasantly close to the strong anthropic principle".

The failing uniqueness, once a central motivation of superstring theory (and a strong link to its bootstrapping past), appeared to be forcing a modification in the way the theory was to be understood. A similar shift was affecting cosmology according to Andre Linde (with a shift to an *inflationary model*, devised by Alan Guth), also writing in 1986:

> At present it seems absolutely improbable that all domains contained in our exponentially large universe are of the same type. On the contrary, all types of mini-universes in which inflation is possible should be produced during the expansion of the universe, and it is unreasonable to expect that our domain is the only possible one or the best one. From this point of view, an enormously large number of possible types of compactification which exist, e.g., in the theories of superstrings should be considered not as a difficulty but as a virtue of these theories, since it increases the probability of existence of mini-universes in which life of our type may appear. ... The old question [of] why our universe is the only possible one now is replaced by the question in which theories [of] the existence of mini-universes of our type [are] possible. This question is still very difficult, but it is much easier than the previous one. From this point of view, an enormously large number of possible types of compactification which exist e.g. in the theories of superstrings should be considered not as a difficulty but as a virtue of these theories, since it increases the probability of existence of mini-universes in which life of our type may appear [43, p. 399].

Hence, Linde finds the existence of a plurality of worlds, predicted by a theory, a *good* thing, rather than a feature to be eliminated by finding appropriate selection or elimination mechanisms. However, the problem is that lack of uniqueness in pinning down the features of our world quite naturally results in a loss of predictive power (presumably one that scales with the departure from uniqueness). As Schwarz writes:

> Ideally, there would be just one consistent theory and it would have a unique stable vacuum. If that were the case then everything would be calculable from first principles. This is certainly the outcome that would be most satisfying. We have no guarantee that this is the way things are, however. At present, it seems at least as likely that there are large classes of stable vacua each characterised by a number of parameters. In this case one would imagine different choices are actually realized in different regions of the universe. Then the fact that a particular vacuum is selected in our little corner of the universe could not ever be understood as a logical necessity except perhaps using an anthropic principle. There would

be no possibility of ever calculating some of the observed phenomenological parameters [59, p. 200].

These considerations form one kind of plurality. The existence of the various *types* of string theory points to another. However, the plurality that is behind the most recent string controversy includes the additional non-perturbative advances discussed in this chapter. In referring to the various vacua as "stable" Schwarz was unaware of (or was sidestepping) a hidden instability in their moduli. The problem is: what holds the compact spaces in place? What *constrains* them, preventing them from decompactifying to large dimensions, like our flat ones? When we take into account the method for correcting this (namely stabilization via flux compactification), the number of solutions becomes incredibly vast, with the standard estimate being around 10^{500}—vast, but certainly more tightly constrained that Narain's infinity of theories. Hence, as the conception of string theory has changed (with new tools and ideas being added, such as Calabi-Yau compactifications initially, and then D-branes), a persistent controversy concerning the theory's predictive power has changed in tandem.

The problematic moduli in question are the Kähler and complex structure moduli parametrizing the size and shape of Calabi-Yau manifolds. These will be free to vary if left unconstrained.[49] Shamit Kachru, Renata Kallosh, Andrei Linde, and Sandip Trivedi (KKLT: [40]) speak of having to "freeze" such moduli, using flux to stabilize moduli in compactification schemes. In this case the flux is quantized so that the moduli values are also quantized in the process. The flux constrains the complex structure moduli, while D-branes have to be introduced to constrain the Kähler moduli.[50] This will determine a countable family of stable Calabi-Yaus. However, the number of possible manifolds, although forming a discrete family, is considerably larger than previous pluralities in string theory. A rough estimate that is often suggested, as above, is 10^{500} possible ground states.[51] This is the contemporary meaning of 'string Landscape' (corresponding to the previous chapter's 'ground state explosion') but, as before (see p. xx), we find once again a split into two ways of viewing the plurality:

---

[49] In the low-energy limit these moduli are like massless scalar fields, and so they can be changed without energy loss. In 1985, Michael Dine and Nathan Seiberg [15] had argued that the size modulus of a Calabi-Yau manifold would indeed decompactify to infinite radius, rendering unusable the whole compactification scheme (on which string phenomenology rested).

[50] In fact, the 'turning on' of fluxes in this way implies that the compact manifold is no longer of Kähler-type. Strominger [68] had already discussed such compactifications in 1986, when he studied compact spaces with torsion. This modifies the usual Cremmer-Scherk 'Cartesian product space' approach to the treatment of the compact and non-compact spaces, since the two lose their autonomy (and one speaks of a 'warped product' instead)—see Becker et al. [4] for an early discussion in the context of the moduli stabilization problem.

[51] A figure computed by assuming that shape moduli are restricted to some integer values, $n = 0, \ldots, 9$ (arising from the flux quantization), and combining this with the maximum possible Euler number for a Calabi-Yau manifold, assumed to be around 500 (based on current theoretical estimates and computer searches). The exact figure is not so important for the purposes of the debate. All that matters is that this number is, as Susskind so nicely puts it, "prodigiously large" [74, p. 285].

1. Treat the landscape's elements as corresponding to dynamical possibilities (once necessary identifications due to dualities have eliminated redundant points).
2. Find some mechanism or principle to break the plurality down to our world.

There are two controversial aspects with option 1: not only does it involve commitment to a gigantic ensemble of unobservable worlds; but in order to make sense of our own world within this ensemble, we must invoke the anthropic principle: we are **here**$_{existence}$ because we are **here**$_{location}$. That is, we find ourselves in this particular ground state (with its Yukawa couplings and particular particle content) because such a ground state (located amongst a plenitude of others) is necessary to support the existence of complex beings like ourselves. Were the values different (corresponding to a different **here**$_{location}$), we would not be **here**$_{existence}$.[52] As Susskind put it in his paper that introduced the terminology of *Landscape*: "The only criteri[on] for choosing a vacuum is utility, i.e. does it have the necessary elements such as galaxy formation and complex chemistry that are needed for life. That together with a cosmology that guarantees a high probability that at least one large patch of space will form with that vacuum structure is all we need" [73, pp. 5–6].[53] That is, no *dynamical* selection mechanism is needed to sift through the possible worlds; nor is any ultimate consistency condition that eliminates all but one possible world.[54]

---

[52] It is useful to compare this with Johannes Kepler's explanation for the planets' specific spacings from one another and from the Sun (as presented in his *Mysteruim Cosmographicum*). Kepler tried to deduce these distances from (geometrical) first principles, using a 'best fit' approach to the nesting of the five Platonic solids within one another, while considering the spheres in which the solids were themselves embedded as grounding the planet's relative distances. On this account, the explanation for the Earth's distance from the Sun, for example, is based on a mathematical scheme involving the regular polyhedra. Of course, the model was soon proved wrong by data, showing previously unknown planets that did not fit Kepler's scheme. The point is, however, that a more natural explanation in this context is simply that had the Earth *not* been at the distance it has (or thereabouts) there wouldn't exist beings such as ourselves capable of posing the question in the first place, since the conditions would not support complex life. Of course, this is over-simplified, and one might question various parts of the anthropic answer, but it clearly shows how an anthropic response might in some cases be a reasonable option. One might think a better response would look not to mathematical principles, but to physical principles: the evolution of galaxies and so on. This latter would perhaps be a closer match to option 2 above.

[53] As Susskind notes, the terminology of "Landscape" came from the study of systems with very many degrees of freedom, in which the metaphor of 'energy landscape' is employed [74, p. 274]. In this context one can find jagged graphs with peaks and valleys, such that the valleys are supposed to represent possible states of the system.

[54] This desire for 'one possible world' coming out of the equations ('one vs. many') might be seen as a throwback to Chew's *frustration* over arbitrariness in physics [10] (see also [11]). It is, perhaps, no accident that Chew's former student, David Gross, is one of the staunchest advocates of the 'uniqueness via selection' option. He was (and likely still is) of the opinion that such forms of reasoning should be "at best. . .the last resort of physical theory" [31, p. 105]. (I will just mention one example of what such a selection rule might look like (due to R. Holman and L. Mersini-Houghton: [32]). Their idea is that decoherence via the backreaction of matter degrees of freedom onto gravitational degrees of freedom can serve to reduce the number of allowed initial states of a universe. Generically, any cosmological model with both matter and gravity will exhibit non-ergodic behavior driven by out-of-equilibrium dynamics so that such universes must satisfy a superselection rule for the initial conditions—they also manage to pull out an explanation of the arrow of time from

Relative to this Type 3 plurality, Susskind adopts a stance more or less aligned with Linde.[55]

The standard argument against such a position is that it demolishes our ability to make predictions. This forms the basis of Lee Smolin's primary objections in his book *The Trouble with Physics*. Smolin strongly distinguishes examples such as the Keplerian one I gave above from the kinds of case involving the universe as a whole and the landscape. To change to Smolin's own example (see [67, p. 163]): why is the Earth so bio-friendly? The puzzle is easily resolved anthropically in this case because we have evidence of billions of other stars (and likely planets) and we will quite naturally find ourselves on a biofriendly one: how could it possibly be otherwise? This is the *weak* anthropic principle: it is generally accepted as valid, though rather trivial reasoning. If we apply the same question to the universe, instead of the Earth, then, the objection goes, we have no evidence of billions of similar universes, and so we cannot run the same argument. As far as the universe is concerned, we have a sample of one: our own.[56]

However, there have been attempts to derive predictions (or, more precisely *accommodations*) using the 'vacua + anthropics' package as a tool. Bousso and Polchinski followed this strategy in 2000 to calculate the value of the cosmological constant [7]. Their approach was simply to find a way to generate a large enough

---

(Footnote 54 continued)
their scheme, based on the fact that the same non-ergodicity lowers the entropy of initial states, thereby allowing one to use the second law of thermodynamics *plus* this low-entropy past.

[55] Alan Guth credits Susskind with being one of the key spokespeople for (the original) inflationary cosmology's good 'public relations' (interview of Alan Guth, by Alan Lightman, September 21, 1987: http://www.aip.org/history/ohilist/34306_1.html). Susskind (together with Sidney Coleman) were in the audience of Guth's first talk in which he introduced the idea. However, when it came to the anthropic people, Guth was on the side of Gross: "I find it hard to believe that anybody would ever use the anthropic principle if he had a better explanation for something." Pointing instead to a future where we have better physics, he says: "I tend to feel that the [physical constants] are determined by physical laws that we can't understand [now], and once we understand those physical laws we can make predictions which are a lot more precise ... my guess is that there really is only one consistent theory of nature, which has no free parameters at all". Of course, some would say that the anthropic Landscape *is* such a theory. (Note that Guth has since switched his allegiance: http://www.iop.org/about/awards/international/lecture09/page_38408.html—it seems clear that in his interview with Lightman, Guth was referring to a *strong* version of the anthropic principle according to which humans are somehow 'special' in the universe. If one has independent reasons to believe in a large enough ensemble of worlds to make worlds like ours likely within it, then one can adopt a weaker version of the principle along the lines of that given above, concerning the Earth's location.

[56] In fact, I think a case might be made for using Smolin's 'reasonable' usage of the anthropic principle (that one can explain away curious, apparent fine-tuning using an ensemble of similar cases) as providing some level of support for the universe-level case. The kind of fine-tuning one finds at the Universe-level is very similar to that at the planetary-level, and so one might reasonably assume that their solutions will be similar (especially so in the absence of any other reasonable alternatives). In other words, so long as one doesn't *restrict* one's evidence to just one finely-tuned universe, but also considers how a similar problem concerning the bio-friendly Earth is resolved (and perhaps other fine-tuning cases of a similar nature in which one has an observable comparison class available, as Smolin insists), then one could begin to mount a defence of the landscape.

space of string vacua to make those possessing a tiny (but non-vanishing) cosmological constant, like our own universe,[57] likely (and thus explain our presence in such
a world). Susskind was following this same path. Michael Douglas also followed
suit, again focusing on "physical questions this [ensemble] might help us resolve"
[17, p. 1]. Douglas' approach was to attempt to better understand the details of the
space of vacua by classifying its elements: "one must simply enumerate string/M
theory vacua and test each one against all constraints inferred from experiment and
observation" (ibid., p. 2). This approach has similarities to one suggested by Kawai,
Lewellen, and Tye in 1987[58] which is clearly doing for the type-2 plurality what
Douglas proposes for the type-3:

> It is also clear that in contrast to the 10-dimensional case the number of 4-dimensional chiral
> models is very large. As yet, a complete classification of all consistent string models is
> unavailable. In this work, we have given a complete treatment of fermionic string models in
> $D \leq 10$ dimensions obtainable from toroidal compactification in the fermionic formulation.
> This subclass of models is already quite large. In the first quantized formalism, all consistent
> string models should be treated on an equal footing. It is plausible that string dynamics may
> select a subset of the first quantized string models (i.e. second quantized vacuum states)
> as locally stable (e.g., by considering solutions of the (as yet unknown) closed string field
> theory). However, even if string dynamics eventually selects a unique ground state, it does
> not necessarily imply that this is the state representing our universe [39, p. 72].

So much by way of setting up the kind of early landscape scenario we have already
seen. They continue:

> A systematic approach to test the string theory would be to completely classify all consistent
> four-dimensional chiral string models and then examine them one by one. We believe such
> a complete classification is a tractable problem and that the relevance of string theory to
> nature can be tested [39, p. 75].

Shortly afterwards, Antoniadis, Bachas, and Kounnas cautioned against such a 'brute
force' approach, stating: "The number of consistent four-dimensional string theories
is so huge that classifying them all would be both impractical and not very illuminat-
ing" [1, p. 104]. It is clear that the additional structure of the type-3 plurality makes
the project more plausible. However, the task is ongoing.

It is often argued that string theory's contributions to mathematics are sufficient to
warrant such inflated levels of support. Such mathematical contributions are impressive on their own merits, but they can often lead to unexpected physical results. John
Ellis expressed this particularly clearly in the paper that introduced the title of 'theory
of everything'[59]:

---

[57] Of course, this tiny non-zero value is fixed by the acceleration of the universe's rate of expansion.

[58] Lerche, Lüst, and Schellekens had remarked earlier [42, p. 505] that one might be able to
completely classify some subclass of the plurality of theories.

[59] As John Schwarz has pointed out, the phrase 'theory of everything' has tended to worsen the
controversy: "The phrase "theory of everything," which has been used in connection with string
theory, is a phrase I don't like myself and have tried to avoid. It was introduced by somebody else.
There are several reasons I don't like it. One reason ... is that it gives other physicists the impression
that people who work in this field feel that their work is more important than what other people are

> Even if many features of [superstring theory] are wrong, new ideas are being brought into particle physics at a rate unequalled since the renaissance of gauge theory in 1971. Our intellects are being mathematically stimulated, and we are thinking of many new types of phenomena that our experimental colleagues can search for. We cannot discover the secrets of nature by pure reason, and must look to an experimental breakthrough. At the very least, the superstring may point us to a previously unmarked stone which, when turned over, may reveal interesting new life beneath [20, p. 597].

Ivan Todorov points out, in response to this kind of *argument from mathematical fertility* that the study of knot invariants was stimulated by the Haag-Kastler operator algebraic approach to local quantum theory[60] yet has not received anything like the kind of support as string theory has [75, p. 158]. Clearly, however, this is too simplistic. I have not seen it suggested that string theory's mathematical achievements *alone* warrant such preferential treatment. Only that it is one component of a case built from very many achievements.

Murray Gell-Mann suggests that string theory might be following an entirely distinct path to that usually followed in the natural sciences:

> My attitude towards pure mathematics has undergone a great change. I no longer regard it as merely a game with rules made up by mathematicians and with rewards going to those who make up the rules with the richest apparent consequences. Despite the fact that many mathematicians spurn the connection with Nature (which led me in the past to say that mathematics bore the same sort of relation to science that masturbation does to sex[61]), they are in fact investigating a real science of their own, with an elusive definition, but one that somehow concerns the rules for all possible systems or structures that Nature might employ. Rich and self-consistent structures are not so easy to come by, and that is why superstring theory, although not discovered by the usual inductive procedure based principally on experimental evidence, may prove to be right anyway [23, p. 208].

There is a sense in which this debate over the fundamentals of the scientific enterprise harks back to a much earlier debate over the same issue, between Thomas Hobbes and Robert Boyle. Hobbes criticised Boyle's experimental method for a variety of reasons, but especially pertinent is his assertion that experiments—Hobbes had in mind those involving the air-pump—are inherently defeasible, with any knowledge

---

(Footnote 59 continued)
doing, and this creates a certain hostility or bad feelings. My personal feeling is that what we're doing is interesting and important but what other people are doing is also interesting and important, and any phraseology that's going to create a wrong impression I think is unfortunate. ... Another is that I think it's misleading, because even if we did solve all the problems we're trying to solve, there would be many things that were not explained—it's not a theory of everything. It's a theory of something—something that's very fundamental and very interesting. But there's a lot more to the world than what you can learn from the basic underlying microscopic physical laws" (Interview with John H. Schwarz, by Sara Lippincott. Pasadena, California, July 21 and 26, 2000. Oral History Project, California Institute of Technology Archives. Retrieved [2nd Jan, 2012] from the World Wide Web: http://resolver.caltech.edu/CaltechOH:OH_Schwarz_J).

[60] Vaughan Jones' work grew out of his studies of subfactors, which was related to the Haag-Kastler approach.

[61] Gell-Mann seems to have been at his most whimsical at this conference. In his lecture he also refers to the other speakers' rapid-fire usage of the overhead projector as like a "tachistoscope," accusing them of engaging in subliminal messaging!

generated from them likewise rendered defeasible (experiments can be rationally compelling though not deductively valid).[62] The experimental approach was victorious in the earlier debate. It remains to be seen whether string theory follows Boyle down the experimental path, or ends up closer to Hobbes. Certainly, in attempting to construct a theory with no free parameters, that explains all forces, all matter, and even their spatiotemporal framework, one is bound to face some difficulties in connecting with everyday experimental science!

10.4 The Future of String Theory

At a meeting on 'Unified String Theory' in 1985, David Gross laid out eight questions and problem areas that needed to be addressed [29]. He revisited these in a 2005 talk [31]:

1. How Many String Theories are There?
2. String Technology.
3. What is the Nature of String Perturbation Theory?
4. String Phenomenology.
5. What is the Nature of High Energy Physics?
6. What Picks the Correct Vacuum?
7. Is there a Measurable, Qualitatively Distinctive, Prediction of String Theory?

He saved an additional question for last: "8. What is String Theory?" Some of these questions become interlinked by 2005, especially the first and last. In addition to the older dualities, giving the "web of theories" (T-duality and the weak-strong coupling dualities), he mentions the AdS/CFT correspondence: this links backgrounds too, but also provides clues as to the question of what string theory is, since it provides a non-perturbative definition of the string theories involved in the duality—for this reason, question 3 is also clearly impacted on.

As we have seen, a persistent stumbling block since it was encountered in the mid-1980s has been the problems posed by the proliferation of string vacua, which has a direct bearing on questions 6 and 7 (and 1), probably the most important from the point of view of string theory's critics. The only known way pointing to some kind of solution it is to invoke the anthropic principle.

The key issue for string theorists is more probably: what *is* string theory? This might seem like a rather ridiculous question to pose after a book devoted to its not inconsiderable history, but there has yet to be presented a *principle* for string theory, and it remains to a large extent a framework of rules of thumb and techniques, albeit an incredibly fruitful and promising one. In the context of his 1985 talk, Gross was concerned that its methods of construction, "often producing, for apparently mysterious reasons, structures that appear miraculous" [31, p. 104] was problematic:

---

[62] See Steven Shapin and Simon Schaffer's book *Leviathan and the Air-Pump* [65] for the locus classics of this debate.

far better to have an well-founded account rather than a miracle, despite the fact that problems were being resolved all the same. As he puts it: "[w]e do not really understand what are the truly fundamental degrees of freedom, what is the underlying dynamical principle and what are the underlying symmetries?" (ibid.). Gross further asks: "how many more string revolutions will be required before we know what string theory is?" [31, p. 104]. We haven't moved so very far in the intervening 10 years: there has been no 'third revolution,' though, as with the AdS/CFT conjecture, one might consider raising the status of the Landscape conjecture to revolutionary status. However, this too is more of an aftereffect of D-branes. Still, it is an aftereffect that reinvigorated the field, coming around a decade since D-branes were understood to be a pivotal concept.[63] In its essential details, however, the landscape is a much older concept in string theory. What changed is that D-branes brought it under greater statistical control. The recent developments on the gauge/gravity duality did truly transform the state of the discipline: whatever string theory is, it's not as it was known prior to 1994/5. Advances have been made.

Much of the most recent work (as of 2013) has been devoted to unpacking the consequences of this duality and pushing it to its limits in order to extract realistic models, instead of QCD-*like* models. With this class of dualities there has also emerged an increased inclination amongst string theorists to engage in debates on the conceptual foundations of string theory, discussing such issues as the emergence of space-time, relational locality and the nature of physical observables (much as had occurred in the mid- to late-1980s. The ability of the AdS/CFT correspondence to provide a potential resolution of the black hole information paradox, allowing unitary condensing and evaporation of black holes (by studying a dual unitary gauge model of the process) is an important event that has to play a role in how string theory is evaluated. The Landscape has blended with some of this machinery, opening up new possibilities for explaining otherwise puzzling features of our universe.

The first revolution was characterised by an obsession with replicating the standard model (especially the fermion generations). The second was concerned more with black hole physics, but also went back to its origins in strong interactions, where it attempted to answer the kinds of strongly coupled problems that other approaches found too difficult. The present era has linked up with cosmology, and is tackling the really big questions about the universe as a whole. Gross latched onto these emerging connections between string theory and other areas such as cosmology:

> Cosmology needs string theory as it tries to push back to the big bang. Inflationary theory needs string theory to justify its sometimes ad hoc or fine-tuned constructions. . . . Conversely string theory needs cosmology. String theorists hope that cosmological observations will enable one to make contact with observation [31, p. 102].

I expect that the next era will focus on pushing these connections to their limits. Solving riddles that appear to be 'out of bounds' appears to be a specialty of string

---

[63] There appear to be roughly decadal cycles (the explosive snores of Terry Gannon's drunk from the preface perhaps?) in which some new big idea transforms string theory: (1974: dual models of everything) → (1984: anomaly cancellation/heterotic strings/Calabi-Yaus) → (1994/5: D-branes and dualities) → (2003/4: the anthropic landscape) → (2014: ?). It seems we are due a new cycle.

theory, and cosmology has these in abundance. Absent direct experiments, such unified puzzle-solving offers a much needed alternative source of empirical support.[64]

To get a better grasp on where string theory has come from, and where it might go in the future, it is instructive to sort its evolution into stages that I will characterise as 'playing with $\mathscr{X}$' (where $\mathscr{X}$ is some particular concept or tool). Different choices for $\mathscr{X}$ will often link up in unexpected and fruitful ways, possibly triggering a new phase of development. We find, for example:

- Playing with the operator formalism.
- Playing with the string picture.
- Playing with limits:

  - zero-slope
  - large-N

- Playing with supersymmetry:

  - worldsheet
  - spacetime

- Playing with compactification:

  - lattices
  - winding
  - orbifolds

- Playing with duality:

  - D-branes
  - black holes

- Playing with the Landscape.

Such phases are themselves characterised by a near-exhaustive approach, examining all possible ways of using, stretching, and thinking about the $\mathscr{X}$ in question, and often mixing in ideas from other phases.[65] I have indicated some possible subdivisions one

---

[64] The style of explanations given by string theory are very much on a par with those in cosmology. Consider: why do there appear to be no magnetic monopoles in the universe? This is a question concerning an empirical fact that we know (it is old evidence, if you like), but that is still in need of an explanation (especially if one believes that the universe began in an extremely hot state). Likewise, the horizon problem: why does there appear to be some kind of conspiracy linking the thermal behaviour of causally disconnected regions of the universe? No theory predicted these features *prior* to our having known about them. However, that inflationary cosmology was able to derive them as consequences (using the same mechanism) is a success of the theory, whether or not they constitute genuine predictions. It is no accident that, like string theory, cosmology often has recourse to anthropic reasoning.

[65] For example, the winding phase became a tool in the orbifold phase. The winding notion was generalised to 'wrapping' once the notion of branes came about. The wrapping and winding were used (in tandem with D-branes) to resolve a problem with stability of compact spaces, which in turn led to the Landscape (with its possibilities for doing statistics of vacua, thanks to the discreteness involved in the winding).

might make, though I'm sure many further subdivisions could be found within each of those I have suggested. I certainly don't mean to suggest that this exhausts the development of string theory. For example, missing from this idea is the analogical reasoning that has permeated all stages of string theory's evolution. In fact, it is almost *always* the case that heavy analogical reasoning is at work in the initial period of these various phases, where they are often pushed until they snap—in which case one will have learned something interesting: a breakdown of the older concepts.

The present phase appears to be based around playing with the *Landscape* and *holography*.[66] This looks set to stay for a while, but it is interesting to speculate on what the next 'playtime' might involve. In most of the cases in the past, however, the new phases have been almost entirely unexpected, which is precisely what leads to the sudden frenetic pace that follows. It is entirely possible that the phase transition will not be a new idea at all, but some confluence of pre-existing ideas (as with D-branes and dualities).

At the second Nobel Symposium in 1986, with a talk possessing the same title as my final section title, John Schwarz writes:

> I was asked recently what is the fundamental equation that we are trying to solve. I found the question somewhat awkward to answer in a few words, because while we know what we are talking about, there does not yet exist a concise and elegant description of string theory [59, p. 197].

At the very same symposium, in the closing talk, Murray Gell-Mann wrote: "there is a hint that the search for the principle underlying superstring theory may bring us back to the vicinity of where we started, the duality version of the bootstrap" [23, p. 205]. Behind this remark lurks a grain of truth: at the root of the belief that there will be a dynamical principle[67] that (non-anthropically) selects the unique configuration describing our world from a bunch of *prima facie* equally qualified configurations, is, I think, a bootstrapper's dream (or hangover). It is a desire to have the world uniquely fall out from the right consistency conditions. For better or for worse, this 'dream' has pushed string theorists on, still searching for the elusive principle while more and more structure is added and the framework is ever more radically altered. Such extra-empirical principles clearly have a role to play in theory-building. Those who adopt the anthropic stance are guided along different channels, and inevitably uncover different aspects of the same structure that is common to both camps, as well as different applications that are not common to both. It isn't at all clear which group the future development of string theory will favour. My guess is that the anthropic stance will succeed partly because it has strong support in aspects of cosmology, but also because the notion of a physical theory that uniquely pins

---

[66] I'm including in this especially the utilization of infrared 'domain walls' to attempt to recover confining gauge theories which I see as conceptually continuous with the earlier constructions involving orbifolds, twisted sectors, and the like, in order to get out certain realistic features.

[67] This belief is quite clearly expressed by one of Chew's students, John Schwarz, when he writes: "There is a widespread belief, which I share, that a beautiful and profound principle lies at the heart of string theory. When elucidated, it should become much clearer why all these miracles have been turning up" [59, p. 198].

down our world seems too strange a prospect. But this is just an opinion. It is more likely that the two stances will continue in parallel, as they appear to have done for some time, defined more by the personalities of those adopting them than by the physics.

10.5 Closing Remarks

I hope to have revealed in this book a little more of the history of string theory than is usually presented, even in professional accounts. The lesson I think emerges from this is that, while the mythological presentations of 'revolutions' and 'dark years' and so on, make for a good story, a more accurate depiction reveals a somewhat less turbulent life story, though no less interesting for it. Though there are indeed curiosities in the history of string theory—preeminent amongst these being the phase of exaptation from hadronic to 'fundamental' strings—for the most part it represents a perfectly rational sequence of events, not so very different locally from any other area of physics. Indeed, I think that in presenting string theory's historical trajectory as a somewhat quirky roller-coaster ride, the proponents of string theory might have shot themselves in the foot! Those that have not studied string theory might be far more willing to give strings a chance if they knew that perfectly ordinary quotidian principles of scientific theory construction lay at its heart. This is the story I have attempted to tell, and it is my hope that it may do a little good in taming some of the hype and hysteria forming the controversy over string theory and its elevated position in the research landscape. I might also add that throughout, the majority of those with an interest in string theory have not been irrationally convinced of its absolute certainty, but rather have seen that the potential payoff is so large that it makes the risk of its being a dead end worth taking: if one has an example of a likely-looking candidate for a unified field theory of all known interactions and elementary particles, then that is surely reason enough to pursue it.

References

1. Antoniadis, I. C., Bachas, P., & Kounnas, C. (1987). Four-dimensional superstrings. *Nuclear Physics,B289*, 87–108.
2. Atick, J. J., & Witten, E. (1988). The Hagedorn Transition and the Number of Degrees of Freedom in String Theory. *Nuclear Physics, B310*, 291–334.
3. Banks, T., Fischler, W., Shenker, S. H., & Susskind, L. (1997). *M* theory as a Matrix Model: A Conjecture. *Physical Review D, 55*(8), 5112–5128.
4. Becker, K., Becker, M., Dasgupta, K., & Green, P. (2003). Compactifications of Heterotic Theory on Non-Kähler Complex Manifolds, I. *Journal of High Energy Physics, 4*, 1–59.
5. Bekenstein, J. (1981). Gravitation, the quantum, and statistical physics. In Y. Ne'eman (ed.), *To fulfil a vision: Jerusalem Einstein centennial symposium on gauge theories and unification of physics forces* (pp. 42–59). Boston: Addison-Wesley Publishing Company Inc.

6. Bogomol'nyi, E. (1976). Stability of classical solutions. *Soviet Journal of Nuclear Physics, 24*, 861–870.
7. Bousso, R., & Polchinski, J. (2000). Quantization of four form fluxes and dynamical neutralization of the cosmological constant. *Journal of High Energy Physics, 6*, 1–25.
8. Callan, C. G., Myers, R. C., & Perry, M. J. (1988). Black holes in string theory. *Nuclear Physics, B311*, 673–698.
9. Callan, C., & Maldacena, J. M. (1996). D-Brane approach to black hole quantum mechanics. *Nuclear Physics, B472*, 591–608.
10. Chew, G. F. (1970). Hadron bootstrap: Triumph or frustration? *Physics Today, 23*(10), 23–28.
11. Cushing, J. T. (1985). Is there just one possible world? Contingency vs the bootstrap. *Studies in the History and Philosophy of Science, 16*(1), 31–48.
12. Dai, J., Leigh, R. G., & Polchinski, J. (1989). New connections between string theories. *Modern Physics Letters A, 4*(21), 2073–2083.
13. Davies, P. C. W., & Brown, J. (Eds.). (1992) Superstrings: A theory of everything? Cambridge: Cambridge University Press.
14. de Vega, H. J., & Sánchez, N. (1987). Quantum dynamics of string in black hole spacetimes. *Nuclear Physics, B309*, 552–576.
15. Dine, M., & Seiberg, N. (1985). Couplings and scales in superstring models. *Physical Review Letters, 55*, 366–369.
16. Dijkgraaf, R., Verlinde, E., & Verlinde, H. (1997). Matrix string theory. *Nuclear Physics, B500*, 43–61.
17. Douglas, M. R. (2003). The statistics of string/M theory vacua. *Journal of High Energy Physics., 2003*, 1–60.
18. Duff, M. J., Howe, P. S., Inami, T., & Stelle, K. S. (1987). Superstrings in $D = 10$ from supermembranes in $D = 11$. *Physics Letters B, 191*(1–2), 70–74.
19. Duff, M. J. (1988). Supermembranes: The first fifteen years. *Classical and Quantum Gravity, 5*, 189–205.
20. Ellis, J. (1986). The superstring: Theory of everything, or of nothing? *Nature, 323*, 595–598.
21. Fairlie, D. (2012). The analogue model for string amplitudes. In A. Capelli et al. (Eds.), *The birth of string theory* (pp. 283–293). Cambridge: Cambridge University Press.
22. Font, A., & Ibáñez, L. E., Lüst, D., & Quevedo, F. (1990). Strong-weak coupling duality and non-perturbative effects in string theory. *Physics Letters B, 249*(1), 35–43.
23. Gell-Mann, M. G. (1987) Superstring theory. Closing talk at the second nobel symposium on particle physics. Marstrand, Sweden, June 7, 1986. In L. Brink et al. (Eds.), *Unification of fundamental interactions* (pp. 202–209). Physica Scripta, The Royal Swedish Academy of Sciences. World Scientific, 1987.
24. Ginsparg, P., & Glashow, S. (1986). Desperately seeking superstrings? *Physics Today*, May 7–8.
25. Ginsparg, P. (1994). First steps toward electronic research communication. *Los Alamos Science, 22*, 156–165.
26. Ginsparg, P. (2011) It was 20 years ago today... arXiv, 1108.2700v2.
27. Greene, B. R., Morrison, D. R., & Polchinski, J. (1998). String theory. *Proceedings of the National Academy of Sciences, 95*, 11039–11040.
28. Green, M. B. (1991). Space-time duality and Dirichlet Sting theory. *Physics Letters B, 266*, 325–336.
29. Gross, D. (1986) Opening questions. In: M. Green & D. Gross (Eds.), *Unified string theories* (pp. 1–2). World Scientific.
30. Gross, D. (2005). The discovery of asymptotic freedom and the emergence of QCD. *Proceedings of the National Academy of Science, 102*(26), 9099–9108.
31. Gross, D. (2005). Where do we stand in fundamental (string) theory. *Physica Scripta, T117*, 102–105.
32. Holman, R., & Mersini-Houghton, L. (2006). Why the universe started from a low entropy state. *Physical Review D, 74*, 123510–8.

33. Hořava, P., & Witten, E. (1996). Heterotic and type I string dynamics from eleven dimensions. *Nuclear Physics, B460*(3), 506–524.

34. Horowitz, G., & Strominger, A. (1991). Black strings and *p*-branes. *Nuclear Physics, B360*, 197–209.

35. Horowitz, G. & Polchinski, J. (2009). Gauge/gravity duality. In: D. Oriti (Ed.), Approaches to quantum gravity: Toward a new understanding of space, time, and matter (pp. 169–186). Cambridge: Cambridge University Press.

36. Huang, K., & Weinberg, S. (1970). Ultimate temperature and the early universe. *Physics Review Letters, 25*(13), 895–897.

37. Hughes, J., Liu, J., & Polchinski, J. (1986). Supermembranes. *Physics Letters B, 180*(4), 370–374.

38. Hull, C. M. & Townsend, P. K. (1995). Unity of superstring dualities. *Nuclear Physics, B438*, 109–137.

39. Kawai, H., Lewellen, D. C., & Henry Tye, S.-H. (1987). Construction of fermionic string models in four dimensions. *Nuclear Physics, B288*, 1–76.

40. Kachru, S., Kallosh, R., Linde, A., & Trivedi, S. P. (2003). de Sitter Vacua in string theory. *Physical Review D, 68*, 046005-1–10.

41. Kikkawa, K., & Yamasaki, M. (1984). Casimir effects in superstring theory. *Physics Letters B, 149*, 357–360.

42. Lerche, W. D., Lüst, & Schellekens, A. N. (1987). Chiral four-dimensional heterotic strings from self-dual lattices. *Nuclear Physics, B287*, 477–507. London: Academic Press.

43. Linde, A. (1986). Eternally existing self-reproducing chaotic inflationary universe. *Physics Letters B, 175*(4), 395–400.

44. Lovelace, C. (1983). Strings in Curved Space. *Physics Letters, 135*(1,2,3): 75–77.

45. Maldacena, J. M. (1998). The large *N* limit of superconformal field theories and supergravity. *Advances in Theoretical and Mathematical Physics, 2*, 231–252.

46. Mermin, N. D. (1992). What's wrong in computopia? *Physics Today, 45*(4), 9.

47. Montonen, C., & Olive, D. I. (1977). Magnetic monopoles as gauge particles? *Physics Letters, 72B*, 117–120.

48. Narain, K. S. (1985). New heterotic string theories in uncompactified dimensions <10. *Physics Letters B, 169*(1), 41–46.

49. Penrose, R. (1999) How computers help - and hurt - scientific research. *Convergence*, Winter, 30.

50. Polchinski, J., & Cai, J. (1994). Consistency of open superstring theories. *Nuclear Physics, B296*(1), 91–128.

51. Polchinski, J. (1990). Strings, quantum gravity, and membranes. In R. Arnowitt et al. (Eds.), *Strings 89: Proceedings of the Superstring Workshop March 13–18, 1989* (pp. 429–439). Texas: Texas A&M University, World Scientific.

52. Polchinski, J. (1994). Combinatorics of boundaries in string theory. *Physical Review Letters, 75*(26), 4724–4727.

53. Polchinski, J. (1995). Dirichlet branes and Ramond-Ramond charges. *Physical Review D, 50*(10), R6041–R6045.

54. Polchinski, J., & Witten, E. (1995). Evidence for heterotic-type I string duality. *Nuclear Physics, B460*, 525–540.

55. Polyakov, A. M. (2012). Quarks, strings, and beyond. In A. Capelli et al. (Eds.), *The birth of string theory* (pp. 544–551). Cambridge: Cambridge University Press.

56. Prasad, M., & Sommerfield, C. (1975). Exact classical solution for 't Hooft monopole and Julia-Zee Dyon. *Physical Review Letters, 35*(12), 760–762.

57. Randall, L., & Sundrum, R. (1999). An alternative to compactification. *Physical Review Letters, 83*(23), 4690–4693.

58. Schrödinger, E. (1939). The proper vibrations of the expanding universe. *Physica, 6*, 899–912.

59. Schwarz, J. H. (1987) The future of string theory. In L. Brink et al. (Eds.), *Unification of fundamental interactions, Physica scripta, the royal swedish academy of sciences* (pp. 197–201). Singapore: World Scientific.

60. Schwarz, J. H., & Sen, A. (1993). Duality symmetries of 4D heterotic strings. *Nuclear Physics*, *312*, 105–114.
61. Schwarz, J. H. (1995). An *SL*(2, ℤ) multiplet of type II superstrings. *Physics Letters B*, *360*(1–2), 13–18.
62. Schwarz, J. H. (1995). The power of M-theory. *Physics Letters B*, *367*, 97–103.
63. Schwarz, J. H. (1995). Superstring Dualities. Caltech Preprint: CALT-68-2019.
64. Sen, A. (1994). Strong-weak coupling duality in four-dimensional string theory. *International Journal of Modern Physics A*, *9*, 3707–3750.
65. Shapin, S., & Schaffer, S. (1985). *Leviathan and the air-pump: Hobbes, Boyle, and the experimental life*. New Jersey: Princeton University Press.
66. Shenker, S. (1991). The strength of nonperturbative effects in string theory. In O. Alvarez, E. Marinari, & P. Windey (Eds.), *Random surfaces and quantum gravity* (pp. 191–200). USA: Springer.
67. Smolin, L. (2006). *The trouble with physics: The rise of string theory, the fall of science, and what comes next*. New York: Houghton Mifflin Books.
68. Strominger, A. (1986). Superstrings with torsion. *Nuclear Physics*, *B274*, 253–284.
69. Strominger, A., Maldacena, J., & Witten, E. (1996). Black hole entropy in *M*-theory. *Journal of High Energy Physics*, *1997*, 1–18.
70. Strominger, A. (2011). The dS/CFT correspondence. *Journal of High Energy Physics*, *10*, 034.
71. Strominger, A., & Vafa, C. (1996). Microscopic origin of the Bekenstein-Hawking entropy. *Physics Letters B*, *379*, 99–104.
72. Susskind, L. (1993). Some speculations about black hole entropy in string theory. In C. Teitelboim & J. Zanelli (Eds.), *The black hole, twenty-five years after* (pp. 118–131). Singapore: World Scientific.
73. Susskind, L. (2003). The anthropic landscape of string theory. arXiv:hep-th/0302219.
74. Susskind, L. (2006). *The cosmic landscape*. New York: Back Bay Books.
75. Todorov, I. (2001). Two-dimensional conformal field theory and beyond. *Letters in Mathematical Physics*, *56*, 151–161.
76. Townsend, P. K. (1995). The eleven-dimensional supermembrane revisited. *Physics Letters B*, *350*, 184–188.
77. Witten, E. (1995). String theory dynamics in various dimensions. *Nuclear Physics*, *B443*, 85–126.
78. Witten, E. (1996). Five-branes and *M*-theory on an orbifold. *Nuclear Physics*, *B463*, 383–397.
79. Witten, E. (1996). Bound states of strings and *p*-branes. *Nuclear Physics*, *B460*, 335–350.
80. Witten, E. (1998). Magic, mystery, and matrix. *Notices Of The AMS*, *45*(9), 1124–1129.

Erratum to: Particle Physics in the Sixties

Erratum to:
Chapter 2 in: D. Rickles, *A Brief History of String Theory*,
The Frontiers Collection, DOI 10.1007/978-3-642-45128-7_2

On page 32: The legend to the Fig. 2.3 is wrong and should rather belong to Fig. 2.5 that is on page 37.

On page 34: The legend to the Fig. 2.4 is wrong and should rather belong to Fig. 2.3 that is on page 32.

On page 37: The legend to the Fig. 2.5 is wrong and should rather belong to Fig. 2.4 that is on page 34.

The online version of the original chapter can be found under
DOI 10.1007/978-3-642-45128-7_2

D. Rickles, *A Brief History of String Theory*, The Frontiers Collection,
DOI: 10.1007/978-3-642-45128-7_11, © Springer-Verlag Berlin Heidelberg 2014

Index

D. Rickles, *A Brief History of String Theory*, The Frontiers Collection,
DOI: 10.1007/978-3-642-45128-7, © Springer-Verlag Berlin Heidelberg 2014

Titles in This Series

Quantum Mechanics and Gravity
By Mendel Sachs

Quantum-Classical Correspondence
Dynamical Quantization and the Classical Limit
By Josef Bolitschek

Knowledge and the World: Challenges Beyond the Science Wars
Ed. by M. Carrier, J. Roggenhofer, G. Küppers and P. Blanchard

Quantum-Classical Analogies
By Daniela Dragoman and Mircea Dragoman

Life: As a Matter of Fat
The Emerging Science of Lipidomics
By Ole G. Mouritsen

Quo Vadis Quantum Mechanics?
Ed. by Avshalom C. Elitzur, Shahar Dolev and Nancy Kolenda

Information and Its Role in Nature
By Juan G. Roederer

Extreme Events in Nature and Society
Ed. by Sergio Albeverio, Volker Jentsch and Holger Kantz

The Thermodynamic Machinery of Life
By Michal Kurzynski

Weak Links
The Universal Key to the Stability of Networks and Complex Systems
By Csermely Peter

The Emerging Physics of Consciousness
Ed. by Jack A. Tuszynski

D. Rickles, *A Brief History of String Theory*, The Frontiers Collection,
DOI: 10.1007/978-3-642-45128-7, © Springer-Verlag Berlin Heidelberg 2014

Extreme States of Matter
on Earth and in the Cosmos
By Vladimir E. Fortov

Searching for Extraterrestrial Intelligence
SETI Past, Present, and Future
Ed. by H. Paul Shuch

Essential Building Blocks of Human Nature
Ed. by Ulrich J. Frey, Charlotte Störmer and Kai P. Willführ

Mindful Universe
Quantum Mechanics and the Participating Observer
By Henry P. Stapp

Principles of Evolution
From the Planck Epoch to Complex Multicellular Life
Ed. by Hildegard Meyer-Ortmanns and Stefan Thurner

The Second Law of Economics
Energy, Entropy, and the Origins of Wealth
By Reiner Kümmel

States of Consciousness
Experimental Insights into Meditation, Waking, Sleep and Dreams
Ed. by Dean Cvetkovic and Irena Cosic

Elegance and Enigma
The Quantum Interviews
Ed. by Maximilian Schlosshauer

Humans on Earth
From Origins to Possible Futures
By Filipe Duarte Santos

Evolution 2.0
Implications of Darwinism in Philosophy and the Social and Natural Sciences
Ed. by Martin Brinkworth and Friedel Weinert

Probability in Physics
Ed. by Yemima Ben-Menahem and Meir Hemmo

Chips 2020
A Guide to the Future of Nanoelectronics
Ed. by Bernd Hoefflinger

From the Web to the Grid and Beyond
Computing Paradigms Driven by High-Energy Physics
Ed. by Rene' Brun, Federico Carminati and Giuliana Galli Carminati

A Brief History of String Theory
From Dual Models to M-Theory
By Dean Rickles

Printed in the United States
By Bookmasters